House Construction for the 1980's

House Construction for the 1980's

by John Capotosto

Student's edition published as
Residential Carpentry for the 1980's

RESTON PUBLISHING COMPANY, INC.
A Prentice-Hall Company
Reston, Virginia

Library of Congress Cataloging in Publication Data

Capotosto, John.
 Residential carpentry for the 1980's.

 Bibliography: p.
 Includes index.
 1. Carpentry. 2. House construction. I. Title.
II. Title: Residential carpentry for the nineteen eighties.
TH5604.C35 1983   694   82-3768
ISBN 0-8359-6648-8   AACR2

10   9   8   7   6   5   4   3   2   1

PRINTED IN THE UNITED STATES OF AMERICA

Contents

Preface   vii

Acknowledgments   ix

1 Fundamentals of Carpentry   1
Hand Tools 1   Power Tools 9   Scaffolds and Ladders 11   Ladders 16
Lumber 19   Fasteners 27   Plans and Specifications 33   Building Code 41
Leveling Instruments 42   The Framing Square 55   The Metric System 60
Ellipses 65   Safety on the Job 66

2 Foundations and Formwork   71
Foundation Layout 71   Form Materials 80   Footings 82   Wall-Form Construction 87   Placing Concrete 96

3 Floor Framing   103
Moisture Content of Wood 104   Types of Framing 105   Termite Protection
110   Sill Construction 111   Beams and Girders 112   Floor Joists 114
Subflooring 128   Glued Floors 131

4 Wall and Ceiling Framing   133
Corner Construction 136   Intersecting Walls 137   Exterior Wall Construction 139   Partitions 145   Wall Sheathing 149   Ceiling Frame 151

5 Roof Framing   157
Roof Styles 157   Roof Frame Members 158   Fundamentals of Roof
Framing 162   Installing the Common Rafter 175   Shortening the Hip Rafter
184   Construction of Various Roof Types 196   Other Roof Elements 202

Intersecting Roofs with Unequal Pitch 206   Roof Trusses 215   Roof Sheathing 220   Plank-and-Beam Framing 225

6 Exterior Finish   238

Door Frames 239   Exterior Windows 241   Cornices 249   Finish Roofing 257   Flashing 263   Applying Shingles 268   Other Roofing 281   Drainage 282   Exterior Wall Finish 285

7 Insulation and Moisture Control   309

Vapor Barriers 310   Thermal Insulation 313   Slab Insulation 319   Weatherstripping 325   Sound Control 326

8 Interior Finish   332

Plastering 332   Gypsum Drywall 337   Wall Paneling 342   Ceiling Tiles 347   Suspended Ceilings 352   Finish Floors 353   Interior Trim 365

9 Stairs   396

Stair Design and Construction 397

10 Cabinetry   418

Cabinet Design and Construction 418

Glossary   437

Bibliography   453

Index   455

Appendix I   461

Appendix II   464

Preface

Residential Carpentry incorporates the latest materials, methods, and techniques of construction in the building industry. It concentrates on the building of the single family dwelling, covering all phases of construction in an orderly, step-by-step fashion. It deals with those phases of special interest and importance to the carpenter, including foundations, walls, roofs, and the finish work. Wherever necessary, the text also includes information on other building trades, and it is profusely illustrated with drawings and photos. These explain and enlarge instructions, making it easier for the reader-user to understand technical matters.

Special attention has been given to insulation because of the serious problem of energy conservation facing our nation. The proper application of insulating and weatherstripping materials is well covered.

Safety is also given ample coverage in discussions of both accident prevention and safety on the job. Safety precautions relating to specific tools and equipment are emphasized.

Throughout, the text teaches a practical approach to carpentry and stresses simplified methods of construction. Formulas, complicated mathematical computations, and special technical terms have their place, of course, but not here. Instead, only essential math and formulas, as they apply to roofing and stair building, are covered. Such terms as "modus of elasticity," "moments of inertia," and "unstable equilibrium" are used in the design of structures and must be thoroughly understood by engineers and architects, but not necessarily by carpenters. Here the carpenter, as a skilled craftsman, trained to work with tools and materials, learns how to build new structures and how to do remodeling, repairs, and maintenance.

Residential Carpentry for the 1980s also provides basic instruction for students in high schools, vocational schools, and colleges. Special emphasis is placed on the fundamentals of carpentry.

The glossary is comprehensive and includes many illustrations which will prove helpful to students as well as those already in the trade.

The author hopes that you will find this edition even more helpful than the previous one.

Acknowledgments

The author wishes to thank the following companies and associations for their cooperation in supplying reference material and photos for use in the preparation of this book.

Acoustical & Insulating Materials Assoc., Park Ridge, Ill.
American Ladder Institute, Chicago, Ill.
American Plywood Association, Tacoma, Wash.
Andersen Corp., Bayport, Minn.
Armstrong Cork Company, Lancaster, Pa.
Asphalt Roofing Manufacturers Assoc., New York, N.Y.
Automated Building Components, Miami, Fla.
Berger Instruments, Boston, Mass.
The Black & Decker Mfg. Co., Towson, Md.
Bostitch Division of Textron, Inc., East Greenwich, R.I.
Bruce, E. L., Co., Memphis, Tenn.
California Redwood Assoc., San Francisco, Calif.
The Celotex Corp., Tampa, Fla.
Certain-Teed Products Corp., Valley Forge, Pa.
Deer Park Lumber Co., Deer Park, N.Y.
Dexter Lock, Grand Rapids, Mich.
Estwing Mfg. Co., Rockford, Ill.
Evans Products Co., Portland, Ore.
Flint, A. W., Co., New Haven, Conn.
Forest Products Laboratory, U.S. Dept. of Agriculture, Madison, Wisc.
Formica Corp., Cincinnati, Ohio
GAF Corp., New York, N.Y.
Georgia-Pacific Corp., Portland, Ore.
Gold Bond Building Products, Buffalo, N.Y.
Gypsum Assoc., Chicago, Ill.
Hand Tools Institute, New York, N.Y.
Keystone Steel & Wire, Peoria, Ill.
Marsh Wall Products, Chicago, Ill.
Masonite Corp., Chicago, Ill.
Modern Materials Corp., Detroit, Mich.
Morgan Co., Oshkosh, Wisc.
National Particleboard Assoc., Silver Springs, Md.

National Assoc. of Home Builders, Washington, D.C.
National Forest Products Assoc., Washington, D.C.
National Oak Flooring Mfgrs. Assoc., Memphis, Tenn.
National Woodwork Mfgrs. Assoc., Chicago, Ill.
Patent Scaffolding Co., Fort Lee, N.J.
Pemko Mfg. Co., Emeryville, Calif.
Pittsburg Corning Corp., Pittsburgh, Pa.
Porter, H. K., Co., Inc., Pittsburgh, Pa.
Proctor Products Co., Inc., Kirkland, Wash.
Red Cedar Shingle & Handsplit Shake Bureau, Seattle, Wash.
Reuten, Fred, Inc., Closter, N.J.
Rockwell International, Pittsburgh, Pa.
Rodman Industries, Marinette, Wisc.
Shakertown Corp., Winlock, Wash.
Simpson Timber Co., Seattle, Wash.
Southern Forest Products Assoc., New Orleans, La.
The Stanley Works, New Britain, Conn.
Steel Scaffolding & Shoring Institute, Cleveland, Ohio
Sumner Rider & Assoc., New York, N.Y.
TECO, Washington, D.C.
U.S. Forest Service, Pacific Northwest Forest & Range Experimental Station, Seattle, Wash.
United States Gypsum Co., Chicago, Ill.
U.S. Plywood, New York, N.Y.
Ventarama, Port Washington, N.Y.
Western Red Cedar Lumber Assoc., Portland, Ore.
Western Wood Moulding & Millwork, Portland, Ore.
Western Wood Products Assoc., Portland, Ore.
Weston Instruments, Newark, N.J.
Weyerhaeuser Co., Tacoma, Wash.
Wild Heerbrugg, Inc., Farmingdale, N.Y.
Wisconsin Knife Works, Inc., Beloit, Wisc.

House Construction for the 1980's

Fundamentals of Carpentry 1

"Jack of all trades, master of none" is an old cliché that does not apply to carpenters. The modern carpenter must be a jack of all trades and a master of one. In addition to being skilled in his own craft, he must have a good working knowledge of various other trades, such as masonry, plumbing, heating and electrical, in order to carry out his own work efficiently. He must coordinate his work with that of the other trades so that all the pieces will come together precisely as in a jigsaw puzzle. Engineers and architects design houses, but it is up to the carpenter to convert these ideas on paper to the finished product—a house that is safe, sound, and comfortable.

There are two kinds of carpentry—rough carpentry and finish carpentry. Unfortunately, the term "rough carpentry" often denotes crude or slipshod work to the uninformed. Of course, this is not the case. The term simply refers to the fact that unfinished (rough) lumber is used in this phase of construction. Rough carpentry deals with the framing of a structure. It involves form building, floor framing, wall framing, and roof framing. It also includes the installation of sheathing, subfloors, and partitions. It must be done with the greatest of care and accuracy. Perhaps it should be called "tough" carpentry, or at the least, "frame" carpentry.

Finish carpentry involves the installation of doors, windows, interior and exterior trim, siding, cornices, paneling, cabinets, built-ins, and many other items, resulting in the completed structure. In some situations, especially on big jobs, carpenters may be assigned to carry out special phases of the work. For example, one person may only hang doors, while another may just install trim, and so on. If possible, this trend should be discouraged, because it restricts the carpenter and limits his overall ability.

Hand Tools

In order to do his job, the carpenter must be able to lay out, cut, fit, and assemble a variety of wood and non-wood materials. To perform this work, he must be familiar with and proficient in the use of many hand and power tools. Although power tools will help do many jobs faster and easier than they can be done by hand, there are many jobs that can be done only with hand tools.

1

The use of hand and power tools is well covered in other texts dealing specifically with the subject. We shall cover only those which pertain strictly to carpentry. Carpenters should follow standard practices when using tools, with the thought of safety and accuracy always in mind. The tools used should be of high quality, and most important, they should always be kept sharp, since dull tools cause accidents and impair the quality of the work.

Cutting Tools.   Many types of handsaws are used by the carpenter. For general cutting, an 8-point crosscut saw—that is, one with 8 teeth per inch —is often used. Ripping is better done with a 7-point saw. Saw blades should be protected when not in use and they should never be placed on the ground, because bits of dirt on the teeth can dull them fast. Most carpenters hang them conveniently on the structure (*Fig. 1–1*). In general, to start a cut with a handsaw, use your thumb to guide the blade. Then move your hand a safe distance away from the line of cut as the sawing progresses. Use a coarse-toothed saw for cutting green (unseasoned) wood.

Figure 1–1 Handsaw hung on framing is within easy reach.

The crosscut saw should be held at a 45-degree angle between saw and work; the ripsaw should be held at 60 degrees (*Fig. 1–2*). Saws can be touched up with a file so they will be kept in good working order; however, when they do become dull, they should be sharpened.

The backsaw has a thin blade with fine teeth, usually 14 to 15 points. The thin blade is kept straight with a reinforcing "backbone." Backsaws

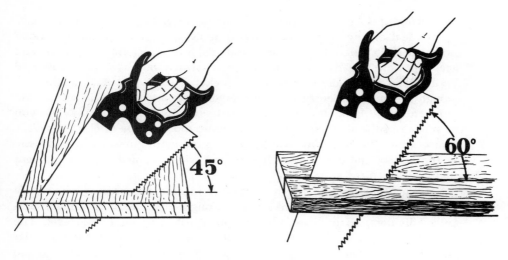

Figure 1–2 Proper method of using the handsaw. *Left:* Cross-cutting. *Right:* Ripping.

are used with wooden miter boxes or in more elaborate metal frames that are adjustable for all angles. When mitering crown moldings in a miter box, hold the work at the mounting angle, as shown in *Fig. 1–3*.

The coping saw is a useful tool that should be included in the carpenter's toolbox. It will cut curved lines with ease. It is often used for coping moldings. The blades of this saw are rather fragile and breakage is common. It is a good idea to keep a supply of blades taped to the back edge of the frame to keep them from being misplaced.

For metal-cutting, the toolbox should also contain a hacksaw and a few good-quality blades.

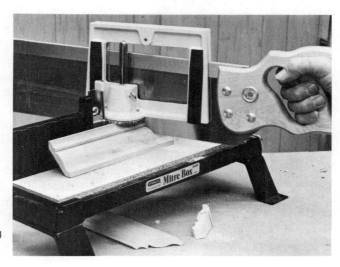

Figure 1–3 Crown molding being cut in miter box.

Measuring and Marking Tools.   The accuracy of carpentry work depends
a great deal on the layout and marking tools used by the carpenter. Good
tools, properly cared for, are essential. Folding rules, flexible tapes, and
steel tapes are equally useful in house construction. The folding rule and
flexible tape are usually carried in the pocket. The 50-foot steel tape is long
enough for most carpentry jobs, but larger tapes are available. Care should
be taken not to kink flexible or steel tapes; this will ruin them. When you
are reeling in the steel tape, wrap a cloth around the tape to clean it as it is
retracted into the case. This will keep grit from getting into the mechanism.
At the end of the day, wipe off the tape with an oiled cloth to prevent
rusting.

Other tools needed by the carpenter include the framing square, try
square, T-bevel, combination square, straightedge, marking gauge, butt
gauge, dividers, chalk lines, plumb bob, spirit level, and line level. The
framing square is useful for laying out rafters, stairs, braces, and angular
cuts. It is discussed further in the section on roof framing. The try square
and combination square are used for squaring and miter-cut layouts. The
sliding head of the combination square makes it useful for gauging depths
and for parallel marking. A square that is off can be a source of trouble for
the carpenter; it should be checked for accuracy periodically. To do this,
place the square on the edge of a straight piece of wood and draw a squaring line on the surface of the board. Flop the square over and draw another
line over the first one. If the lines coincide, the square is true. If the lines
taper, the tool is off and should be repaired, if possible, or discarded (*Fig.
1–4*).

The blade of the T-bevel can be set to any desired angle. It can be
used for transferring or laying out angles. One important use is in laying
out rafter cuts. When not in use, the pointed edge of the blade should be
placed in the recess in the handle.

The straightedge is used for marking lines longer than 24 inches. It is
also used in conjunction with a spirit level to increase its span for greater
accuracy. It is often used when door frames are being hung. Many carpenters make their own by laminating several pieces of straight-grained wood.
A length of heavy-gauge aluminum also makes an excellent straightedge. It
is lightweight, will not bend or warp, and can also be used as a guide for
power saws.

The level and plumb bob enable the carpenter to lay out and check
vertical and horizontal lines. Levels are made from a number of materials in
various sizes. Long levels eliminate the need for straightedges. For this reason, many carpenters use a 6-foot level. You can check the accuracy of a
level by placing it on a level surface with the bubble perfectly centered.
Then, swing it around 180 degrees and note the position of the bubble (*Fig.
1–5*). It should remain centered. If the bubble moves off-center, look for
the cause, which may be dents or foreign matter along the edge of the
frame.

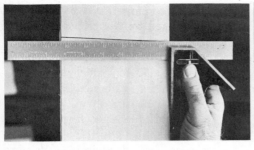 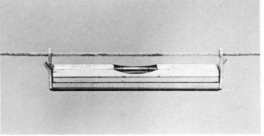

Figure 1–4 Checking accuracy of a square.   **Figure 1–5** An accurately centered bubble.

The plumb bob is a weighted device pointed at one end and suspended from a line. In use, the pointed end and line will hang in a vertical plane. Some bobs are filled with mercury to give them added weight without increasing the size. The slimmer the profile of a bob, the greater its resistance to wind.

Hand Finishing Tools.   Planes and chisels head the list of sharp-edged cutting tools used by the carpenter for finishing wood surfaces. Planes are used to smooth rough surfaces and to bring surfaces to size. The carpenter usually has several planes in his collection, each plane serving a special purpose. In all, there are about a dozen different planes available.

The block plane is very short and is suited for planing end-grained wood. The smooth plane, jack plane, and jointer plane are used for removing stock and producing finished smooth surfaces. The jointer plane is the longest of all planes and is especially suited for planing door edges.

Wood chisels have cutting edges that range in size from ⅛ inch to 2 inches. The popular sizes for carpenters are ⅜-inch, ½-inch, ¾-inch, and 1-inch widths. The parts of a chisel are shown in *Fig. 1–6.*

Chisels must be kept sharp and their edges protected when not in use. In use, always keep both hands well back of the cutting edge.

Figure 1–6 A typical wood chisel.

Hand Boring Tools.   The hand drill is used for making small holes up to $^{11}/_{64}$ inch in diameter. Holes ⅜ inch in diameter and larger are made with the bit brace. Both types are shown in *Fig. 1–7.* Auger bits are used for holes up to 1½ inch in diameter. Larger holes are bored with expansive

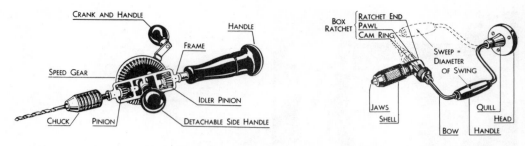

Figure 1–7 Two hand tools used for boring holes. *Left:* Hand drill. *Right:* Hand brace.

bits which have a range from ⅞ inch to 3 inches in diameter. For cabinet work where small-size holes are needed for hinges and catches, many carpenters prefer to use the push drill, which uses fluted bits (*Fig. 1–8*).

Figure 1–8 Push drill.

Assembly Tools.   Hammers are perhaps the most commonly used of the assembly tools. They are made in a variety of sizes and styles, each with a specific purpose (*Fig. 1–9*). The two types generally used by carpenters are the straight-claw and curved-claw. The straight-claw is used mainly for framing; its claw is handy for prying pieces apart (*Fig. 1–10*). The hammer head should be of forged steel and tightly fitted to the handle. Basic rules should be followed when you are hammering. Grasp the hammer handle near the end and swing it so it strikes the nail or other surface squarely (*Fig. 1–11*). Glancing blows can cause nails to fly. Hardened steel cut and masonry nails should never be driven with a nail hammer. Glancing blows can cause these nails to shatter. Use only a heavy hammer with a large face (*Fig. 1–12*), and be sure to wear safety goggles (*Fig. 1–13*).

Screwdrivers are of two basic types, Phillips and slotted. Lengths and sizes vary considerably. The length of a screwdriver is measured from the tip to the ferrule. *Fig. 1–14* shows the various parts of a screwdriver. To function properly, the tip of the screwdriver should fit the slot of the screw

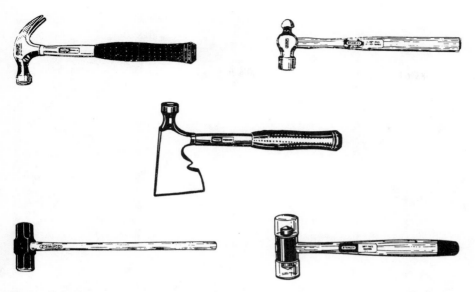

Figure 1–9 Various hammers. (a) curved-claw, (b) ball-peen, (c) hatchet, (d) engineer's, (e) soft-face.

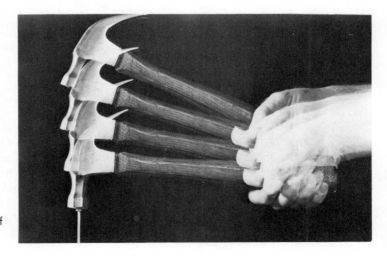

Figure 1–10 Straight-claw framing hammer.

Figure 1–11 Proper use of the hammer is important.

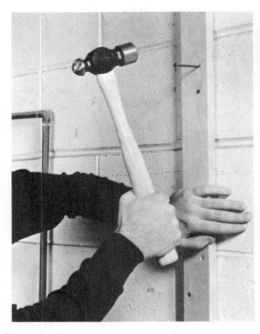

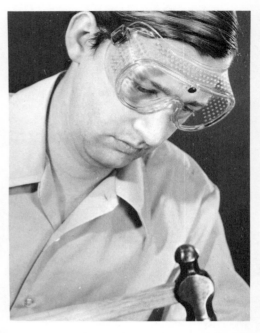

Figure 1–12 Masonry nails should be struck only with a wide-faced heavy hammer.

Figure 1–13 Safety goggles are highly recommended when masonry nails are used.

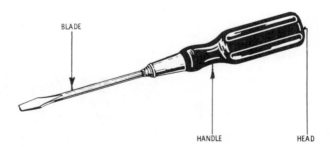

Figure 1–14 Parts of a screwdriver.

snugly. A good selection of screwdrivers for the carpenter would include 4-, 6- and 8-inch sizes with both slotted and Phillips tips.

Offset screwdrivers are used for work in tight areas where a regular screwdriver will not fit. They are made with slotted and Phillips tips, with or without ratchets.

Other tools used by carpenters include files, wrenches, pliers, clamps, nail sets, hatchets, and staplers.

SAFETY RULES FOR HAND TOOLS

- Never use files without handles.
- Before use, always be sure hammer heads are securely fastened to their handles.
- Always keep cutting tools sharp; dull ones may slip.
- Discard striking tools if they are excessively worn.
- Do not allow chisel and punch heads to "mushroom."
- Do not hold nails or tacks in your mouth while working.
- Always wear safety goggles when using hammers and other striking tools.
- When using sharp-edged tools, always strike, chop, chisel, or carve with the cutting action away from yourself.

Power Tools

The power tools used by the carpenter are mostly of the portable type. They are conveniently held in the hand and brought to the work. Stationary power tools are those in which the work is brought to the tool. This does not mean that stationary tools cannot be used on the jobsite. Many carpenters use small transportable stationary tools such as the radial-arm saw and the cutoff saw (*Fig. 1–15*).

Figure 1–15 Radial-arm saw in use at jobsite.

The portable power tools used by the carpenter include the portable saw, saber saw, drill, and router. The portable saw (also called a circular saw or builder's saw) is excellent for straight cuts; it is fast and efficient and can be fitted with various blades to handle different materials. The larger the tool, the greater its capacity. The size of the saw is determined by the blade diameter; thus a 6-inch saw has a 6-inch blade. For carpentry work the saw should be large enough to cut through 2-inch-thick stock with the blade tilted at 45 degrees (*Fig. 1–16*). The saw should have a substantial blade guard and a clutch. The clutch will prevent dangerous kickbacks which can happen if the blade binds during a cut. Dull blades can also be the cause of kickbacks.

Figure 1–16 The portable saw is ideally suited for making compound cuts.

Never force the tool. If it starts to labor, back it off and try to find the cause of the problem. Often a saw will bind when the kerf closes. This can happen when you are cutting warped boards, but more often it occurs because the work is improperly supported. The line of cut should always be outside the supports for cutoff work. When you are ripping panels, use a support, such as a 2 × 4, parallel to the cutting line. This will prevent buckling and binding.

Because the blades of these saws cut on the upstroke, keep the good or face side of the work down. The blade should project about ¼ inch below the work.

Trim saws (small-size portable saws) are useful for cutting paneling and for other light-duty work. Lightweight and easy to use, they have the same features as the larger saws except that the blades are only 4 inches in diameter.

SAFETY RULES FOR PORTABLE SAWS

- Do not stand in water.
- Rest the saw base on the work firmly and be sure that the blade is not touching the work when you start a cut.
- Do not allow the work to bind against the blade when cutting, since this can "throw" the saw. Support the work properly.
- Do not force the saw. A tool is "forced" when it is fed through the work too fast. This causes the blade to slow down and could damage the bearings and motor.
- Be sure the switch is off when the power plug is inserted.
- Do not cut short pieces with the portable saw. The piece, together with your thumb, can be turned into the blade by uneven thrust.

- Do not stand in the "line of cut." If the saw binds, it can kick back and severely cut your legs.
- Use caution when cutting loose knots; they can cause the saw to kick.
- Keep both hands on the saw. If that is not practical, keep your free hand well away from the blade.
- Do not talk to anyone or look away when operating the saw.
- Do not make plunge cuts (internal cuts) with the portable saw.
- Adjust the saw blade so that it projects about ¼ inch through the bottom of the work.
- Unplug the cord when changing blades.

The saber saw (*Fig. 1–17*) can be used for making straight or decorative cuts in wood and other materials. When fitted with the proper blade it can also be used on metal. It is ideal for paneling work, for cutting to scribed lines, for making fixture and window openings, and for irregular cutting. The saber saw and the portable saw are more widely used by carpenters than any other power tools, but all have a place in the carpenter's toolbox. The important concern is that the carpenter have a good understanding of the various tools and that he use them properly and safely.

Figure 1–17 The saber saw.

Scaffolds

Scaffolds are platforms used to support workers and materials, enabling them to reach and work on the structure as it rises above ground level. Scaffolds are temporary structures and may be made of wood or metal, or a combination of both. Many commercially made types are available. Often, however, carpenters build their own scaffolds, using lumber that is available at the jobsite. Regardless of whether they are complex or simple, commercially made or "homemade," they must be structurally sound and safe for use. The carpenter must be familiar with national and local safety codes regarding the erection and use of scaffolding.

Metal Scaffolds.   Tubular metal scaffolds made of steel or aluminum are very popular. They are portable, strong, and easy to assemble and disassemble. A type generally used by carpenters is shown in *Fig. 1–18*. The units can be used side by side and stacked. The scaffold members must be free of rust and all locking devices must be in good working order. If the unit is fitted with casters, the brakes must be in good working order. When you are erecting the scaffold, be sure that each leg is fully supported. Check the frame to make certain it is plumb and level. For planks, use only lumber that is graded for plank use. Also, secure the plank to the scaffolding when necessary.

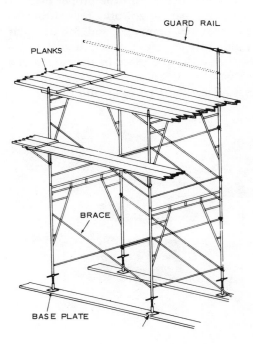

GUARD RAIL

PLANKS

BRACE

BASE PLATE

Figure 1–18 A sturdy metal scaffold.

Wood Scaffolds.   There are two basic types of wood scaffolds: the single-pole and the double-pole. They may be classified as light-duty or heavy-duty. The light-duty scaffold must be capable of supporting a working load of 25 pounds per square foot. The heavy-duty scaffold must support a working load of 75 pounds per square foot. Otherwise, they are similarly made.

   The poles or uprights for wood scaffolds must be of clear, straight 2 × 4s. Ledgers consist of 2 × 6s, braces of 1 × 6s, and guardrails of 2 ×

4s. Ledgers must extend past the poles at least 4 inches. The lower ends of the poles must rest on blocking so they will not sink into the ground. Lumber for planks must be 2 × 10 or heavier. The planks should extend beyond the ledgers at each end by at least 6 inches and not more than 12 inches. The edges of the planks should be placed close enough to prevent tools or materials from falling through. They can be spiked to the ledgers so they will not shift.

The double-pole scaffold is shown in *Fig. 1–19*. It is freestanding and diagonally braced. The single-pole scaffold shown in *Fig. 1–20* is unsafe and should never be used. Note that it lacks a guardrail and it is poorly designed.

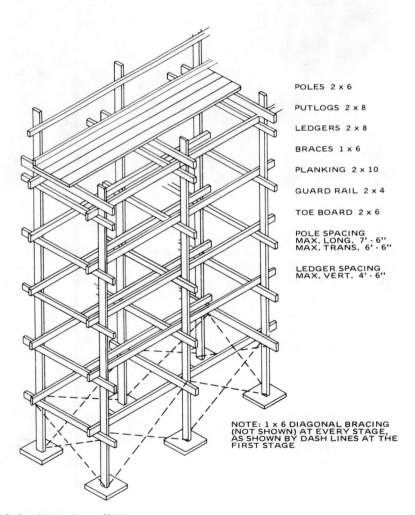

POLES 2 x 6

PUTLOGS 2 x 8

LEDGERS 2 x 8

BRACES 1 x 6

PLANKING 2 x 10

GUARD RAIL 2 x 4

TOE BOARD 2 x 6

POLE SPACING
MAX. LONG. 7' - 6"
MAX. TRANS. 6' - 6"

LEDGER SPACING
MAX. VERT. 4' - 6"

NOTE: 1 x 6 DIAGONAL BRACING
(NOT SHOWN) AT EVERY STAGE,
AS SHOWN BY DASH LINES AT THE
FIRST STAGE

Figure 1–19 Double-pole scaffold.

Figure 1–20 A poorly designed and unsafe scaffold.

Bracket Scaffolds.   Wall-mounted bracket scaffolds are often used by carpenters for light-duty work. They are lightweight and easily transportable. *Fig. 1–21* shows several of many types available. Various methods are used to fasten the brackets to the walls. Some are nailed or bolted, while others are made to hook onto studding. The safest type are those which are bolted. Bolts must pass through and be fastened to a 2 × 6 block laid across the inside of two studs, as shown in *Fig. 1–22*.

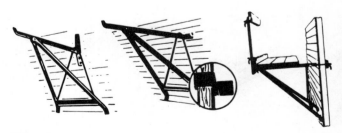

Figure 1–21 Various wall brackets.

Figure 1–22 Securing wall bracket.

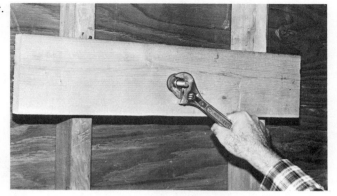

Other Scaffolds. Trestle jacks are generally used for indoor work. Made of steel, most are adjustable, with working heights of from 7 to 12 feet. Special handle nuts lock all height adjustments.

Post jacks consist of a foot-operated ratchet mechanism. They are supported by 4 × 4-inch braced posts. They are widely used for shingling side walls (*Fig. 1–23*).

Roofing brackets are useful on steep slopes. The unit shown in *Fig. 1–24* has a notched extension that hooks over nails fastened to the roof frame. The adjustable plank support can be leveled regardless of roof slope.

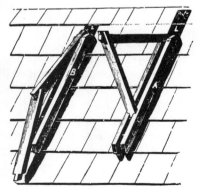

Figure 1–23 Foot-operated post jack.   Figure 1–24 Adjustable roofing bracket.

Ladder brackets are used on rung ladders. They form a support for the plank. Some types are adjustable, as shown in *Fig. 1–25*.

Figure 1–25 Adjustable ladder bracket. In use, the ladder must lean at a safe angle.

SAFETY RULES FOR SCAFFOLDING

- Inspect scaffolds daily before use. Never use damaged or worn equipment.
- Follow all codes pertaining to scaffolding.
- Use adequate sills or pads under scaffold posts.
- Plumb and level the scaffold as it is being erected.
- Fasten all braces securely.
- Do not climb cross-braces.
- Use guardrails and toeboards if required.
- Use caution when you work near power lines. Consult power company for advice.
- Do not use ladders or other makeshift devices to increase the height of a scaffold.
- Never overload scaffolds.
- For planks, use lumber that is graded as scaffold plank.
- Make sure planks overlap ends by 12 inches and extend 6 inches beyond the scaffold supports.
- Secure planks to scaffolding when necessary.
- Do not ride a rolling scaffold.
- Remove all material before moving a rolling scaffold.
- Keep caster brakes applied except when moving the scaffold.

Ladders

Various types of ladders are used by carpenters. Some are shown in *Fig. 1–26*. The principal materials used in the manufacture of ladders are wood, aluminum, magnesium, and fiberglass.

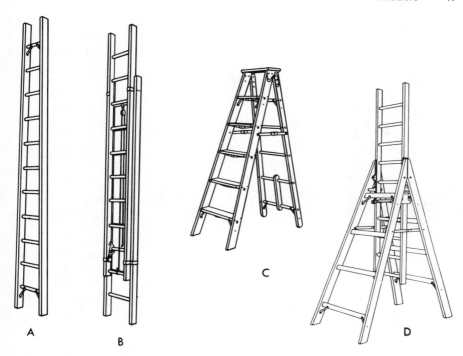

Figure 1–26 Various ladders used by carpenters: A. Single ladder. B. Extension ladder. C. Stepladder. D. Trestle ladder.

Wood ladders are sturdy and bend very little for the designed loads. They weigh more than metal ladders and in large sizes are difficult to handle. Because they do not conduct electricity (when dry), they are safe to use with power tools and near electrical circuits.

Metal ladders are not affected by moisture or sunlight and they do not decay. They weigh less than wood ladders but are most costly. They should not be used around electrical circuits because they conduct electricity, and can be a shock hazard.

Fiberglass is used to make the side rails of straight and stepladders. Such ladders are light, strong, corrosion-resistant, and nonconductive. They will not absorb moisture nor will they dry-rot. They are tougher than wood or metal ladders and more costly.

Stepladders are portable and self-supporting. They usually fold flat for storage. They have broad steps which should be at least ¾ inch thick and 3½ inches deep. The are made in heights up to 20 feet.

Straight ladders have rungs and are made in lengths up to 30 feet. Some carpenters make their own on the job, but it is wiser to use approved commercial ladders.

Extension ladders consist of two straight sections made to slide past each other. Some are made to extend up to 60 feet. Various types of safety "shoes" are made for use at the base of the ladder. These prevent slipping (*Fig. 1–27*).

Safety shoe

Figure 1–27 Versatile safety shoe for ladder. *Left:* Used on hard surface. *Right:* Used on soil.

Ladders should never be painted, because the paint may conceal defects. Instead, use clear finishes such as lacquer, linseed oil, or varnish. When you lean a straight ladder against a wall, make sure the base is one-quarter of its working length away from the wall. If the ladder is placed with the base too far out, the ladder can break or slip. If too close, it may fall backward. Observe the following ladder safety rules.

SAFETY RULES FOR LADDERS

- Inspect all ladders carefully before each use.
- Keep all fastenings tight and oil the moving parts often.
- Keep rungs free of dirt, grease, and oil.
- Be sure the ladder stands on firm ground. When necessary, use non-slip bases.
- Place the ladder so that the distance from its base to the wall is one-fourth of the ladder's height.
- If positioned for climbing onto a roof, a ladder must extend 3 feet above the roof.
- Never place a ladder in front of a door or similar opening unless proper precautions are taken.
- Overlap extension ladders at least 3 feet for 36-foot lengths, 4 feet for 48-foot lengths, and 5 feet for 60-foot lengths. Install the sections so that the upper section is outermost.
- Be sure that the locks on extension ladders are hooked securely before you climb. Never extend a ladder while standing on it.
- Always face the ladder when ascending or descending. Place the ladder close enough to the work so that you won't have to over-reach.
- Never stand on the top three rungs of straight or extension ladders.

- Never stand on the top pail-rest or rear rungs of stepladders.
- Use extreme care when working near electric wires and equipment.
- Never overload a ladder; and remember that standard ladders are made to support only one person. Special ladders are available for supporting two people.
- Never use a ladder in the horizontal position, on saw horses to serve as a scaffold.
- Store ladders properly. A cool, dry, ventilated area is best. Hang the ladder so it won't sag.

Lumber

When the early settlers in this country built houses, there were no lumber-yards or mills where they could purchase the required lumber. They simply felled trees and used primitive methods to produce the lumber needed (*Fig. 1–28*). As the country began to grow, lumber mills were established and more and better lumber became available. Lumber, which was once rough and not too uniform in size or grades, gradually improved through the years. Today, lumber is produced in modern computerized mills and is available in standard sizes and grades.

Figure 1–28 Early method of producing lumber.

At present, lumber is sold and specified by its nominal size—that is, its *original* cutting size. When sold to the user, it has been reduced in size owing to seasoning and dressing. A piece of 2 × 4 lumber (nominal size) actually measures 1½ × 3½ inches (dressed size). Plans and specifications list lumber in the nominal size. The dimension differences must be taken into consideration by the carpenter.

Not all lumber is surfaced or planed smooth. The number of faces surfaced is designated by a imple code. For example, S4S means that the lumber has been surfaced on all four sides. S2S indicates that two sides have been surfaced. A board with S1S2E designation means that it has been surfaced on one side and two edges. A chart with the standard nominal and dressed sizes is shown in *Fig. 1–29* on pages 22 and 23.

Lumber grading has been established to set quality control standards among mills manufacturing the same or similar woods. This assures the carpenter or other user that a piece of lumber with a particular grade will be the same whether it is bought in New York or in Oshkosh. Official grade stamps on the lumber guarantee the assigned grade. The participating mills are identified by a number assigned to each mill. This number is found at the upper left of the grade stamp, shown in *Fig. 1–30*. The symbol at the lower left identifies the sponsoring association, Western Wood Products in this case. The grade (C-select) is shown at the top. The moisture content is at the bottom center, and the species of wood (sugar pine) is at the lower right.

The grading systems of hardwoods and softwoods differ. Softwood is divided into three main classes, according to its intended use: yard lumber, structural lumber, and factory and shop lumber. Yard lumber is widely used for construction and general building purposes. It is subdivided into select and common classifications. Select lumber is of good appearance and is available in grades A to D. Grade A is suitable for natural finishes and is practically clear—that is, free from knots and blemishes. Grade B can also take transparent finishes and is generally clear. Grade C will take high-quality paint finishes. Grade D is suitable for paint finishes between higher finishing grades and common grades.

Common lumber, which is suitable for general construction and utility purposes, is graded in numbers from 1 to 5. The number 1 common is suitable for use without waste; it is sound and tight-knotted. Number 2 common is less restricted in quality than number 1 but of like quality. It is used for framing, sheathing, and other structural forms where the stress is not too great. Number 3 common permits some waste, with grade characteristics less than number 2. It is used for footings and rough flooring. Number 4 common permits waste, for it is of low quality and may have coarse features such as holes and decay. It is used for sheathing, subfloors, and cheaper types of construction. Its main use is in the manufacture of boxes and crates. Number 5 common is not produced in all species. Its only requirement is that it be usable. It is utilized for boxes and crates.

The building frame supports the finished members of the structure. It includes posts, beams, joists, subfloor, plates, studs, and rafters. Yard lumber, usually of softwood, is utilized for these members. It is cut into standard sizes including 2 × 4, 2 × 6, 2 × 8, 2 × 10, 2 × 12, and other sizes required for the framework.

Structural lumber is used for those members of the frame in which the smallest cross-section dimension is 5 inches or greater.

The most widely used grade of yard lumber for residential construction is number 2 common. When the lumber size needed in a particular grade is not available, it can usually be built up as required. Lumber sizes are standardized for convenience. Softwoods run 8, 10, 12, 14, 16, 18, and 20 feet in length; 2, 4, 6, 8, 10, and 12 inches in width; and 1, 2, and 4 inches in thickness. Hardwoods do not have standard lengths or widths. They run ¼, ½, 1, 1¼, 1½, 2, 2½, 3, and 4 inches in thickness.

Plywood is available in widths of 36, 48, and 60 inches and in lengths of 60 to 144 inches. Thicknesses range from ⅛ inch to over 1¼ inches.

Lumber is priced and quantities are estimated in board feet, except for moldings and furring, which are priced and sold by the linear foot. A board foot is a unit equivalent in size to a board measuring 12 inches by 12 inches by 1 inch thick, nominal size. A board measuring 6 inches wide by 24 inches long and 1 inch thick is also equal to 1 board foot. Board footage can be determined mathematically, by the use of tables, or by means of framing square tables.

Mathematically, a simple formula is used to determine the number of board feet required:

$$\frac{P \times T \times W \times L}{12} = \text{board feet}$$

P = Number of pieces
T = Thickness of wood in inches
W = Width of wood in inches
L = Length of wood in feet

For example: Find the board feet in a piece of lumber measuring 2 inches thick, 10 inches wide, and 6 feet long.

Solution:

$$\frac{1 \times 2 \times 10 \times 6}{12} = \frac{120}{12} = 10 \text{ board feet}$$

Some carpenters find it easier to convert all dimensions to inches, then divide by 144 instead of 12, as follows:

$$\frac{1 \times 2 \times 10 \times 72}{144} = \frac{1440}{144} = 10 \text{ board feet}$$

Product	Description[a]	Nominal Size		Dressed Dimensions		
		Thickness in Inches	Width in Inches	Thicknesses & Widths[b] in Inches		Lengths in Feet
				Surfaced Dry	Surfaced Unseasoned	
DIMENSION LUMBER	S4S	2 3 4	2 3 4 6 8 10 12 over 12	1-1/2″ 2-1/2″ 3-1/2″ 5-1/2″ 7-1/4″ 9-1/4″ 11-1/4″ off 3/4″	1-9/16″ 2-9/16″ 3-9/16″ 5-5/8″ 7-1/2″ 9-1/2″ 11-1/2″ off 1/2″	6 ft. and longer in multiples of 1′
SCAFFOLD	Rough Full Sawn or S4S	1-1/4 & Thicker	8 and Wider	Same	Same	6 ft. and longer in multiples of 1′
TIMBERS	Rough or S4S	5 and Larger		Thickness In. Width In. 1/2 Off Nominal		6 ft. and longer in multiples of 1′

		Nominal Size		Dressed Dimensions		
		Thickness in Inches	Width in Inches	Thickness in Inches	Width in Inches	Lengths in Feet
DECKING[b] Decking is usually surfaced to single T&G in 2″ thickness and double T&G in 3″ and 4″ thicknesses	2″ Single T&G	2	6 8 10 12	1-1/2″	5″ 6-3/4″ 8-3/4″ 10-3/4″	6 ft. and longer in multiples of 1′
	3″ and 4″ Double T&G	3 4	6	2-1/2″ 3-1/2″	5-1/4″	
FLOORING	(D & M), (S2S & CM)	3/8 1/2 5/8 1 1-1/4 1-1/2	2 3 4 5 6	5/16″ 7/16″ 9/16″ 3/4″ 1 1-1/4″	1-1/8″ 2-1/8″ 3-1/8″ 4-1/8″ 5-1/8″	4 ft. and longer in multiples of 1′
CEILING & PARTITION	(S2S & CM)	3/8 1/2 5/8 3/4	3 4 5 6	5/16″ 7/16″ 9/16″ 11/16″	2-1/8″ 3-1/8″ 4-1/8″ 5-1/8″	4 ft. and longer in multiples of 1′
FACTORY & SHOP LUMBER	S2S	1 (4/4) 1-1/4 (5/4) 1-1/2 (6/4) 1-3/4 (7/4) 2 (8/4) 2-1/2 (10/4) 3 (12/4) 4 (16/4)	5 and wider (4″ wider in 4/4 No. 1 Shop & 4/4 No. 2 Shop)	25/32 (4/4) 1-5/32 (5/4) 1-13/32 (6/4) 1-10/32 (7/4) 1-13/16 (8/4) 2-3/8 (10/4) 2-3/4 (12/4) 3-3/4 (16/4)	Usually sold random width	4 ft. and longer in multiples of 1′

[a] S1S—Surfaced one side.
  S2S—Surfaced two sides.
  S4S—Surfaced four sides.
  S1S1E—Surfaced one side, one edge.
  S1S2E—Surfaced one side, two edges.
  CM—Center matched.
  D & M—Dressed and matched.
  T & G—Tongue and grooved.
  EV1S—Edge vee on one side.
  S1E—Surfaced one edge.

Figure 1-29 Lumber sizes.

Product	Description	Nominal Size		Dressed Dimensions		
		Thickness in Inches	Width in Inches	Thickness in Inches	Width in Inches	Lengths in Feet
SELECTS & COMMONS S-DRY	S1S, S2S, S4S, S1S1E, S1S2E	4/4 5/4 6/4 7/4 8/4 9/4 10/4 11/4 12/4 16/4	2 3 4 5 6 7 8 and wider	3/4 1-5/32 1-13/32 1-19/32 1-13/16 2-3/32 2-3/8 2-9/16 2-3/4 3-3/4	1-1/2 2-1/2 3-1/2 4-1/2 5-1/2 6-1/2 3/4 Off nominal	6 ft. and longer in multiples of 1'
FINISH & BOARDS S-DRY	S1S, S2S, S4S, S1S1E, S1S2E	3/8 1/2 5/8 3/4 1 1-1/4 1-1/2 1-3/4 2 2-1/2 3 3-1/2 4	2 3 4 5 6 7 8 and wider	5/16 7/16 9/16 5/8 3/4 1 1-1/4 1-3/8 1-1/2 2 2-1/2 3 3-1/2	1-1/2 2-1/2 3-1/2 4-1/2 5-1/2 6-1/2 3/4 Off nominal	3' and longer. In Superior grade, 3% of 3' and 4' and 7% of 5' and 6' are permitted. In Prime grade, 20% of 3' to 6' is permitted.
RUSTIC & DROP SIDING	(D & M) If 3/8" or 1/2" T & G specified, same over-all widths apply. (Shiplapped, 3/8-in. or 1/2-in. lap)	1	6 8 10 12	23/32	5-3/8 7-1/8 9-1/8 11-1/8	4 ft. and longer in multiples of I'
PANELING & SIDING	T&G or Shiplap	1	6 8 10 12	23/32	5-7/16 7-1/8 9-1/8 11-1/8	4 ft. and longer in multiples of I'
CEILING & PARTITION	T&G	5/8 1	4 6	9/16 23/32	3-3/8 5-3/8	4 ft. and longer in multiples of I'
BEVEL SIDING	Bevel or Bungalow Siding Western Red Cedar Bevel Siding available in 1/2", 5/8", 3/4" nominal thickness. Corresponding thick edge is 15/32", 9/16" and 3/4". Widths for 8" and wider, 1/2" off nominal.	1/2 3/4	4 5 6 8 10 12	15/32 butt, 3/16 tip 3/4 butt, 3/16 tip	3-1/2 4-1/2 5-1/2 7-1/4 9-1/4 11-1/4	3 ft. and longer in multiples of I' 3 ft. and longer in multiples of I'

Product	Description	Thickness in Inches	Width in Inches	Thickness Surfaced Dry	Thickness Surfaced Green	Width Surfaced Dry	Width Surfaced Green	Lengths in Feet
STRESS RATED BOARDS	S1S, S2S, S4S, S1S1E, S1S2E	1 1-1/4 1-1/2	2 3 4 5 6 7 8 and Wider	3/4 1 1-1/4	25/32 1-1/32 1-9/32	1-1/2 2-1/2 3-1/2 4-1/2 5-1/2 6-1/2 Off 3/4	1-9/16 2-9/16 3-9/16 4-5/8 5-5/8 6-5/8 Off 1/2	6 ft. and longer in multiples of I'

bMINIMUM ROUGH SIZES   Thicknesses and Widths Dry or Unseasoned All Lumber (S1E, S2E, S1S, S2S)
80% of the pieces in a shipment shall be at least 1/8" thicker than the standard surfaced size, the remaining 20% at least 3/32" thicker than the surfaced size. Widths shall be at least 1/8" wider than standard surfaced widths.

When specified to be full sawn, lumber may not be manufactured to a size less than the size specified.

Figure 1-29 *continued*

Figure 1–30 Lumber grade stamp.

Lumber less than 1 inch in thickness is figured as 1 inch. Stock thicker than 1 inch is figured by the nominal size. If the size is in fractions, such as $1\frac{1}{4}$, change it to an improper fraction ($\frac{5}{4}$), placing the numerator above the line and the denominator below.

For example: Find the board footage in a piece of lumber measuring $1\frac{1}{4}$ inch $\times$ 10 inches $\times$ 10 feet.

Solution:

$$\frac{1 \times 5 \times 10 \times 10}{4 \times 12} = \frac{500}{48} = 10^{5}/_{12} \text{ board feet}$$

To calculate the board footage by tables, simply locate the nominal size in inches and the actual length. The number of board feet can be read directly from the chart. A table used for rapid calculations is shown in *Fig. 1–31*.

The Essex board measure table found on the back of the framing square is an easy way to compute the board feet in a piece of lumber. For details, refer to the section on the framing square at the end of this chapter.

The moisture content of lumber is important. All lumber retains a certain amount of moisture after cutting. This moisture must be reduced to a satisfactory level in order for the lumber to be suitable for use commercially. Generally, a moisture content of 15 percent to 19 percent is considered average and acceptable for framing and exterior use. The moisture content is reduced by seasoning; that is, stacking the lumber to let it air-dry. This is a slow process. Kiln-drying produces the same results but in a fraction of the time. The lumber is placed in huge ovens where it is heated to drive out the moisture.

Some shrinkage will always take place in structures; therefore, the carpenter makes allowances if such shrinkage will affect the structure. Most shrinkage will occur during the first year after construction. The method of cutting logs into boards affects the shrinkage also. Wood that has been cut tangent to the growth rings of the log will undergo much greater shrinkage than wood that has been cut crosswise against the annual rings. There is very little shrinkage along the length of the log.

The proper storage of lumber is essential; it protects the lumber against fungi and insects, prevents defects caused by alternate wetting and drying, and helps maintain its appearance and dimensional stability.

Rapid Calculation of Board Measure.

Width	Thickness	Board feet
3"	1" or less	1/4 of the length
4"	1" or less	1/3 of the length
6"	1" or less	1/2 of the length
9"	1" or less	3/4 of the length
12"	1" or less	Same as the length
15"	1" or less	1 1/4 of the length

Board Feet

Nominal size (in.)	Actual length in feet								
	8	10	12	14	16	18	20	22	24
1 x 2		1 2/3	2	2 1/3	2 2/3	3	3 1/2	3 2/3	4
1 x 3		2 1/2	3	3 1/2	4	4 1/2	5	5 1/2	6
1 x 4	2 3/4	3 1/3	4	4 2/3	5 1/3	6	6 2/3	7 1/3	8
1 x 5		4 1/6	5	5 5/6	6 2/3	7 1/2	8 1/3	9 1/6	10
1 x 6	4	5	6	7	8	9	10	11	12
1 x 7		5 5/8	7	8 1/6	9 1/3	10 1/2	11 2/3	12 5/6	14
1 x 8	5 1/3	6 2/3	8	9 1/3	10 2/3	12	13 1/3	14 2/3	16
1 x 10	6 2/3	8 1/3	10	11 2/3	13 1/3	15	16 2/3	18 1/3	20
1 x 12	8	10	12	14	16	18	20	22	24
1 1/4 x 4		4 1/6	5	5 5/6	6 2/3	7 1/2	8 1/3	9 1/6	10
1 1/4 x 6		6 1/4	7 1/2	8 3/4	10	11 1/4	12 1/2	13 3/4	15
1 1/4 x 8		8 1/3	10	11 2/3	13 1/3	15	16 2/3	18 1/3	20
1 1/4 x 10		10 5/12	12 1/2	14 7/12	16 2/3	18 3/4	20 5/6	22 11/12	25
1 1/4 x 12		12 1/2	15	17 1/2	20	22 1/2	25	27 1/2	30
1 1/2 x 4	4	5	6	7	8	9	10	11	12
1 1/2 x 6	6	7 1/2	9	10 1/2	12	13 1/2	15	16 1/2	18
1 1/2 x 8	8	10	12	14	16	18	20	22	24
1 1/2 x 10	10	12 1/2	15	17 1/2	20	22 1/2	25	27 1/2	30
1 1/2 x 12	12	15	18	21	24	27	30	33	36
2 x 4	5 1/3	6 2/3	8	9 1/3	10 1/3	12	13 1/3	14 2/3	16
2 x 6	8	10	12	14	16	18	20	22	24
2 x 8	10 2/3	13 1/3	16	18 2/3	21 1/3	24	26 2/3	29 1/3	32
2 x 10	13 1/3	16 2/3	20	23 1/3	26 2/3	30	33 1/3	36 2/3	40
2 x 12	16	20	24	28	32	36	40	44	48
3 x 6	12	15	18	21	24	27	30	33	36
3 x 8	16	20	24	28	32	36	40	44	48
3 x 10	20	25	30	35	40	45	50	55	60
3 x 12	24	30	36	42	48	54	60	66	72
4 x 4	10 2/3	13 1/3	16	18 2/3	21 1/3	24	26 2/3	29 1/3	32
4 x 6	16	20	24	28	32	36	40	44	48
4 x 8	21 1/3	26 2/3	32	37 1/3	42 2/3	48	53 1/3	58 2/3	64
4 x 10	26 2/3	33 1/3	40	46 2/3	53 1/3	60	66 2/3	73 1/3	80
4 x 12	32	40	48	56	64	72	80	88	96

Figure 1–31 Board measure table.

RULES FOR STORING LUMBER

- The lumber should be unloaded in a dry place, not in water or a muddy area.
- Untreated lumber should not be in direct contact with the ground; it should be elevated on stringers to allow air circulation.
- Lumber stored in an open yard should be covered with a porous material that will allow moisture to escape. Polyethylene covers are not suitable because they are nonporous.

- Additional protection is provided by having lumber delivered in paper-wrapped packages that have been treated with a weather-protective coating.
- Framing lumber should be enclosed as soon as possible for protection against the elements.
- Lumber should be used in the order in which it is received.
- Exterior siding and finish should be stored in a closed, unheated area.
- Interior flooring and millwork should be stored in a closed area where heat can be applied during damp weather to maintain the desired moisture content.

Fire Safety.   Since lumber and other flammable materials are used regularly in carpentry and other construction work, everyone on the jobsite should observe the fundamental rules for fire safety and fire prevention. Each worker should have a basic understanding of the various types of fire and a working knowledge of how to put them out. If a fire occurs, the first thing you should do is to alert all workers, then turn in a fire alarm. If the fire is small and easily controllable, you should of course try to put it out yourself. If it looks like a bad or dangerous one, it is advisable to leave it for the fire department to handle.

Fires are classified in four distinct types, designated as class A, B, C, or D. Class A fires occur in wood, cloth, paper, and similar items. These can be extinguished with water or with a carbon dioxide extinguisher. The fire extinguisher for this type of fire is marked A, in a green triangle.

Class B fires are those involving flammable liquids such as oil, gasoline, grease paint, thinners, and so forth. These cannot be extinguished by water. Such fires must be put out by eliminating the air that feeds them. The agent used in these extinguishers smothers the fire with either foam or carbon dioxide (CO_2). The extinguisher is marked B, in a red square.

Class C fires occur in electrical equipment. The extinguishing agent must be non–electricity-conducting and, as in class B fires, a smothering effect is required. Carbon dioxide may be used. The extinguisher is marked C, in a blue circle.

Class D fires occur in combustible metals such as powdered aluminum, magnesium, potassium, zinc, titanium, and lithium. These are put out with a dry powdered compound. The extinguisher symbol is D, in a yellow star.

Some extinguishers are designed to extinguish more than one type of fire. This is noted by the markings on the label. The extinguisher shown in *Fig. 1–32* is suitable for class A, B, and C fires.

Extinguishers should be prominently displayed and all workers on the job should be familiar with their location and operation.

Figure 1–32 Markings indicate that this extinguisher is suitable for Class A, B, and C types of fire.

Fasteners

The method of fastening the various wood members of a frame house is important. Among the fasteners used, nails are in the majority; however, other fasteners such as screws, hangers, and reinforcing plates are also utilized.

The fasteners have the primary function of providing rigidity and strength to the joined parts. The proper choice and use of fasteners will enable the structure to withstand high winds and other forces.

Nails. The most common fastener used in frame construction is the nail. It is made in many sizes and types to fill every need. It may be bright, blued, cement-coated, galvanized, or painted. Various head and point shapes are available. Heads may be flat, round, oval, slotted, or countersunk, and shanks may be smooth or threaded, as shown in *Fig. 1–33.*

The holding power of a nail depends on the size and type of nail as well as the material being nailed. The amount of penetration also influences the holding power of the nail.

The gauge of a nail refers to the wire size used in its manufacture. The lower the gauge number, the greater the diameter of the nail.

The length of nails is expressed in "d" or penny size. Each increase in penny size is equal to ¼ inch. A 2d or 2-penny nail is 1 inch long. A 4d nail is 1½ inches long. The various sizes of common nails are shown in *Fig. 1–34.*

Some of the most frequently used nails are briefly described here.

Annular thread nails have good holding power but are not good for lateral loads.

Box nails have a thinner shank than common nails and are used primarily for toenailing because the thinner shank is less likely to split the wood.

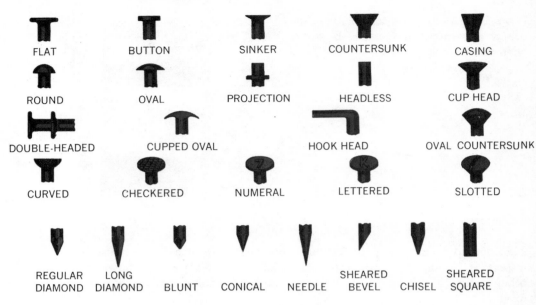

Figure 1–33 Head and point styles of various nails.

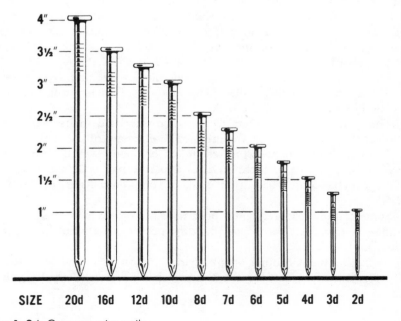

Figure 1–34 Common wire nails.

Casing nails are made of finer wire than common nails. The casing nails are similar to finishing nails, but they have a conical head, which gives them better holding power.

Cement-coated nails have a rosin coating which increases their holding power.

Common nails are the most widely used in construction. They have a heavy cross section and flat heads. They range in size from 2d to 60d (*Fig. 1–35*).

Figure 1–35 Common bright diamond-point nail.

Double-headed nails are used primarily for formwork. The double head permits easy removal.

Finishing nails are made of finer wire than common nails and have a smaller head, which permits sinking.

Hot-dipped galvanized nails are steel nails coated with zinc in molten form. They are corrosion- and stain-resistant.

Knurled nails are designed for use in masonry. They are extremely hard and have a tendency to chip when struck with a hammer. *They must not be used without safety goggles.*

Painted nails are used mostly for matching panels in finish work.

Screw-thread nails have excellent holding power for all loads. However, they are not suitable for end-grain nailing.

Shingle nails are corrosion-resistant and are available in colors to match stained shingles.

Spiral-thread nails have excellent holding power and can be used where end-grain nailing is required.

Fig. 1–36 shows other special-purpose nails.

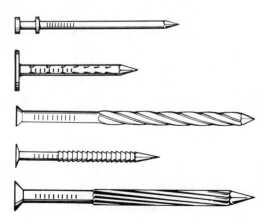

Figure 1–36 Special-purpose nails. From top down: duplex head, roofing, flooring, drywall, and masonry.

Generally, nails should be three times longer than the total thickness of the material being fastened, so that two-thirds of the nail is anchored into the second piece. For greater holding power, nails should be angled slightly toward each other. Common nails are driven flush with the work. Finishing and casing nails are driven with the head protruding slightly. A nail set is used to sink the heads below the surface.

When it is necessary to nail close to the end of a piece of wood, the nail should be driven into a prebored hole slightly smaller in diameter than the nail shank. This will decrease the possibility of the nail splitting the wood, and will increase its holding power. Otherwise, use a blunt-pointed nail or blunt the point of a standard nail before using it.

Paneling, trim, and siding should be nailed to the framing members, not to the sheathing alone. This is especially important when you are using composition sheathing or insulation board, because these have lower nail-holding power than wood sheathing.

When siding or paneling is butt-jointed, the joint should be made over a stud and the nails should be driven at a 45-degree angle through prebored holes into the joint. When you are nailing exterior siding and trim where the wood will be exposed to moisture, be sure to use only noncorrosive nails such as the hot-dipped galvanized type. For any construction work, be sure to select the correct size and type of nails for the job at hand and to follow the recommended nailing procedure of the manufacturer.

Screws.   Although screws have greater holding power than nails, their use in construction is usually limited to the installation of hardware and some interior work. They are made in many head styles, the most common being the flathead, roundhead, and oval head. Heads may be slotted or cross-slotted (Phillips) (*Fig. 1–37*). Screws are identified by their length, gauge, and head style. Lengths vary from ¼ to 6 inches in steel and ¼ to 3½ inches in brass. Gauge diameters vary from 0 to 24. The larger the gauge number, the greater the diameter. Each larger gauge number represents an increment in diameter of .013 inch. The usual wood screw is plain steel; however, stainless, brass, and plated screws are also available.

Figure 1–37 Slotted and cross-slotted screw heads.

Screws require shank and pilot holes in order to be driven properly. *Fig. 1–38* shows the proper method of installing a flathead screw. Roundhead screws are not countersunk. If the screw is to penetrate hardwood, a small amount of soap or wax on the thread will make driving easier.

When you are preboring, note that the diameter of the shank hole

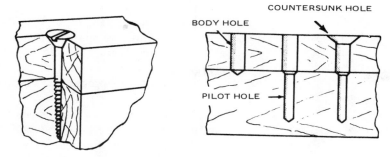

Figure 1-38 Proper method of sinking a screw.

should be about ⅞ the diameter of the screw shank and the hole for the threaded portion should be about ⅞ the diameter of the screw at the root of the threads. Special tools are available which drill all three holes in one operation: the pilot hole, the shank clearance hole, and the countersink for the screw head.

Lag screws have a conical point and square or hexagonal head. They have great holding power and are used mainly for heavy timber construction. Sizes range from 1 to 16 inches, in diameters from ¼ inch to 1 inch.

Sheet-metal screws are similar in appearance to wood screws but they have a full thread and a flat root diameter (wood-screw threads are V-shaped) (see *Fig. 1-39*). Designed primarily for sheet metal, they may also be used for joining sheet metal to wood.

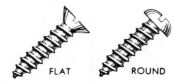 

Figure 1-39 Various sheet-metal screws.

Bolts and lag screws are used for heavy construction work. They are sometimes used for formwork. They have great holding power and are available in sizes ranging from 2 to 16 inches.

Metal framing anchors are widely used in structural framework. They eliminate the need for toenailing, drilling, and notching and they strengthen connections between framing members. The illustration, *Fig. 1-40*, shows some of the most common types used in residential construction. One popular anchor, used when it is desirable to keep the joists and headers or beams in one level, is the sling-type joist hanger shown in *Fig. 1-41*. This type straddles the joist and the beam as well.

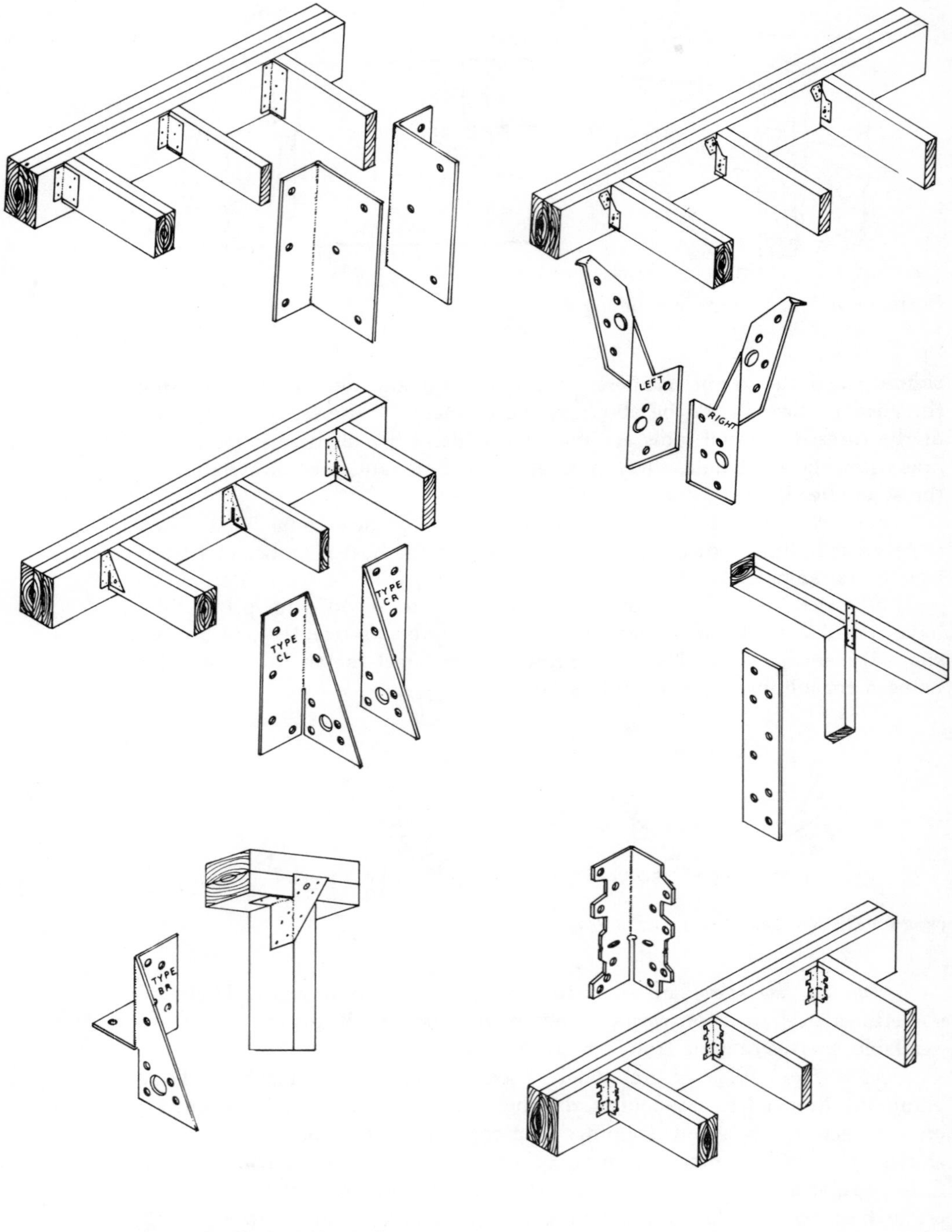

Figure 1–40 Framing anchors.

Figure 1–41 Sling-type joist hanger.

Powder-Driven Fasteners. An explosive charge is used in a specially designed tool to sink various fasteners into a wide variety of materials. The depth of penetration is controlled by the use of color-coded cartridges. The tools resemble guns and operate in much the same manner (*Fig. 1–42*). When the trigger is depressed, a firing pin strikes the rear end of a cartridge, causing it to explode. The fastener, acting as a projectile, penetrates the target, which may be a stud, concrete wall, or similar object. The fasteners are very powerful and will even penetrate a 1-inch-thick steel plate.

Great care must be exercised when you use these tools. In some localities a special permit is required for their use.

Figure 1–42 A powder-driven fastener in use.

Plans and Specifications

The carpenter uses a set of plans or drawings to build a structure. The plans contain graphic and written information which indicates such things as the type of framing, roof slope, type of finish materials, and all other necessary data needed by the people working on the structure. In effect, they tell what, where, and how to use the various materials of construction. They are the main source of information for those responsible for the actual work.

A complete set of plans includes the site plan, elevations, foundation plans, floor plans, detail drawings, and specifications.

Site Plan. The site or plot plan shows property lines and locations, contours and profiles, building lines, existing structures, approaches, grades, and utilities. A typical site plan is shown in *Fig. 1–43*.

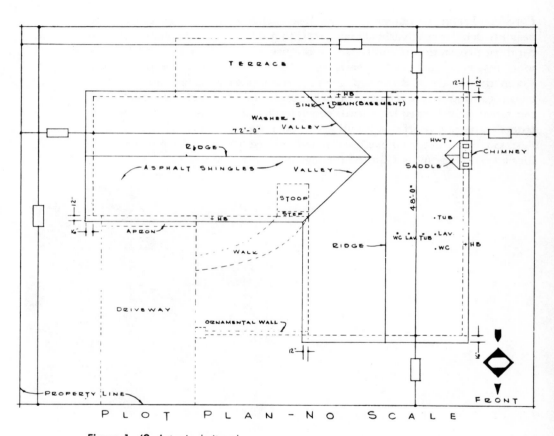

Figure 1–43 A typical site plan.

Elevations. The elevations show the front, rear, and side views of the structure (*Fig. 1–44*). Construction materials may be indicated on the elevation. Section views show how the structure looks when cut vertically by a cutting plane, and may be shown on the elevation sheet or separately. Wall-section views extend from the base of the foundation to the roof.

Floor Plans. Floor and foundation plans are horizontal cross sections through a building. They show all openings, regardless of their height from the floor. They show the size and shape of the building; layout, size, and shape of rooms; length and thickness of walls; and type of materials for construction. They also include the size and type of doors and windows and location of stairs, together with the number of risers. Types and locations of utilities are also shown (*Fig. 1–45*).

Details.   Features that cannot be shown clearly in the scale drawings appear on detail drawings, and may include such items as doors, windows, cornices, fireplaces, and built-ins (*Fig. 1–46*).

Specifications.   The specifications consist of written information supplementing the working drawings. Such matters as legal responsibility, guarantees of quality and performance, permits, and inspections are clearly spelled out. The specifications describe the materials required by the various trades and the methods of application. The work of each trade is described in detail, with a breakdown of the kinds and grade of materials to be used. Nothing is left to guesswork. Here is a typical list of specification headings:

1. Agreements and conditions
2. Excavating, filling, and grading
3. Foundation
4. Rough carpentry
5. Roofing
6. Finish carpentry
7. Lath and plaster
8. Cabinets
9. Stairs
10. Plumbing
11. Heating
12. Electrical
13. Appliances
14. Landscaping

Each of the headings in the list is fully detailed, enabling builders, tradesmen, and suppliers to calculate the amounts and cost of the necessary materials.

Architectural Drawings.   The carpenter must be thoroughly familiar with the working drawings, or blueprints, as they are sometimes called. He must be able to look at the drawings and visualize the structure clearly in his mind. He must understand the symbols, abbreviations, and notations. Typical architectural symbols and abbreviations are shown in *Fig. 1–47*. Although he does not perform the functions of other trades, the carpenter must have some knowledge of them in order to carry out his work successfully. The carpenter does not install heating ducts, for example, but he must be able to identify them on the drawings so that he can make allowances for them in the framing. It would be difficult for the carpenter to put up even a simple partition if he did not know how to read dimensions.

Figure 1-44 Elevation views of a single-family dwelling.

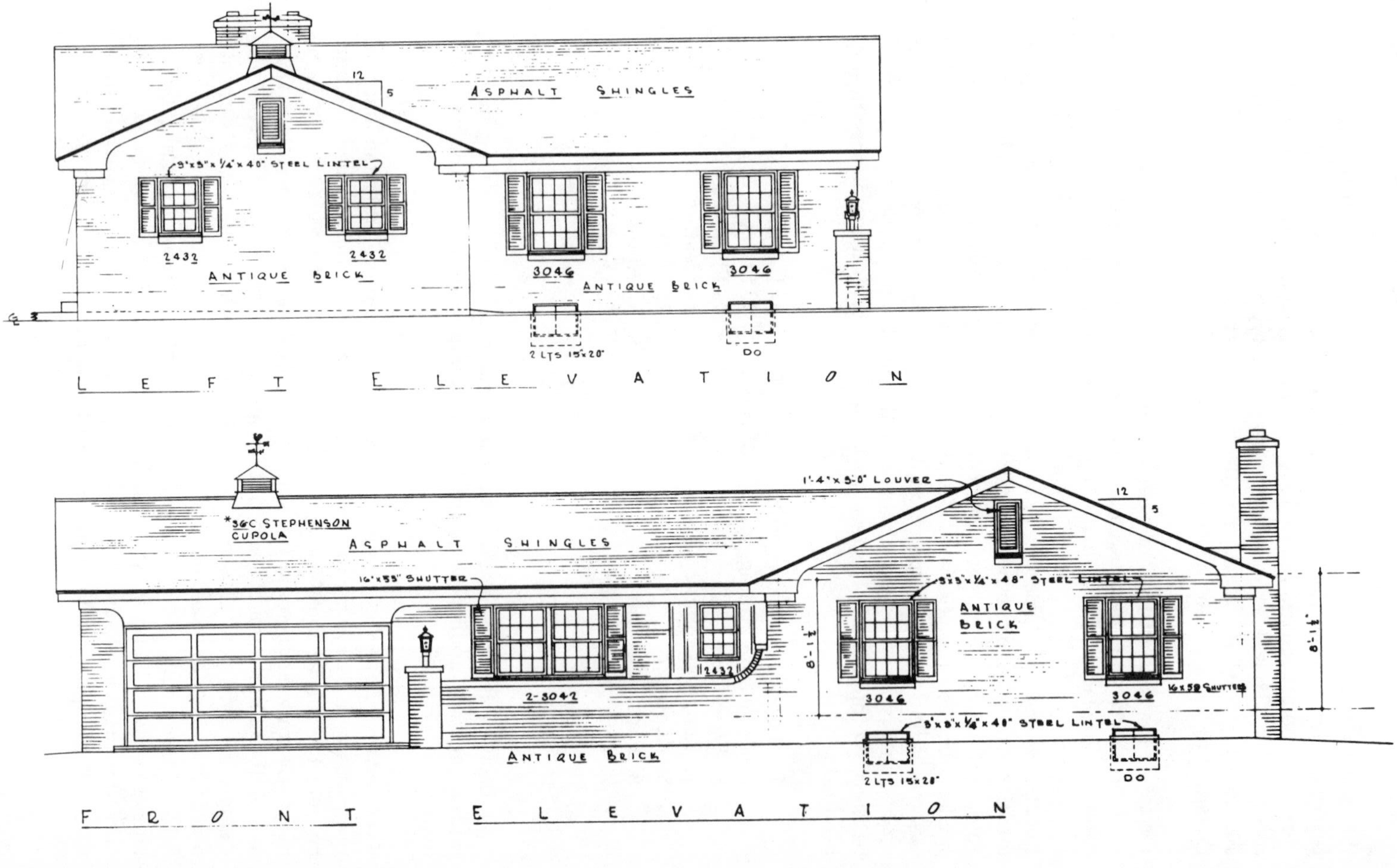

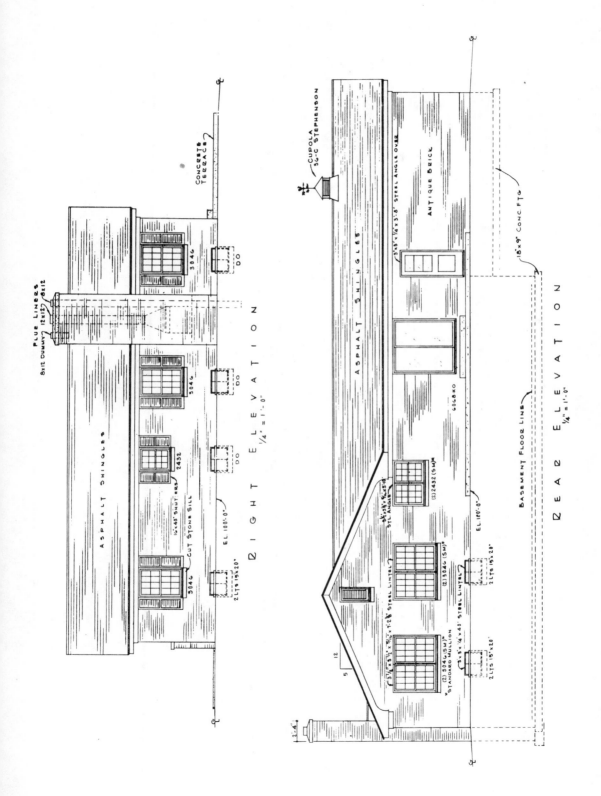

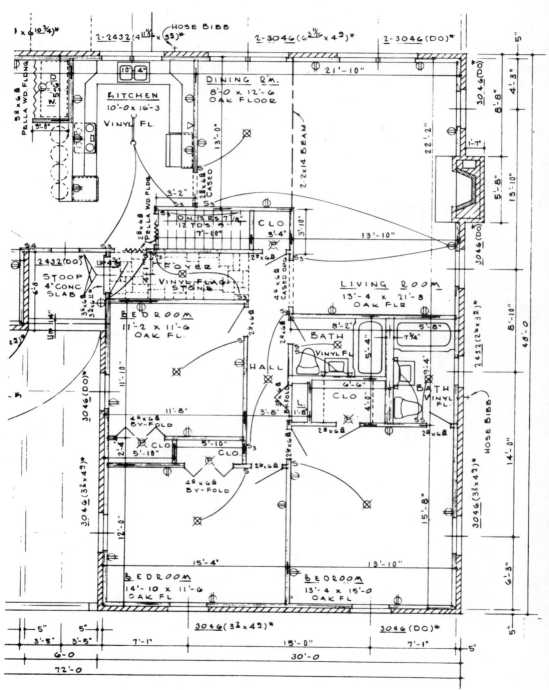

Figure 1–45 Partial section of a floor plan.

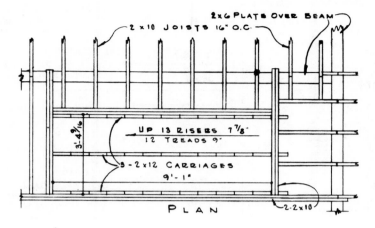

PLAN

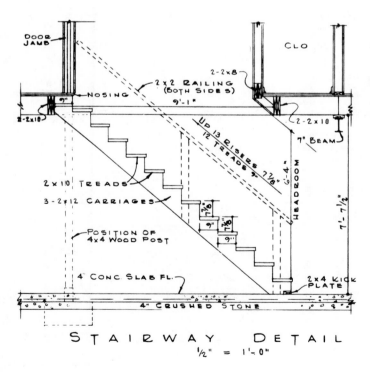

STAIRWAY   DETAIL
½" = 1'-0"

Figure 1–46 Stairway detail.

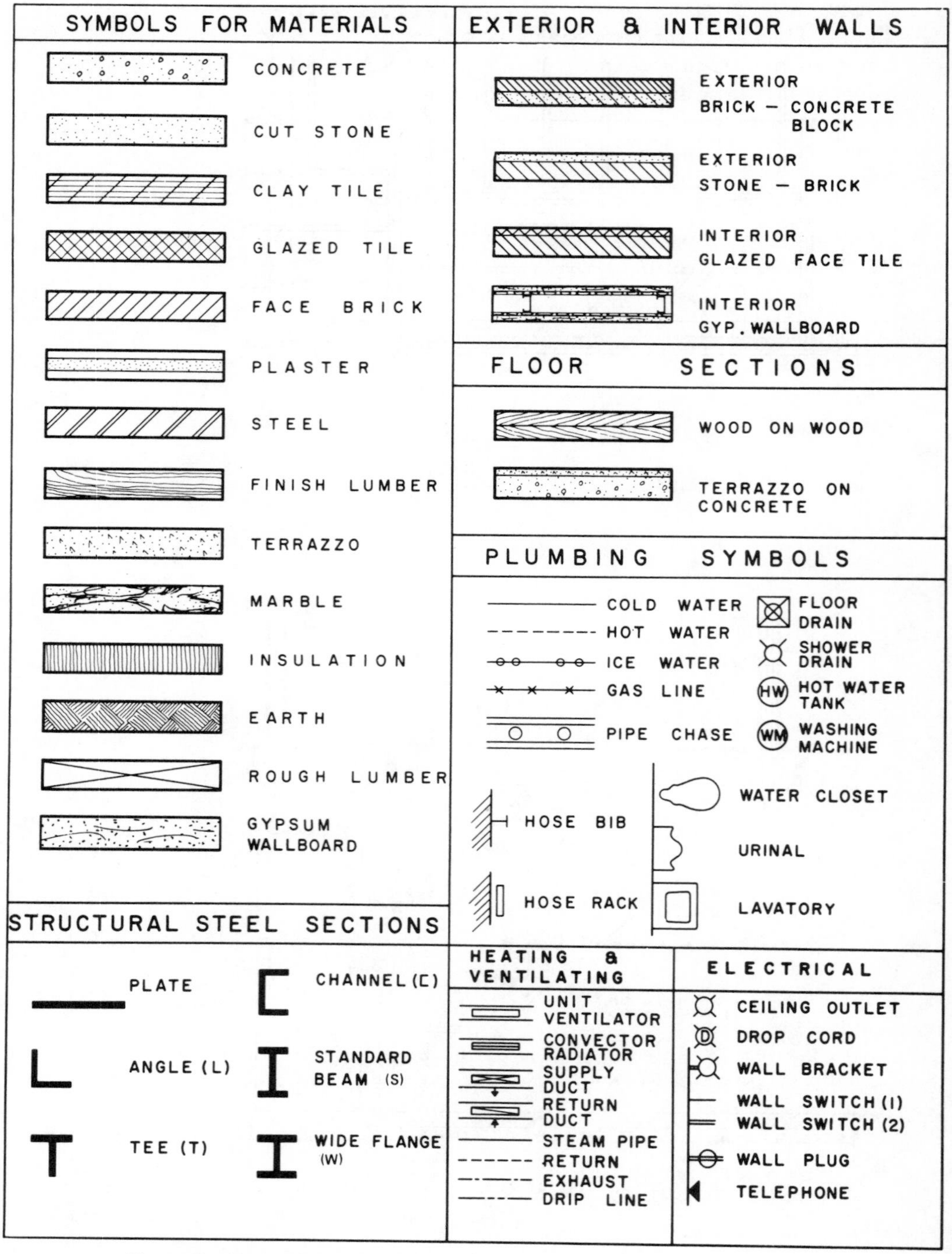

SYMBOLS FOR MATERIALS	EXTERIOR & INTERIOR WALLS
CONCRETE	EXTERIOR BRICK — CONCRETE BLOCK
CUT STONE	EXTERIOR STONE — BRICK
CLAY TILE	INTERIOR GLAZED FACE TILE
GLAZED TILE	INTERIOR GYP. WALLBOARD
FACE BRICK	**FLOOR SECTIONS**
PLASTER	WOOD ON WOOD
STEEL	TERRAZZO ON CONCRETE
FINISH LUMBER	**PLUMBING SYMBOLS**
TERRAZZO	COLD WATER — FLOOR DRAIN
MARBLE	HOT WATER — SHOWER DRAIN
INSULATION	ICE WATER — HOT WATER TANK (HW)
EARTH	GAS LINE — WASHING MACHINE (WM)
ROUGH LUMBER	PIPE CHASE
GYPSUM WALLBOARD	HOSE BIB — WATER CLOSET
	HOSE RACK — URINAL
	LAVATORY

STRUCTURAL STEEL SECTIONS

PLATE	CHANNEL (C)
ANGLE (L)	STANDARD BEAM (S)
TEE (T)	WIDE FLANGE (W)

HEATING & VENTILATING	ELECTRICAL
UNIT VENTILATOR	CEILING OUTLET
CONVECTOR RADIATOR	DROP CORD
SUPPLY DUCT	WALL BRACKET
RETURN DUCT	WALL SWITCH (1)
STEAM PIPE	WALL SWITCH (2)
RETURN	WALL PLUG
EXHAUST	TELEPHONE
DRIP LINE	

Figure 1–47 Architectural symbols.

Architectural plans are drawn to scale; size is reduced but proportions are retained. Thus it is usually possible to determine a size even when a dimension is missing. A careful study of the drawings may show the missing dimension in another view. Often it is possible to find the missing dimension by adding the given measurements, then subtracting them from the overall dimensions. Accuracy is important.

Working drawings are usually made to a scale of ⅛ inch or ¼ inch per foot. This means that for a ¼-inch scale, every foot of measurement in a structure will be shown as ¼ inch on the drawing (*Fig. 1–48*). When a drawing is not to scale, this fact is noted on the drawing, usually with a phrase such as "NOT TO SCALE" or "DO NOT SCALE."

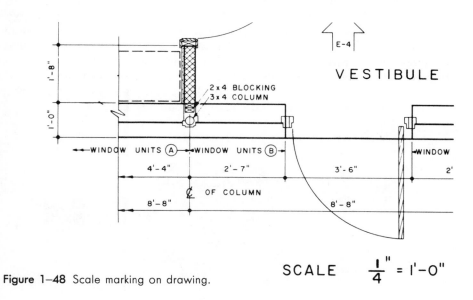

Figure 1–48 Scale marking on drawing.

SCALE $\frac{1}{4}'' = 1'-0''$

Building Code

The building code establishes regulations for the construction of buildings. Its purpose is to safeguard the health, safety, and welfare of the public. The code contains standards of performance and specifications for materials, structural requirements, fire resistance, ventilation, room sizes, lighting, and other considerations. Most communities write their own codes, while others may adopt state codes. Unfortunately, most codes are not up to date. New and better materials and methods are constantly being developed, but regardless of their merit, if they are not specifically mentioned in the code, their use is not permitted. This usually results in increased costs.

Model Codes. Codes are constantly being revised, but this is both costly and difficult. One step in the right direction is the adoption of model codes. Four major organizations have developed such suggested building codes, which have been accepted by many communities. The organizations that prepare these codes also continuously update them to include new materials and methods.

Leveling Instruments

Builder's Level and Transit Level. In order to lay out grade levels, building lines, and other work involving accurate measurements, the modern carpenter makes use of two surveying instruments, the builder's level and the transit level. Both are optical instruments which simplify the many measuring tasks involved in residential construction. The builder's level consists of a telescope affixed to a horizontal circle (*Fig. 1–49*). It contains leveling screws, a scale, index pointer, and a spirit level or bubble. The telescope can be turned sideways for measuring horizontal angles. It can be rotated over a 360-degree scale so that angles can be measured and read. The builder's level is used primarily for determining elevations, measuring angles, running straight lines, and laying out building lines.

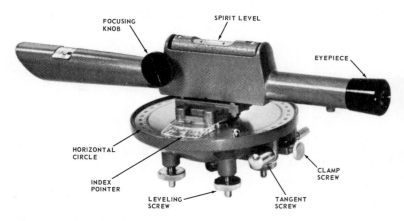

Figure 1–49 Builder's level.

The transit level (*Fig. 1–50*) is basically the same as the builder's level, but in addition to performing the functions of the builder's level, it can also measure vertical angles. It is useful for setting stakes and for plumbing walls. (Note: In carpentry "plumb" always refers to the vertical plane and "level" refers to the horizontal plane.)

Both instruments are mounted on sturdy tripods. Some have legs that can be adjusted for use on sloping grounds (*Fig. 1–51*). Other items used in conjunction with the levels include measuring tapes and the leveling rod.

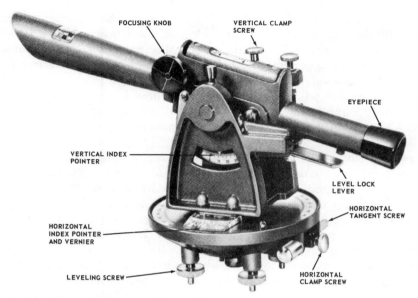

FOCUSING KNOB

VERTICAL CLAMP
SCREW

EYEPIECE

VERTICAL INDEX
POINTER

LEVEL LOCK
LEVER

HORIZONTAL
TANGENT SCREW

HORIZONTAL
INDEX POINTER
AND VERNIER

LEVELING SCREW

HORIZONTAL
CLAMP SCREW

Figure 1–50 Transit level.

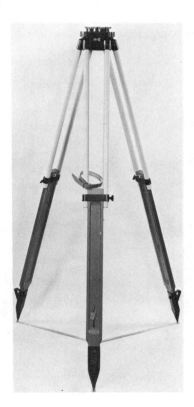

Figure 1–51 Level tripod with adjustable legs.

Because of their extreme accuracy, laser levels and transits are frequently used in construction sites. Most of these instruments emit a low-level laser beam which sweeps around like a beacon. This type is not harmful to your eyesight. However, some laser instruments emit a powerful highly focused beam that can cause serious eye injury or even blindness. Therefore, when you are on sites where lasers are being used, be extremely careful and never look directly into the laser beam. Wherever such equipment is in use, appropriate warning signs should be posted.

Leveling Rod.   The leveling rod is useful for sighting over long distances. It is used to measure vertical differences in elevation. Basically, the rod consists of an upright graduated in inches or hundredths of a foot.

The level operator can use a self-reading rod and read the figures through his telescope. For small home-construction projects, where short distances are involved, a folding rule may be used instead of the rod. When long distances are involved, a rod with target may be used. The target is usually operated and read by a rodman. One type of target with a vernier scale is shown in *Fig. 1–52*. The rod operator usually carries a spirit level, which he uses to make sure that he is holding the rod straight up and down. Elevations are usually given in feet and hundredths of a foot instead of feet and inches. The table shown in *Fig. 1–53* can be used to convert inches to hundredths of a foot.

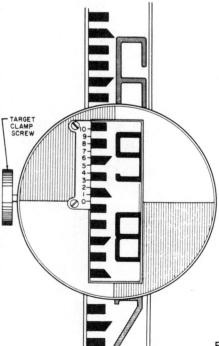

TARGET
CLAMP
SCREW

Figure 1–52 Leveling rod with vernier scale.

In.	Ft.	In.	Ft.	In.	Ft.
1	.08	5	.42	9	.75
2	.17	6	.50	10	.83
3	.25	7	.58	11	.92
4	.33	8	.67	12	1.00

Figure 1–53 Table for converting inches to hundredths of a foot.

Tapes are useful when layouts involve long distances. They are available in various lengths and graduations. Several types are shown in *Fig. 1–54*. A dirty line should be wiped clean before rewinding to prevent a collection of dirt in the case.

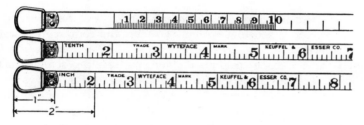

Figure 1–54 Various steel tape graduations. *Top:* Metric. *Center:* Feet and tenths of a foot. *Bottom:* Feet and inches.

The telescopes of the level and transit level are similar. They are designed to magnify the image and appear to bring it up close. A 15-power telescope will make the image appear 15 times closer than when viewed by the unaided eye. Intersecting cross-lines mark the line of sight and permit the target or object to be centered accurately. The line of sight through a telescope is continuous and perfectly straight. Therefore, if the telescope is level, any point along the line of sight is exactly level with any other point.

The leveling vial or bubble is in alignment with the telescope. Thus, when the bubble of the instrument is level, it can be assumed that the telescope will also be level. Leveling screws are rotated to either raise or lower the base of the instrument until the bubble remains level, regardless of the position of the telescope.

Setting Up the Instrument. Mount the instrument to the tripod head, screwing it down carefully. Place the tripod legs firmly on the ground and spread the legs about 3 feet apart. If the ground slopes, place the third leg uphill, as in *Fig. 1–55*.

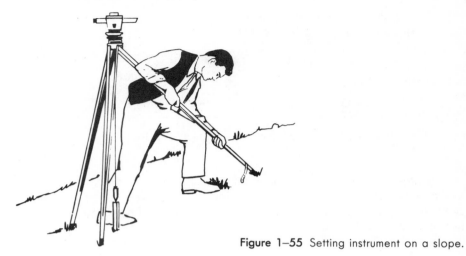

Figure 1–55 Setting instrument on a slope.

Some instruments have four leveling screws, but some of the newer ones use a two-screw leveling system. To level the four-screw instrument, proceed as follows:

Swing the telescope to align it over two leveling screws. Grasp the two screws and, while observing the bubble, rotate the screws either toward or away from each other (*Fig. 1–56*). Adjust the screws to center the bubble. Swing the telescope 90 degrees over the second set of leveling screws, and again adjust the screws to center the bubble. After centering the bubble the second time, swing back to the first position and adjust again. Refine the level by repeating the procedure as many times as necessary to center the bubble in both positions. When the telescope is perfectly level, you will be able to swing the telescope in all directions without a change in the bubble.

Figure 1–56 Leveling a four-screw instrument.

In the two-screw system, the screws press against a shifting plate, causing the instrument to pivot on one fixed leveling screw. To level such instruments, position the telescope as in *Fig. 1–57*. Rotate the screw marked B to center the bubble. Note that the bubble moves in the direction of rotation. Now position the telescope as in *Fig. 1–58* and repeat the procedure using the screw marked C. Refine the leveling by repeating both steps.

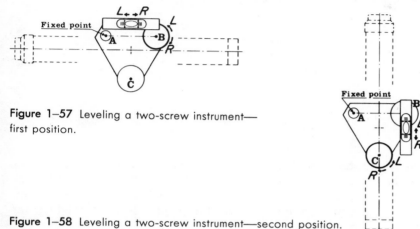

Figure 1–57 Leveling a two-screw instrument—first position.

Figure 1–58 Leveling a two-screw instrument—second position.

Setting Up Over a Point.   To set the instrument up over a stake point, a plumb bob is required. Pass the plumb-bob cord over a plumb-bob hook assembly under the tripod head (*Fig. 1–59*). Then tie the cord with a slip-knot (*Fig. 1–60*). This will permit the bob to be raised or lowered when you slide the knot along the cord.

PLUMB BOB HOOK

Figure 1–59 Plumb-bob hook is attached to shifting clamp screw.

Figure 1–60 Knotting the plumb-bob cord.

Plumb Bob
Hook

The tripod should be firmly set into the ground and positioned so the bob is within ¼ inch of the point and about ⅛ inch above (*Fig. 1–61*).

Loosen the shifting screw beneath the tripod head and shift the instrument to bring the bob directly over the point (*Fig. 1–62*). Tighten the screw and level the instrument.

Figure 1–61 Setting up near a point.   **Figure 1–62** Final adjustment over a point.

Sighting the Instrument. Aim the telescope at the object and rough-sight by using the barrel or V-groove at the top of the telescope. Then, viewing through the eyepiece, focus the instrument. The crosslines will appear sharp against the object. *Fig. 1–63* shows a view of the target as seen through the telescope. The builder's level can only be moved in a horizontal plane, but the transit level can be moved vertically to plumb columns, align points, or to establish slopes (*Fig. 1–64*).

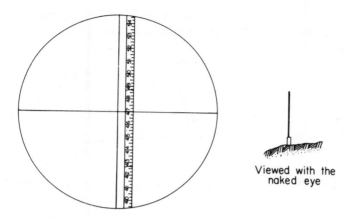

Viewed with the
naked eye

Figure 1–63 Target as sighted through the telescope.

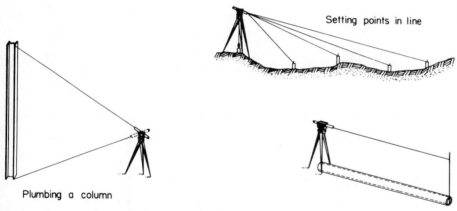

Setting points in line

Plumbing a column

Figure 1–64 Various applications of the transit level.   Establishing a slope

Leveling.   The level or transit can be used to find the difference in grade between two or more points. This procedure is called leveling. A simple example is shown in *Fig. 1–65*, where the difference in elevation between point A and point B is to be measured. The instrument (builder's level or transit level) is set up and leveled at a convenient location. A reading is taken with the rod at point A and another with the rod at point B. The difference in grade is the difference in the readings of the two points.

The instrument can also be set up between two points. Again, a reading is made first with the rod at point A, then with the rod shifted to point

B (*Fig. 1–66*). When the grade is too steep to permit sighting from one point, the instrument may have to be moved several times and the readings computed as shown in *Fig. 1–67*.

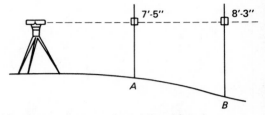

Figure 1–65 The difference in elevation between points A and B is 10 inches.

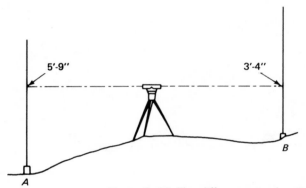

Figure 1–66 The difference in elevation between points A and B is 29 inches.

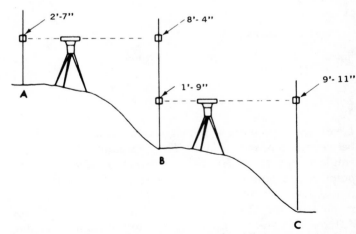

Figure 1–67 The instrument may have to be moved when the difference in elevation is too great to be taken from one position.

To set grade stakes, place the instrument centrally within the building lines, as shown in *Fig. 1–68*. Drive stakes at each corner so they are equally set below the line of sight. Use a rod or rule to establish the proper depth for the first stake.

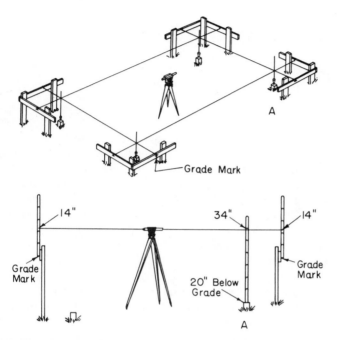

Figure 1–68 Setting a grade mark.

Setting Marks in a Line.   This can be done using either a builder's level or transit level. To set marks in a straight line between points A and B (*Fig. 1–69*), using the builder's level, set the instrument over point A, then sight a plumb-bob cord held over point B. Lock the instrument, then align the bob successively along the intermediate stakes, marking each point.

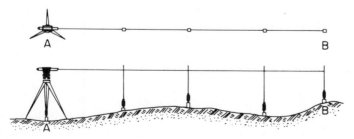

Figure 1–69 Setting marks in a line with the builder's level.

Using the transit level, set it up over point A, then tilt the telescope downward, sighting onto point B. Lock the horizontal clamp screw, then sight and mark the intermediate stakes (*Fig. 1–70*).

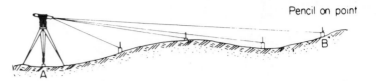

Pencil on point

Figure 1–70 Setting marks in a line with the transit level.

Plumbing a Column. Set the transit level conveniently so you can sight the column along its top and bottom. Level the instrument, then sight the base of the column, placing the vertical crosswire in line with the edge of the column. Lock the horizontal clamp, then elevate the telescope to the top of the column. The vertical wire should remain on the edge of the column. If it is off, have the column adjusted until the two coincide. Select another position for the instrument, 90 degrees from the first, and repeat the procedure. Make the necessary adjustments. When the readings hold, the column will be plumb.

Reading the Vernier. The horizontal circle of the instrument (level or transit level) and its vernier must be used to read a horizontal angle. The horizontal circle is a complete circle, marked in degrees in four quadrants (*Fig. 1–71*). The circle contains 360 degrees and each degree has 60 minutes. The circle moves independently of the telescope. It can be rotated by hand, but it does not move when the telescope is turned. The vernier scale moves around the inside of the horizontal scale as the telescope is turned right or left. The zero mark on the vernier is the index and the position of the index gives the angle reading of the horizontal circle. The graduations on either side of the vernier index permit measurements in minutes as well as degrees. An enlarged view of the vernier index (*Fig. 1–72*) reveals that it contains twelve graduations from 0 to 60 on either side of the zero. The left-hand set of graduations is used to read clockwise angles; the right-hand set is used to read counterclockwise angles. Each graduation on the vernier is $1/12$ narrower than the graduations on the horizontal circle; since each marking on the horizontal circle represents one degree, or 60 minutes, each marking on the vernier represents $1/12$ of one degree, or 5 minutes.

If the index of the vernier is aligned with the zero of the horizontal circle, no other mark on the vernier will coincide with the marks on the horizontal circle. If the instrument is turned clockwise until the first graduation to the left of the vernier index coincides with the one-degree mark to the left of the zero on the horizontal circle, the telescope will have been

Figure 1–71 The horizontal circle is used to set and measure horizontal angles.

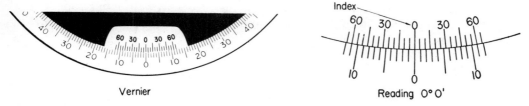

Vernier

Index

Reading 0° 0'

Figure 1–72 Details for horizontal circle and vernier.

turned on an angle of 5 minutes clockwise. The angle reading will be 0°5′ clockwise (*Fig. 1–73*). If the telescope is moved a little more, so that the second mark of the vernier aligns with the second mark of the horizontal circle, the reading will be 10 minutes (each vernier mark being 5 minutes), and so on.

To read an angle to the nearest 5 minutes, read the last degree mark on the horizontal circle that was passed by the vernier index. Then find the 5-minute mark on the vernier that most nearly lines up with a mark on the horizontal circle. Read the minutes from this mark on the vernier. Remem-

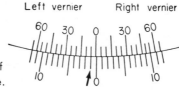

Left vernier   Right vernier

Figure 1–73 When the first graduations to the left of the zero coincide, the angle measures 0°5′ clockwise.

ber that when the degree numbers increase to the left, clockwise angles are read and the left vernier is used. The opposite is true when the degree numbers increase to the right. *Fig. 1–74* shows a reading of 23°50′ counterclockwise.

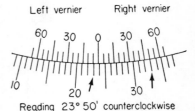

Reading 23° 50′ counterclockwise

Figure 1–74 The right vernier is used to read counterclockwise angles.

Measuring Horizontal Angles.   The builder's level or transit level may be used to lay out or measure horizontal angles. To measure the angle BAC (*Fig. 1–75*), set the instrument over point A, then, using the vertical crossline, sight along line AB. Set the zero mark of the horizontal circle to align with the index mark of the vernier. Now swing the telescope to align with point C. Read the angle on the horizontal circle against the vernier index.

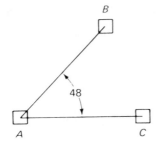

Figure 1–75 Using the level to measure a horizontal angle.

Measuring Vertical Angles.   The transit level must be used to measure vertical angles. When the angle measured is above a horizontal line, it is said to be a plus vertical angle. When below a horizontal line, it is said to be a minus vertical angle. Vertical angles are read using the vertical arc and vernier in the same manner as for horizontal angles. The right-hand vernier is used for plus angles; the left vernier for minus angles. An example of measuring a vertical angle is shown in *Fig. 1–76*. The angle is read from the vertical arc and vernier.

Laying Out a Right Angle.   The right-angle layout is used when establishing building lines. The level or transit level can be used for this procedure. Set up the instrument over point A, then sight point B. Zero the horizontal circle with the vernier index, then swing the instrument 90 degrees and set stake C. The angle BAC will equal 90 degrees. This procedure is repeated to lay out the other corners of the building lines (*Fig. 1–77*).

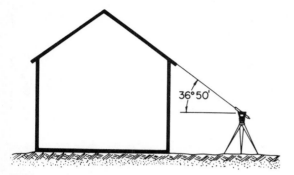

Figure 1–76 Measuring a roof slope with a transit level.   Figure 1–77 Laying out building lines.

The Framing Square

It would be difficult to build a house without the aid of the framing square. This is one of the most useful of the carpenter's tools, especially when used for roof framing, where it is used to determine lengths of common, hip, valley, and jack rafters. It is also used for laying out the top, bottom, and side cuts of rafters. It has a table that enables the carpenter to quickly tell the contents in board measure of any size board or timber. The framing square is also useful for finding the lengths of common braces, for stair layout, and for laying out octagons, among other things.

The square is made in the form of a right angle, but it contains two legs of different lengths and widths. These are called the body and tongue. The body (also called the blade) is the longer and wider leg, and is usually 24 inches long and 2 inches wide. The tongue is the shorter and narrower leg, and is usually 16 inches long and 1½ inches wide (*Fig. 1–78*). If you hold the square with the body in your left hand, the tongue in your right hand, and the corner pointing away from you, the top surface will be the face (*Fig. 1–79*). The face has a rafter table stamped on the body and an octagon scale on the tongue, and usually bears the manufacturer's trademark. The back of the square is the side opposite the face. It contains a board-measure table on the body and brace-measure table on the tongue. The heel is the point or corner at the outside where the legs of the square meet. The inside corner is sometimes also referred to as the heel.

The scales along the inner and outer edges are inch graduations divided into eighths, tenths, twelfths, and sixteenths.

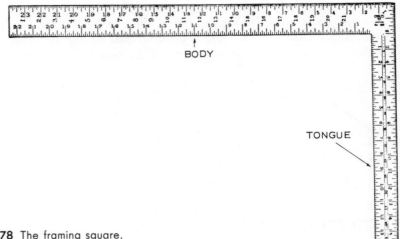

Figure 1–78 The framing square.

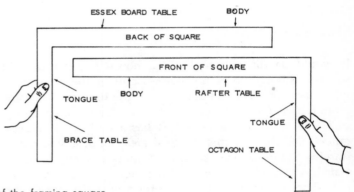

Figure 1–79 Parts of the framing square.

A hundredths scale located on the back near the heel consists of a 1-inch line divided into 100 parts. The longer lines indicate 25 hundredths, while the next shorter ones indicate 5 hundredths. The individual lines, of course, are hundredths of an inch. A sixteenths scale is located directly below the hundredth graduations (*Fig. 1–80*). This enables the carpenter to easily convert hundredths to sixteenths, and is especially helpful when utilizing the rafter tables, where dimensions are given in hundredths.

The octagon or "eight-square" scale is located along the center of the face of the tongue. It is used for laying out octagons from squares of even dimensions. For example, suppose you want to lay out an 8-inch octagon (8 inches flat to flat). Proceed as follows: Lay out an 8-inch square, then draw vertical and horizontal center lines (*Fig. 1–81A*). Next, set a pair of dividers from the first dot to the eighth dot on the octagon scale (*Fig. 1–81B*).

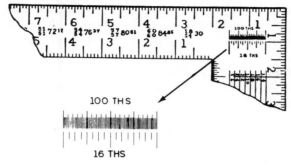

Figure 1–80 The hundredth scale on the framing square.

(Note that each dot represents the size of the square in inches. If the square size were 12 inches, you would set the divider to span 12 dots.) Lay off the divider setting to each side of the center lines, then connect the points with the diagonal lines to form the octagon (*Fig. 1–81C*).

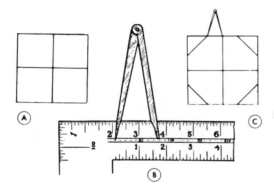

Figure 1–81 Using the framing square to lay out 8-inch octagon: A. Lay out an 8-inch square with vertical and horizontal center lines. B. Set divider to span 8 points on the octagon scale of the framing square. C. Transfer the divider measurement to each side of the center lines and connect these points to form the octagon.

The Essex board-measure table found on the back of the framing square is an easy way to compute the board feet in a piece of lumber. All computations are based on lumber that is 1 inch thick. The inch markings on the edge of the square (*Fig. 1–82*) represent the width of a board 1 inch thick. The length is provided in the vertical column under the 12-inch mark. To find the number of board feet in a piece of lumber that is 4 inches thick, 8 inches wide, and 14 feet long, locate the number 14 under the 12-inch mark on the blade. Follow this line across the blade to the inch mark representing the width of the board, in our case the 8-inch mark. At the meeting point, you will find the figures 9 and 4. The figure 9 on the left is the number of feet and the figure 4 on the right is the number of inches. Thus the board measure for a 1-inch-thick piece of lumber that is 8 inches wide and 14 feet long is 9 feet and 4 inches, or $9\frac{1}{3}$ board feet. Since the board in our example is 4 inches thick (not 1), multiply the $9\frac{1}{3}$ by 4 to get the total board feet. Our final answer is $37\frac{1}{3}$ board feet. (See also the discussion, earlier in this chapter, of finding board measure by mathematical equation.)

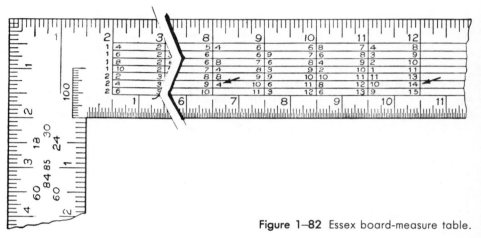

Figure 1–82 Essex board-measure table.

Brace Table. This table is found at the back of the tongue along its center (*Fig. 1–83*). It gives the brace size for equal runs and rises in intervals of three units from 24 inches to 60 inches. Actually, it is the hypotenuse measurements of various right triangles or common braces. For example, if you want to know the length of a brace where the rise on a post is 54 inches and the run on the beam is 54 inches, find the figure 54/54 on the table. Next to this figure you will find the number 76.37 (or $76\frac{3}{8}$), which is the length of the brace.

For sizes not shown on the table, you can add, subtract, or divide from the figures listed to find the others. Note that the figure 60/60 is twice that of 30/30 on the table. Note also that the brace for the 60/60 is 84.85 or double that for the 30/30, which is 42.42. Therefore, you can find the length of braces longer or shorter than those shown on the table by halving or doubling the runs on the table. For example, to find the brace length for a rise and run of 96 inches, find one-half the length or 48 inches first, and then multiply by 2. The figure for 48/48 is 67.88, and 67.88 multiplied by 2 equals 135.76 or 11 feet $3\frac{3}{4}$ inches. Suppose you want to know the brace length where the run and rise are each 8 inches. Take the

Figure 1–83 The lengths of common braces are found in the brace table.

brace measure for the 24/24 rise and run, which is 33.94. Divide this number by 3, because 24 is three times greater than 8. The answer is 11.31. You could also have taken the 48/48 measure and divided by 6, since 48 is six times greater than 8. The 48/48 figure is 67.88 and divided by 6 equals 11.31

To make the brace, mark off the length on the stock, then draw 45-degree lines to these marks and cut diagonally, as shown in *Fig. 1–84*.

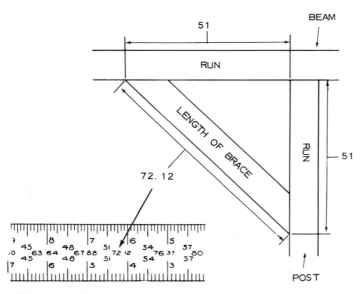

Figure 1–84 Using the table on the square to lay out braces.

At the right-hand end of the table near the heel, you will find the figure 18/24–30, which means that when the run or rise is 18 inches and the other is 24 inches, the brace will be 30 inches long. These figures can be used in finding brace measures for unequal runs, provided they are in the proportions of 3:4:5. For example, a run of 6 feet and a rise of 8 feet will require a brace of 10 feet. Likewise, a 9-foot run and 12-foot rise will require a 15-foot brace.

Some squares have polygon tables that give the square settings for forming the required angle cuts. Referring to the table (*Fig. 1–85*), you will note that it is marked "Angle Cuts for Polygons," followed by the number of sides and two figures. For example, for an octagon cut the reading is "8 sides 18–7½." This means that for an eight-sided polygon you must apply the square to the work with 18 set on the blade and 7½ set on the tongue. The last figure is always the side cut as shown in *Fig. 1–86*.

The rafter table and its use is covered fully in the section on roof framing.

Figure 1–85 The polygon table.

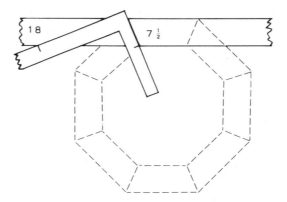

Figure 1–86 To cut eight pieces to form an octagon, take the first figure from the table (18 for octagon) and set blade to 18. The second figure (7½) is set on the tongue and gives the side cut.

The Metric System

Primitive societies based their measurements on parts of the body. Lengths were measured with the forearm, palm, finger, foot, and so forth. Capacities of containers were determined by the amount of seeds or stones required to fill them. Thus the "carat," used today as a unit of gem measure, was derived from the carob seed.

As societies developed and multiplied and trade and commerce flourished, greater accuracy in measurements was required. The invention of the numbering system and the science of mathematics made new measuring systems possible, but they still varied among different nations. The early colonists in America used the English system of measurement based on the inch, foot, and yard. This system is still the most commonly used system of weights and measures in the United States today, but we are in the process of a much delayed and needed change.

In the year 1790, France developed the metric system, which is a base-10 or "decimal" system. The base unit of length for this system is the meter, the base unit of mass is the gram, and the base unit of volume is the liter. Larger and smaller versions of each base unit are derived by multiplying or dividing by 10 and its powers. In 1840 the metric system was made compulsory in France.

Other nations had long recognized that there was a need for an international coordinated system. They gradually adopted the metric system, because it was found to be simpler to use and better adapted to scientific and engineering work. By 1900, thirty-five nations had officially adopted the metric system.

In 1960, a simplified modernized metric system was adopted by the General Conference on Weights and Measures, an international conference of diplomats, and was called the International System of Units, abbreviated as SI. Further improvements in this system have since been made.

It was not until December 1975 that the U.S. Congress enacted the Metric Conversion Act to develop a coordinated national program for the voluntary conversion to the metric system. The United States is the last major nonmetric nation in the world, but conversion is inevitable. More and more industries, such as the photographic, liquor, and medical/pharmaceutical, are slowly converting to metrics. The construction industries have been slow to make any changes, but in 1974 the Construction Industries Coordinating Committee (CICC) was formed to make plans for the eventual conversion to the SI system. They proposed that building dimensions and building product sizes should be based on an international building module of 100 mm and multiples of this basic module.

The advantages of the metric system are many. It is the international language of measurement. The United States is surrounded by a metric world and must accept it. The decimal system is simpler to use and better than the English system because it simplifies calculations, estimates, and measurements. Temporarily living with two systems may prove costly; the sooner we convert the better. Schools and teachers must realize the importance of teaching students how to use the metric system and the conversion tables. Carpentry students may find themselves unable to ply their trade without such knowledge.

Certain rules must be observed in writing metric quantities. All symbols are in small letters—e.g., mm or cm—unless derived from a proper name. The symbols are not to be used as abbreviations, and are not followed by a period unless at the end of a sentence.

Note that in metrics, a number never begins with a decimal point. A zero should be added before the decimal point—e.g., 0.48 instead of .48—for all numbers less than 1.

Also, spaces should be used instead of commas in numbers greater than three digits—for example, 2 051 366 mm instead of 2,051,366 mm.

At present, two spellings are in use for liter (*litre*), meter (*metre*), and gram (*gramme*), but it is anticipated that these will also be standardized. In the United States, the first spelling is most commonly used: liter, meter, and gram.

There is no confusion in terminology in the metric system. The unit denotes liquid or dry measure. In the English system, however, an ounce might denote an ounce of weight or an ounce of volume; a ton might denote 2,000 pounds or a nautical ton of 100 cubic feet (cargo capacity).

In the metric system, everything is in multiples or divisions of 10. There are no fractions used; decimals are used instead. This makes it very easy to learn and use; calculations are faster and more accurate. During the changeover period, using the conversion tables to change measurements to equivalent metric units will greatly facilitate calculations. If the tables are not handy, using these two factors will greatly simplify calculations. For example,

$$\text{no. of inches} \times 2.54 = \text{no. of centimeters}$$
$$\text{no. of feet} \times 0.3048 = \text{no. of meters}$$

Under the SI (International System of Metric Units), there are seven base units, forty-three derived units, two supplementary units, and twenty accepted units. For most everyday purposes, the seven-base units are the most important.

Base Units.   These have only one characteristic and are dimensionally independent, as shown in *Fig. 1–87.*

Derived Units.   These are metric units with at least two characteristics, such as length and width or length and time. They are formed by (1) multiplying a base unit by itself or (2) multiplying a base unit by one or more of the other units. For example, m² is the symbol for an area unit of a square meter (1 meter × 1 meter). An example of the second is the unit C, which is the symbol for coulomb, the unit of quantity of electricity (C = 1 second × 1 ampere).

Supplementary Units.   These are two purely geometrical units, the radian

Quantity	Base Unit	Symbol
Length	Meter	m
Mass	Kilogram	kg
Time	Second	s
Electric current	Ampere	A
Temperature	Kelvin	K
	Degree Celsius	°C
Luminous intensity	Candela	cd
Amount of substance	Mole	mol

Figure 1–87 Table of base units.

and steradian units, for measurement of angles; they have not yet been fully categorized.

SI-Accepted Units. These are units outside the SI which are so widely used that they have come to be accepted by the GCPM (General Conference on Weights & Measures). For example, *degree, hour, liter,* and *knot* are not metric units but are accepted units. Note that in volume measurement, the liter represents the liquid capacity of a cubic decimeter (cube measuring 10 cm long × 10 cm wide × 10 cm high) and is an accepted unit but not a derived unit, since it is not based on the meter. The unit of volume, based on the meter, is a cubic meter.

The following tables, shown in *Fig. 1–88* will assist the student in his calculations:

Lumber sizes are not expected to change much in the metric system. Nominal sizes will most likely disappear. All measurements are based on the meter, which measures 39.37 inches. One inch is equivalent to 25.4 millimeters. The common stud spacing of 16 inches equals 406.4 mm, which is rounded off to 400 mm when the metric system is used in building. Rafter spacing of 24 inches rounds out to 600 mm in the same manner. A nominal 2 × 4 (actual size 1½ × 3½) equals about 40 mm × 90 mm in the metric system.

In constructing roofs, unit rise will be expressed as millimeters of rise per meter run instead of the inches rise per foot of run now used. Manufacturers are now producing metric framing squares and short and long tape rules with metric graduations.

When it is applied to the framing square, the metric system actually simplifies framing calculations. There is no need to convert feet to inches and fractions to decimals. Millimeters are one-thousandth part of a meter. Therefore, no further conversion is necessary. To change millimeters to meters, simply move the decimal three places to the left.

LENGTH MEASUREMENT

Metric Unit	Value	Symbol
kilometer	1000 meters	km
meter	base unit	m
decimeter	0.1 meter	dm
centimeter	0.01 meter	cm
millimeter	0.001 meter	mm

VOLUME MEASUREMENT

Metric Units	Value	Symbol
cubic meter	derived unit	m^3
cubic decimeter	0.1 cubic meter	dm^3
cubic centimeter	0.01 cubic meter	cm^3
cubic millimeter	0.001 cubic meter	mm^3

MASS MEASUREMENT

Metric Unit	Value	Symbol
kilogram	base unit	kg
gram	0.001 kg	g
decigram	0.1 g	dg
centigram	0.01 g	cg
milligram	0.001 g	mg

Figure 1–88 Table of metric measurements.

In the conventional foot and inch system, the common stud spacing of 16 inches would be rounded out to 400 mm, which is close enough to the actual figure of 406.4 mm. Likewise, 24-inch rafter spacings become 600 mm in the metric system.

A metric framing square is shown in *Fig. 1–89*. The tongue, which is the shorter and narrower part, measures 40 mm wide and 400 mm long. The body, which is wider and longer than the tongue, measures 50 mm wide and 600 mm long. Compare these measurements with those of the square shown in *Fig. 1–78*.

An informative booklet entitled *Metric Steel Square* is available free to students. Write to Ad Services, The Stanley Works, Box 1800, New Britain, CT 06050.

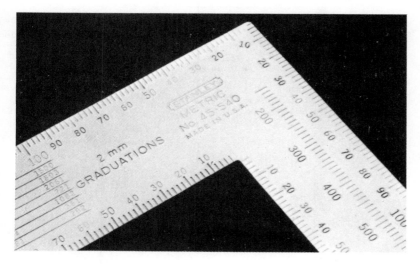

Figure 1–89 A metric framing square.

When the metric conversion is complete, architects will design structures with coordinated metric building dimensions and product sizes based on the basic module of 100 mm and multiples of this unit. The simplicity of the system is apparent when the millimeter is used in building design. For example, a modular dimension of 6,400 mm is exactly 64 modules (of 100 mm each) and is easily divisible by 400 mm (standard stud spacing). In the present system of feet and inches, the multiplier is hidden in modular dimensions. For instance, you cannot tell at a glance that 21'4" represents 64 modules of 4 inches or 16 units of 16-inch stud spacings.

Cost computations are also greatly simplified. For example, to find the cost of material to cover an area of 2.8 meters by 5.6 meters, at $16 per square meter, simply multiply 2.8 × 5.6 × 16. This equals $250.88. Now try to find the cost of an area 9'4" by 18'5" at a cost of $1.50 per square foot. See the difference?

The changeover to metrication of construction industries in other countries did not take too long, nor was it a period of difficult adjustment for people. Likewise, the transition in this country should go smoothly.

Ellipses

The carpenter is often called upon to cut an opening in a pitched roof to allow the passing of a vent pipe or similar item (*Fig. 1–90*). When a round pipe passes through a sloping roof, the resulting hole is elliptical in shape, having a major and minor axis. The following method is used to lay out the ellipse:

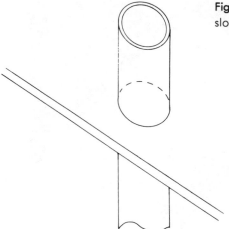

Figure 1–90 A round pipe, passing through a sloping surface, requires an elliptical opening.

1. Place the framing square on a board to the cut of the roof, taking the unit run (12 inches) on the blade and the unit rise on the tongue. Draw diagonal line 1 (*Fig. 1–91*). This represents the roof angle.
2. Next, lay out two lines, indicated as 2 and 3 in *Fig. 1–92*, perpendicular to the edge of the board. The distance between these two lines should be equal to the diameter of the pipe. The diagonal measurement between the two lines (2 and 3) is equal to the length of the ellipse (major axis). The width of the ellipse (the minor axis) is equal to the diameter of the pipe.
3. Lay out the ellipse on the roof by locating the center of the pipe (P) (*Fig. 1–93*). Draw the center lines AB and CD. Line AB is the major axis, parallel to the roof rafters, and line CD is the minor axis, perpendicular to line AB. A divider is set at one-half the major diameter. Place one leg of the divider at point C, then strike the intersecting arcs X and Y (*Fig. 1–94*) on line AB.
4. Place nails at points X, Y, and C. Tie a string tautly around the three nails; then remove nail C and replace with pencil held vertically at the same location. Then, keeping string taut at all times, grasp pencil and move it around the board, marking a perfect ellipse (*Fig. 1–95*).
5. Make the cutout with a saber saw. Be sure to allow clearance around the pipe—about ½ inch for vent pipes, about 1 inch for flue pipes, and about 2 inches for chimney pipes.

Safety on the Job

It is the responsibility of an employer to provide safe working conditions

Figure 1–91 Place the square on the board to the cut of the roof and draw a diagonal line.

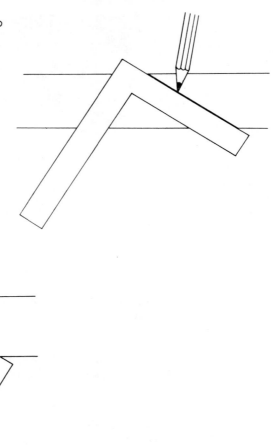

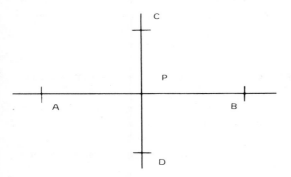

Figure 1–92 Method of finding length of elliptical hole for pipe.

Figure 1–93 Markings indicate major and minor axes of ellipse. Pipe center is indicated by letter P.

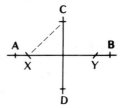

Figure 1–94 Set a divider to one-half the major diameter, then strike arcs at X and Y.

Figure 1–95 Laying out the ellipse with pencil and string.

and proper equipment for his workers. It is equally essential that the workers heed and observe all safety rules, precautions, and procedures in the daily practice of their trade to insure the health and safety of themselves and others around them. Besides the safety rules previously noted in this chapter for various specific hand and power tools, here is an additional list of safety precautions to be observed:

- Do not remove safety guards from tools.
- Do not work with wet hands or feet. When working in a wet area, use rubber-soled shoes and rubber gloves.
- Be sure the power is off and the cord is disconnected before changing blades on a power tool.
- When plugging in a power tool, be sure the switch is off.
- Before use, check circular saw blades for fractures.
- Inspect pneumatic hoses for defects or signs of wear, and use only if in good condition.
- Make sure that the safety valve on a compressor is operational before you use it.
- Do not use a compressor unless you have had ample instructions in its use.
- Always wear eye protection when using pneumatic tools.
- Use the recommended air pressure for the tool you are using.
- Disconnect the air supply before doing maintenance on tools.
- Keep ropes, chains, and cables in good condition. If any are worn, they should be removed from service.
- Do not use knotted ropes; knots weaken the rope considerably. Use knots only for the attachment of materials. *Fig. 1–96* illustrates sev-

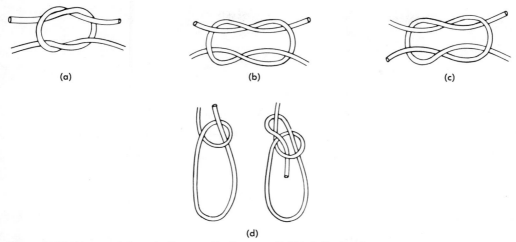

(a) (b) (c)

(d)

Figure 1–96 Various knots: A. Square. B. Granny. C. Thief. D. Bowline.

eral common knots. Study the drawings carefully and practice making the square and bowline knots until you become proficient. The granny and the thief knot, also shown in the illustration, are unsafe and must never be used. Note how they resemble the square knot, which is safe.

Foundations and Formwork 2

A sound foundation is essential in the construction of a structure, whether it be a single-family dwelling or a skyscraper. Certain basic principles apply in the construction of footings and foundations. *Fig. 2–1* shows a foundation under construction. The foundation provides a stable base for the building. It distributes the weight of the building and applied loads over the bed area to prevent unequal settling of the building. When soil conditions are poor and do not provide good bearing, footings should be spread over a greater area. The foundation must be laid out accurately and must be of proper design to support the load above. The design of the foundation is the responsibility of the architect, but the layout and actual construction of the formwork is the responsibility of the carpenter.

Figure 2–1 A foundation under construction.

Foundation Layout

The first step involved in laying out building lines is to locate them on the property. This may be done by taking measurements from lot markers, curbs, property lines, or existing buildings. The procedure is called *staking out*. Before the first stake is driven, the carpenter must be certain that all legal matters have been resolved. Setbacks and sideyards must meet code requirements, for example.

If possible, the building lines that locate the outer walls of the structure should be laid out with instruments, but other methods may be used instead. After the corners are established, batter boards are installed and used as an aid in locating and retaining building lines during construction. *Fig. 2–2* shows a typical building-line layout. Since the building has an abutment, it will require additional batters as shown.

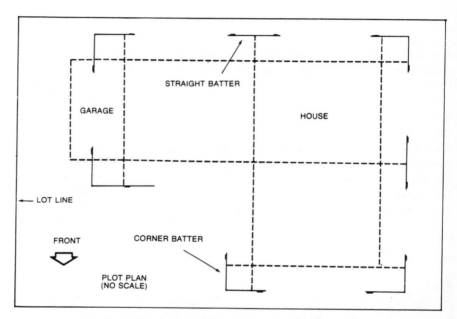

Figure 2–2 Building-line layout. Place batter boards away from excavating area.

A corner stake and one side of the building line must be established. The rest of the stakes are added to form the outline of the building. All corners must be perfect right angles. If leveling instruments are not used, the layout may be made or checked by means of the 3:4:5 method, which is based on the Pythagorean theorem. According to this famous mathematical proposition, a triangle with sides of 3, 4, and 5 units, or multiples of these figures, will form a right triangle. Conversely, if one side of a right triangle measures 3 feet, and the side meeting it at a right angle measures 4 feet, then the diagonal, or hypotenuse, will measure 5 feet (*Fig. 2–3*). For greater accuracy, however, multiples of these numbers are used, such as 6 feet, 8 feet, and 10 feet or 9, 12, and 15 feet. To put this theorem into practice, drive a stake A at one corner of the building line. Drive a nail into the center of this stake. Loop two lines and hook them over the nail in the stake. Run one line AB along the established building line. Run line AC at an approximate right angle to line AB, positioning it by eye. Place a mark

Figure 2–3 The right-angle triangle is useful in layout work. If the base is 3 feet and the altitude 4 feet, the hypotenuse will be 5 feet. Multiples of these figures may be substituted.

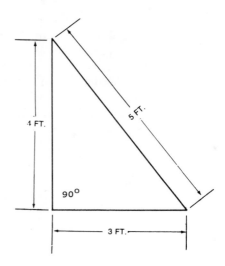

Y on this line AC at a distance 8 feet from the corner A. Now drive a stake X along line AB, exactly 6 feet from corner A. Drive a nail at the center of this stake, then hook a tape measure over it. Stretch the tape along line XY. Now manipulate the tape and line AC so that the 8-foot mark on line AC crosses the 10-foot mark on the tape. Fasten line AC securely. Lines AB and AC will form a perfect right angle (*Fig. 2–4*).

Another application of the 3:4:5 method utilizes a square made in these proportions. Use two pieces of straight 1 × 6 lumber 12 feet long.

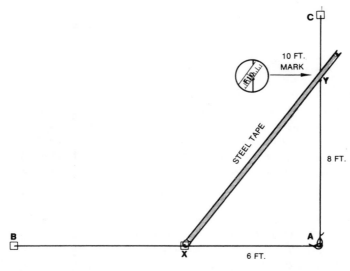

Figure 2–4 Application of the 3:4:5 method of laying out a right-angle corner using steel tape.

Join them in an approximate right angle, using one nail close to the corner. Measuring from the corner, place a mark 6 feet along the edge of one leg, and another 8 feet along the edge of the other leg. Lay a third piece of lumber, with two marks 10 feet apart, diagonally across the legs. Manipulate the legs so the marks on the legs and the diagonal coincide. Add more nails to the corner and fasten the diagonal securely. The result will be a large layout square with a 90-degree corner. *Fig. 2–5* shows the square in use.

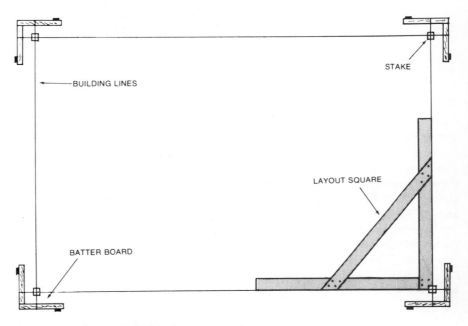

Figure 2–5 Layout of building lines using a large square.

When the corners of the building have been accurately located, drive a stake at each corner and mark the stake carefully with a nail at the top to precisely pinpoint the corner. When the four corner stakes are in place, check the building lines for squareness. Do this by measuring the diagonals. If the layout is perfectly square, the diagonals will be equal in length (*Fig. 2–6*).

Batter Boards. Batter boards consist of stakes and ledgers placed at least 3 feet outside the building line, as a guide to locating and retaining the building lines during construction. They must be accurately positioned and leveled with each other.

Use 1 × 6 boards for the ledgers and 2 × 2s or 2 × 4s for the stakes. Point one end of each of the stakes, and make them long enough so they

Figure 2–6 The carpenter can check layout for squareness by measuring diagonals. If diagonals are equal, the layout is square.

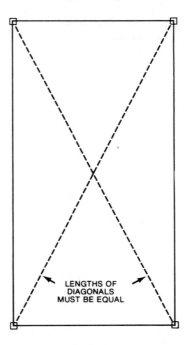

LENGTHS OF
DIAGONALS
MUST BE EQUAL

can penetrate into firm soil and still extend sufficiently above ground. The batter boards may be made up single or double, as in *Fig. 2–7*.

Stretch taut lines across opposite corners and position the strings on the batter boards so they cross precisely over the corner stakes.

Use a plumb bob for exact placement of these lines. Mark the top of the ledger where the lines cross, then cut a small kerf and pass the line through the kerf. When all the lines are in place, check again for accuracy

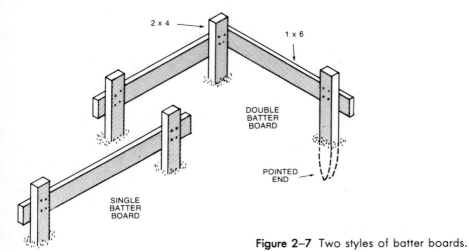

2 x 4

1 x 6

DOUBLE
BATTER
BOARD

POINTED
END

SINGLE
BATTER
BOARD

Figure 2–7 Two styles of batter boards.

by measuring the diagonals as explained above. Make a second kerf cut to indicate the excavation line.

The batter boards should be placed far enough from the building corner to allow room for excavating, as shown in *Fig. 2–8*. Some carpenters install the batter boards so that their height is the same as the foundation wall.

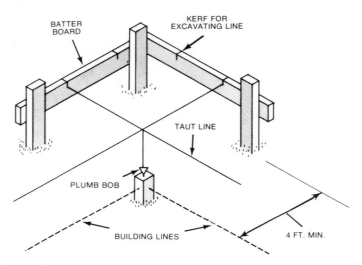

Figure 2–8 Method of staking and laying out foundation lines.

Excavation. The width and depth of the excavation are determined by the size of the foundation. It must be wide enough to permit working space for form-building, laying of drain tile, and waterproofing (*Fig. 2–9*). The amount of back slope is dependent on the soil. Soft, sandy soils require considerable backslope. For stable soils, the backslope can be almost straight.

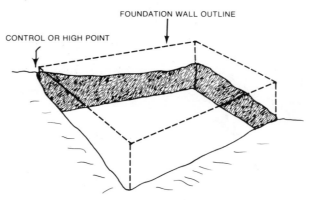

Figure 2–9 Procedure for establishing depth of excavation.

Measuring Grade Elevations. Grade elevations are usually related to a reference point on the job called the benchmark, which is assigned an arbitrary elevation of +100 feet. Other elevations around the jobsite are determined by comparison with this reference point. Higher points will have elevations greater than +100 feet and lower points less than +100 feet. The benchmark must be stable and stationary. Usually one corner of the building is made the benchmark ´against which grade elevations are checked.

Height of Foundation. Generally, the height of the foundation wall, and thus the depth of the excavation, is established by using the highest elevation around the perimeter as a control (*Fig. 2–10*). The foundation wall should extend a minimum of 8 inches above the finished grade. This will assure good drainage and moisture protection for the framing members. It also provides indirect termite control. By maintaining a space between grade surface and the framing members, termite tunnels can be observed if they occur. Proper steps may then be taken to eliminate the termites. If rough grading is required because of steep terrain, the topsoil should be removed and piled separately so that it can be reused later. *Fig. 2–11* shows how corners are established for excavation and footings.

Concrete Foundations. The type of foundation depends on the kind of structure as well as its location. Weather and climate are factors that can determine the type of foundation to be used. In the North, where freezing weather occurs, the footings must extend below the frost line. In the warmer climates, where freezing is not a problem, houses may be built on

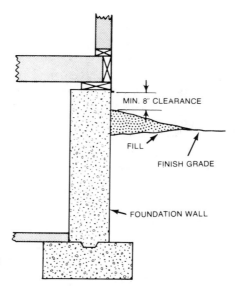

Figure 2–10 Foundation wall must be a minimum of 8 inches above the finish grade.

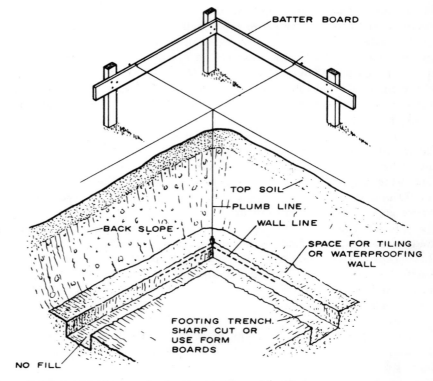

Figure 2–11 Establishing corners for excavation and footing.

slabs or with low foundation walls. No portion of the footing should bear on freshly filled ground.

It is possible to build foundations of material other than concrete; however, our concern here is strictly with concrete footings and foundations. These require some type of form, made of earth, steel, wood, or other material, which serves as a mold into which freshly mixed concrete is poured. When the concrete has hardened, the form is usually removed and in most cases is reused. Some forms are left in place permanently. These are generally round column forms made of fiber or waterproof corrugated paper. The forms most often used in the construction of small homes are made of wood or a combination of wood and steel. These may be built on the jobsite but usually they are made in standard-size panels and taken from job to job.

To properly build a form, some knowledge of concrete is essential. Concrete is a combination of varying amounts of cement, sand, gravel, and water and weighs about 150 pounds per cubic foot. When it is poured into a form in its plastic state, it exerts hydrostatic pressure. This pressure acts on the side walls of the form with a tendency to force them apart. As the concrete sets, the hydrostatic pressure diminishes; and as the pour continues, the concrete in the lower part of the form may start to set, thus becoming self-supporting. If the rate of pour is too fast, the entire form may be filled before the concrete starts to set. The temperature of the concrete at the time of pouring also has an effect on the pressure in the form. At warmer temperatures, the concrete sets faster. At lower temperatures, the entire form may be poured before any of the concrete starts to set. Thus, at rapid pour rates or lower temperatures, greater pressure is exerted on the form walls. Retardants that may be added to the concrete in hot weather must also be considered when building forms. Their use has the same effect as lower temperatures.

The ideal temperature for pouring concrete is 60 degrees Fahrenheit. Although it is impossible to control the weather, certain steps may be taken to keep the concrete within the temperature bracket of about 50 to 75 degrees. In hot weather, the mixing water should be cold. If necessary, slush ice may be added to the water to cool it. It is possible to cool the aggregate by sprinkling it with water. If possible, materials should be kept in the shade. Forms can be sprinkled with water just before the concrete is poured. *Fig. 2–12* shows a ready-mix truck preparing to pour into foundation forms.

Figure 2–12 Ready-mix truck preparing to pour concrete into foundation forms.

Cold weather also poses problems in concrete work. The fresh concrete must be protected from freezing during the placing and curing. After pouring, the concrete should be covered so that it will retain its moisture. This will also protect it from freezing. When pouring is done in very cold weather, it may be necessary to heat the concrete materials. This involves steps quite the opposite of those taken when the temperatures are too high. For example, the mixing water and even the aggregates may have to be heated.

Form Materials

The materials used in formwork include lumber, plywood, hardboard, steel, aluminum, and reinforced plastic. In residential construction, lumber and plywood are commonly used.

The type of lumber used in formwork usually depends on availability and cost. Many species of lumber are suitable. The important requirement is that the lumber be straight, sound, and strong. It should not be fully dried, nor should it be green. Partially seasoned stock is ideal. Fully dried lumber will swell excessively when wet, and green material will have a tendency to dry out and warp in hot weather.

The kinds of lumber used for formwork include Douglas fir, Western hemlock, California redwood, Southern yellow pine, Idaho white pine, Northern white pine, sugar pine, and ponderosa pine. Redwood is not recommended for architectural forms because of its tendency to stain. However, it is excellent for structural forms.

Boards used for form sheathing may be square-cut or tongue-and-grooved. Square-cut boards are easier to install and more economical than matched tongue-and-groove stock, but a rough finish surface results. The tongue-and-groove sheathing results in a smoother and flatter surface. Also, there is no loss of mortar through the joints, as is the case when plain boards are used.

Plywood Formwork. Plywood is widely used for form sheathing because it has many advantages over the use of boards. It is available in large sheets and in varying thicknesses. It is made of layers of veneer cross-banded and glued. The number of layers is always odd—3, 5, 7, and so on. The grain direction of alternate layers is always laid at right angles. This equalizes stresses and keeps warpage to a minimum. The glue used for form boards must be waterproof.

The most common species of plywood used for formwork is Douglas fir, but others are also used. Thicknesses of $\frac{1}{2}$, $\frac{5}{8}$, and $\frac{3}{4}$ inch are the most common. Sizes range from the standard 4 $\times$ 8-foot sheet to 5 $\times$ 12 feet. Other thicknesses from $\frac{1}{4}$ to $1\frac{1}{8}$ inches are also available.

In some formwork, the specifications may call for curved surfaces.

The table in *Fig. 2–13* shows the various bends possible with different thicknesses of plywood.

Concrete-form–grade plywood is available edge-sealed and mill-oiled. The edge-sealing keeps out moisture and mill-oiling prolongs the life of the panel. The mill-oiled panel must still be oiled on the job, however. Some form coatings will not work with mill-oiled plywood. In such cases, be sure to specify unoiled plywood.

Panel thickness in.	Curved across grain	Curved parallel to grain
1/4	24 in.	5 ft
3/8	36 in.	8 ft
1/2	6 ft	12 ft
5/8	8 ft	16 ft
3/4	12 ft	20 ft

These radii apply for millrun panels carefully bent. Select pieces with clear, straigt grain can be bent to smaller radii.

Figure 2–13 Bending radii of various plywood panels.

Overlaid Plywood. Overlaid plywood is similar to exterior plywood but is made with a resin-treated fiber facing. This produces a smooth surface suitable for many uses. When this material is used for formwork, its lack of grain patterns or knots produces a smooth, grainless surface in the concrete. Overlaid plywood is made in medium density (MDO) and high density (HDO). Both are suitable for formwork, but the HDO has greater resistance to abrasion and water penetration.

Hardboard Forms. Hardboard is made of refined wood fibers that have been formed into sheets under heat and pressure. Two basic types are made—tempered and untempered. The tempered material contains oils and resins which make it very hard and moisture-resistant. For concrete formwork, special additional processing is required.

Hardboard for forms is only ¼ inch thick and is not structurally capable of supporting concrete by itself. It is used only as a facing and must be supported with 4-inch or wider lumber (*Fig. 2–14*). The hardboard is installed with a small space between abutting edges.

Hardboard is available in sheets 4 feet wide and from 4 to 16 feet in length. Form-grade hardboard can be bent cold to a radius of 25 inches. Smaller radii are possible, but only by ordering the material prebent by the manufacturer.

Figure 2–14 Form side faced with hardboard.

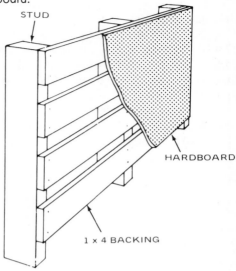

STUD

HARDBOARD

1 x 4 BACKING

Other Materials. Steel, aluminum, and plastic are seldom used in residential formwork. Their use is usually limited to special-purpose forms. However, some form panels are manufactured with a combination of steel and plywood.

Footings

Wall footings support the foundation wall. They must be designed to carry the dead weight of the foundation as well as the live weights and dead weights placed on the wall. The footing achieves this by spreading the force loads over a large area of ground. The live and dead loads are carried by the walls and columns to the footings. Normally, one-half of the dead and live loads on the joists spanning from the wall to the nearest beam is carried by the wall. The beam carries the other half of the load. Truss loads are carried by the walls supporting the trusses. The size and shape of the footing depends on the condition of the soil and the loads it is to carry (*Fig. 2–15*).

Normally, the depth of the footing is equal to the wall thickness, and the width is generally twice that amount (*Fig. 2–16*).

In climates subject to freezing, the footings must be placed below the frost line. This is usually 4 feet; however, building codes should be consulted since they may differ. Footings must never be placed on loose or replaced soil. In good cohesive soils, they may be poured directly into a trench without the need of forms (*Fig. 2–17*).

Type of soil	Tons per square foot
Soft clay	1
Firm clay or fine sand	2
Compact fine or loose coarse sand	3
Loose gravel or compact coarse sand	4
Compact sand - gravel mixture	6

Figure 2–15 Table of load-carrying capacities of soils.

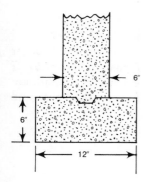

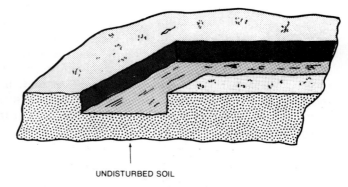

UNDISTURBED SOIL

Figure 2–16 Footing details.

Figure 2–17 Trench cut in firm soil can be used as a footing form.

Keyed Footings. Footings for block walls are left flat at the top; however, for concrete walls they are usually keyed. The key is a recess placed at the center of the footing by various means. It provides a strong joint between the wall and the footing and will prevent lateral movement of the wall. The recess or key can be made by pressing oiled 2 × 4s into the concrete before it sets. The 2 × 4s should be beveled (*Fig. 2–18*). The 2 × 4s are removed when the concrete is sufficiently hard. Another method is to use 2 × 2s placed diagonally.

Bricks or steel rods are also useful in strengthening the joint between the wall and the footing. These are inserted into the concrete and left in permanently.

Stepped Footings. In areas where the ground slopes, it is often desirable to step the footings in order to save on excavation and materials. The bottom of each step should be level, and the footings should be cast in one

Figure 2–18 Beveled 2 × 4 is pressed into
concrete to form key in footing.

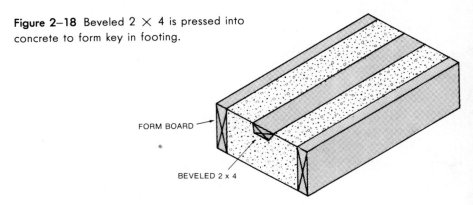

FORM BOARD

BEVELED 2 x 4

piece. To eliminate cracking, the underside should be free of sharp corners.
Forms for stepped-down footings can be made, as shown in *Fig. 2–19*.

The dimensions of the steps in wall footings depend on the type of
soil prevalent. In granular soil the steps may be designed in the ratio of
two horizontal units to one vertical unit. In cohesive soils, it is common
practice to make the depth of the step equal to the footing thickness and
the length of the step 1½ times the footing thickness.

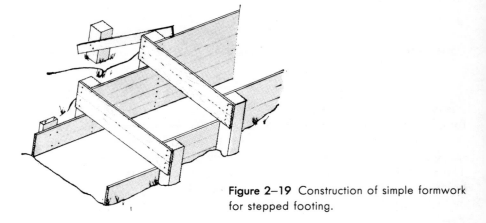

Figure 2–19 Construction of simple formwork
for stepped footing.

Column Footings.   The footings for columns and piers can be square or
sloped. Usually they are made with square sides (*Fig. 2–20*), or with a pedestal if they are to support a wood post (*Fig. 2–21*). The pedestal keeps the
post off the floor, where it might be subject to moisture. The pin anchors
the post. *Fig. 2–22* shows how the form for a column is made.

If a steel post is to rest on a footing, bolts for the bottom plate are embedded in the concrete. Alternatively, the post may be set in place on the footing and the concrete floor is poured around it.

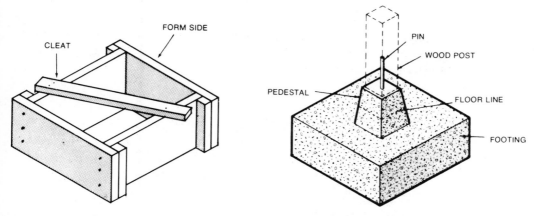

Figure 2–20 Typical column footing form.   Figure 2–21 Typical post footing.

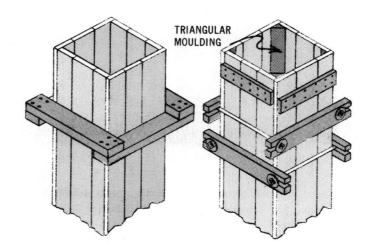

Figure 2–22 Column forms.

Erecting Footing Forms.   The footing forms are erected when excavating has been completed. Lines are restrung on the batter boards. At the intersection of the lines at each corner, a plumb line is dropped. To indicate the outer perimeter of the building, drive stakes at each corner and mark the corner. From these corners, locate the outer edges of the footing and proceed to install the form boards. Drive the stakes at intervals of about 3 feet if 1-inch form boards are used. The stakes should be driven so the tops are slightly below the surface of the form boards to facilitate striking off the concrete. Because they are so shallow, the fresh concrete exerts little pressure on footing forms and they are therefore made with a minimum of bracing (*Fig. 2–23*).

Forms are assembled with double-headed nails driven from the stake side, not the inside. Take measurements from the outer form boards to locate the inner boards. Use 1 × 2 spacers to keep the forms spread evenly. These spacers are removed as the concrete is poured.

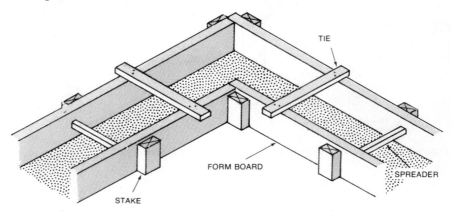

Figure 2–23 Place the stakes in the footing form below the form boards.

Monolithic Forms. Separate forms are usually required for the footings and the foundation walls when a house is to have a basement. However, when the foundation wall is very low, a monolithic form may be used for the single pouring of concrete to form a combined footing and wall (*Fig. 2–24*).

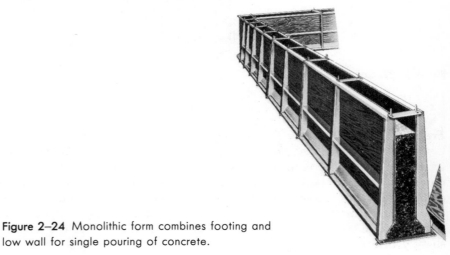

Figure 2–24 Monolithic form combines footing and low wall for single pouring of concrete.

Wall-Form Construction

The thickness of the foundation wall depends on the vertical loads above, the lateral pressure of the soil, and the unsupported height and length of the wall itself. For most small buildings, concrete foundation walls are 8, 10, or 12 inches thick. Foundation walls that are 8 inches thick may be used when the wall does not extend more than 4 feet into the earth. If the soil is poor or very wet, the foundation wall should be at least 12 inches thick.

Forms are constructed for each wall face. They consist of boards, plywood, or other material, spaced the desired distance apart. To withstand the pressure of the fluid concrete, form ties are utilized. These ties are used in conjunction with spreaders. The spreaders keep the form faces properly spaced and the ties prevent the outward pressure of the concrete from bulging the form. Ties and spreaders can be made on the job by the carpenter. The spreader consists of a piece of wood cut to a length that corresponds to the wall thickness.

The ties are lengths of steel wire that extend through the form and are secured around studs or wales on each side of the form (*Fig. 2–25*). The ties are twisted with a piece of wood or other object to pull the form boards tightly against the spreaders. Many manufactured devices are available which combine ties and spreaders. These are superior to the stick-and-wire system and are widely used today. One type, called a snap tie, is shown in *Fig. 2–26*. These are made to securely hold the forms at the proper spacing. They have weakened points called "break backs" which will

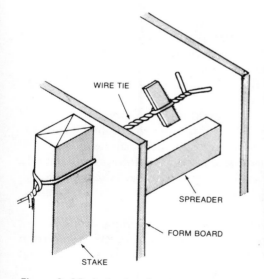

Figure 2–25 A simple wire tie.

Figure 2–26 Installing a snap tie.

break about 1 inch inside the wall after the forms are removed. Wedge-type clamps are used to tighten the tie rods. The small holes left in the concrete wall after the ties are broken are patched with mortar. The three-section tie shown in *Fig. 2–27* consists of inner and outer threaded rods. These are connected by the cone nut at each end of the inner rod. Clamps are used to tighten the assembly. When the form is stripped, the outer rods and cones are removed, leaving a smooth, tapered recess that is easily filled with mortar.

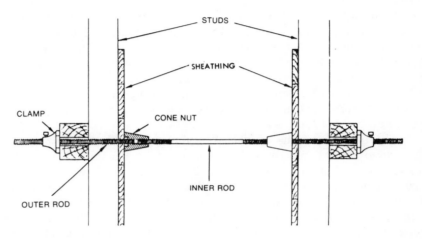

Figure 2–27 Three-section tie.

Another type of tie is the coil tie (*Fig. 2–28*), which will take a working load of 6,000 to 9,000 pounds. Coil ties consist of helix coils which are electrically welded to longitudinal struts of the ties, and into which special coil bolts are threaded. Some coil ties are made to take cones that must be removed with a special wrench when the forms are stripped from the concrete. After the concrete has set, the bolts are removed and reused. The spreader, in this instance, also serves as the tie.

If wood spreaders are used, they should always be removed so they are not buried in the concrete. *Fig. 2–29* shows one simple method of removing wood spreaders that have been combined with snap ties.

The snap ties are machine-made and provide uniform and accurately dimensioned ties with specified safe working-load capacity. Some heavy-duty form ties have safe working-load capacity of 12,000 to 50,000 pounds.

Form Panels. The basic wall-form parts are the sheathing, studs, wales, braces, ties, and spreaders. The sheathing retains the concrete in the form until it hardens. The studs support the sheathing. Wales are horizontal members which support the studs and align the forms. The braces support the forms against construction and wind loads (*Fig. 2–30*).

COIL TIE / SPREADER

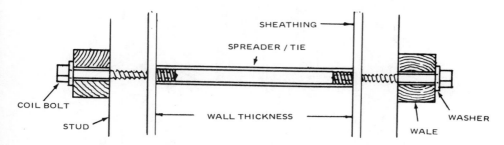

SHEATHING

SPREADER / TIE

COIL BOLT

WALL THICKNESS

STUD

WASHER

WALE

Figure 2–28 Details of combination coil tie and spreader.

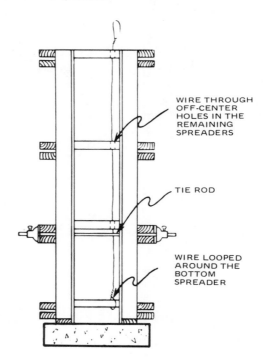

WIRE THROUGH OFF-CENTER HOLES IN THE REMAINING SPREADERS

TIE ROD

WIRE LOOPED AROUND THE BOTTOM SPREADER

Figure 2–29 Wall form showing removable spreaders.

The forms may be built on the job, but most carpenters prefer to use prefabricated form panels. These may be factory-built with steel frames and plywood faces. Some carpenters prefer to build their own (*Fig. 2–31*). In either case, the use of prefabricated panels reduces labor, and since they may be reused many times, they are far more economical than the "one-

Figure 2–30 Form for poured concrete wall.

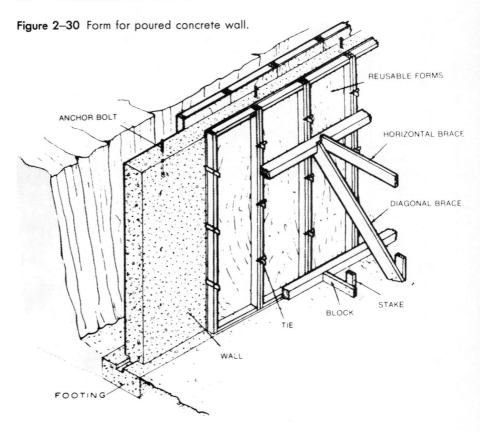

Figure 2–31 A simple plywood form section.

shot," built-in-place forms. The panels are made in convenient sizes, usual-ly 4 × 8 feet, but some carpenters prefer to use 2 × 8-foot panels. For basement foundations, 2 × 4s are used for form frames. The facing materi-al is generally ¾-inch plywood.

Erecting Form Panels.   The form panels must be erected accurately on the footings. Mark a line on the footing to indicate the inner and outer faces of the wall. Then place the panels accordingly. The adjacent panels are fastened with double-headed nails, as shown in *Fig. 2–32.* Carriage bolts may be used instead of nails. Wales, braces, and ties are added to align and strengthen the form.

After the form is erected, it mut be checked carefully to make certain that it is square and plumb. Nail markers are then installed around the inner perimeter of the form. These are placed in a level plane at the height of the foundation wall (*Fig. 2–33*). The concrete is then poured until it reaches these markers.

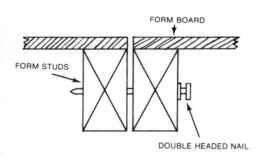

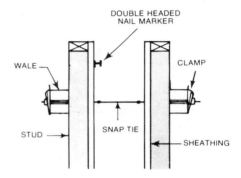

Figure 2–32 Assemble form sections with double-headed nails or bolts.

Figure 2–33 Use nail markers to indicate top of foundation wall.

Low Walls.   Forms for foundation walls up to 3 feet high can be made without the use of wales. Place the studs 2 feet apart and brace as shown in *Fig. 2–34.* Use wood ties at the top. Another method for low walls makes use of steel stirrups. These consist of three pieces of 2 × 2 angle iron welded to form an inverted U. They eliminate the use of ties and braces.

Corners.   The corners of wall forms must be made carefully. It is important that the joint be very tight to prevent the possibility of concrete leakage. One method used for making tight corners is shown in *Fig. 2–35.* This is a special corner bracket which fits between the wales. The corner may also be tightened by the use of double-headed nails or carriage bolts.

When pilasters are required to strengthen a wall, the form is made with an offset. Form construction details for a pilaster are shown in *Fig. 2–36.* A form with a horizontal offset is used when the wall is to contain a ledge for brick veneer (*Fig. 2–37*).

Openings in Foundation Walls.   Openings in the foundation wall for

Figure 2–34 Wales can be eliminated in a low foundation wall form.

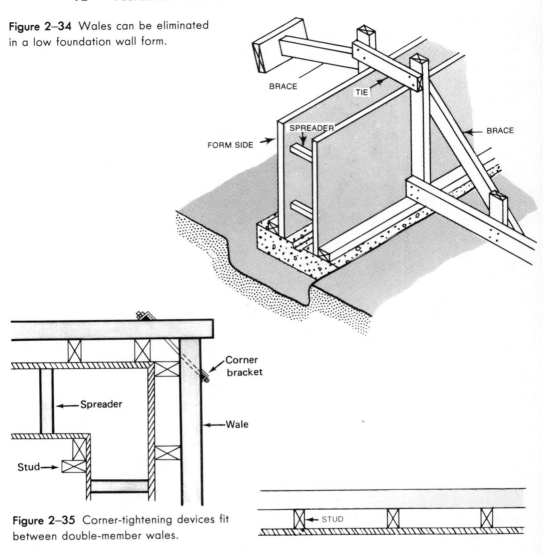

Figure 2–35 Corner-tightening devices fit between double-member wales.

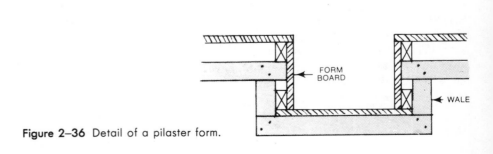

Figure 2–36 Detail of a pilaster form.

Figure 2–37 Detail of form with offset for brick ledge. Tie rods are not shown.

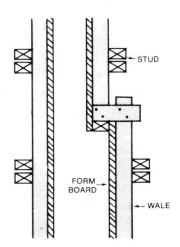

STUD

FORM
BOARD

WALE

doors and windows are built into the form. There are several methods employed, depending on the type of door or window to be used. Generally, boxlike frames called *bucks* are fastened between the walls of the form. They are removed after the concrete has been poured and hardened. These bucks may be made with nailing blocks or strips, which remain in the wall after the forms and frames have been removed. They are beveled so that they will be locked permanently into the concrete (*Fig. 2–38*). The strips are held with double-headed nails or screws driven from the opening side. The

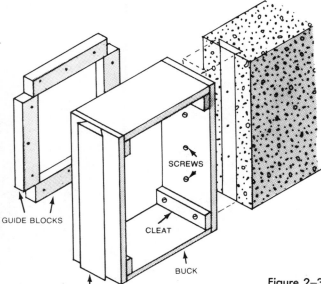

SCREWS

GUIDE BLOCKS

CLEAT

BUCK

BEVELED NAILER

Figure 2–38 Details of buck frames for forming window openings in foundation wall.

fasteners are removed before the form is dismantled. This permits removal of the buck, while the nailers remain in the wall.

Guide blocks made of 2 × 4s are nailed to the inside face of the form. They aid in locating the frame in the wall and also serve as braces for the frame.

When steel basement windows are to be installed after pouring, the frame is made with beveled strips at the sides to facilitate removal. The form surface must be smooth and well oiled so that it can be easily removed. The recess left by the strip receives the steel flange of the window. Mortar is then used to seal the window frame permanently (*Fig. 2–39*).

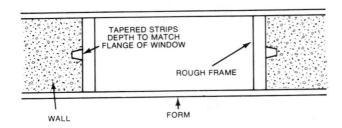

Figure 2–39 Rough form for steel basement window. Tapered strips facilitate removal.

Some window and door frames are permanently cast into the concrete. Key strips made of 1 × 2-inch stock are fitted to wood window and door frames if they are to be installed before pouring. The key holds the frame firmly in place. *Fig. 2–40* shows a basement frame in a concrete form. The front form board has been omitted in the drawing. Care must be taken to make the door and window openings in the form rigid, to prevent distortion from concrete pressure. For large openings, use ample bracing.

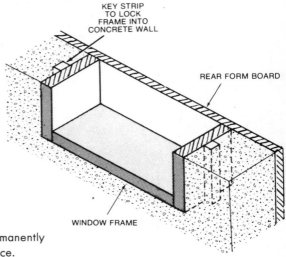

Figure 2–40 The basement frame is permanently cast in the wall. The key holds frame in place.

Lintels.   When an opening occurs below the top of a concrete or block wall, reinforced horizontal supports must be used to carry the load above. These supports, called lintels, may take various forms. In poured concrete walls they consist of steel reinforcing rods embedded in the concrete (*Fig. 2–41*). The rods are wired in place at least 2 inches above the opening. In concrete-block construction, the lintels may consist of steel angle or special lintel blocks in which steel rods are embedded.

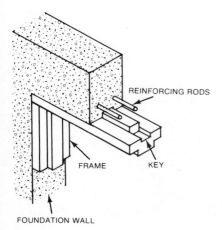

Figure 2–41 Reinforcing rods are used over door and window openings.

Notches for Beams.   Provision must be made in the framework for beams and girders. *Fig. 2–42* shows how a wall is notched over for a beam. The notch is made with a clearance of ½ inch at the end and sides. This provides sufficient ventilation so that moisture can escape.

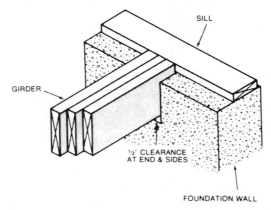

Figure 2–42 Girder framing into masonry or concrete wall.

Placing Concrete

Before concrete is placed, all formwork must be inspected for accuracy and stability. Use the level to determine that nail markers are at the exact height. Oil or coat the forms so that they will release easily from the hardened concrete.

Pour the concrete as close as possible to where it is needed. Do not allow it to flow around the form, because this will cause the aggregates to collect. Instead, pour the concrete in layers of about 12 inches successively around the perimeter of the form. To prevent honeycombing, the concrete should be spaded (*Fig. 2–43*).

Figure 2–43 Pouring concrete into form.

Sill Anchors.   Sill anchors are used to fasten the sill plate firmly to the foundation. Anchor bolts used for this purpose are made in several styles. The most common is the L-shaped anchor, which has a ½-inch thread at one end. The other end is bent at right angles. Anchors are spaced about 6 feet apart and are embedded in the concrete while it is still plastic (*Fig. 2–44*).

Slab Construction.   In homes without a basement, the foundation walls rest on a concrete slab laid directly on the ground. Poured properly, a slab makes a satisfactory floor. It is important that the ground around the slab

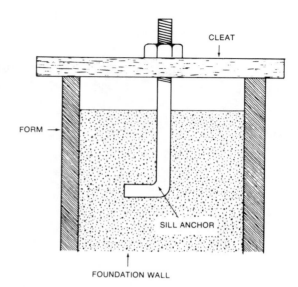

Figure 2–44 Cleat holds anchor in position while concrete sets.

be graded to slope away from the floor on all sides, at least 1 foot in every 25. The slab itself should be at least 8 inches above the surrounding grade.

The soil beneath the slab (the subgrade) should be cleared of roots, grass, and debris and given a granular fill such as gravel or crushed stone. The fill should be well compacted and at least 4 inches thick. This subbase usually assures a dry floor slab, but even so, a vapor barrier must be used between the subbase and the slab. The barrier may be made from 6-mil polyethylene, 55-pound roll roofing, or various asphaltic materials manufactured for the purpose.

Perimeter insulation is installed where the slab meets the foundation wall. In warm climates, the insulation prevents excessive heat buildup in the living quarters; in cold climates, it reduces heat loss to the outside. The insulation must meet certain specifications. It must be rigid enough to support the weight of the slab; it must be waterproof and vermin-proof; and it should not absorb moisture. Note that the insulation is placed above the vapor barrier and between the slab and foundation wall (*Fig. 2–45*).

Slab pouring is timed so that all necessary apertures or connections for utilities have been either provided for or already installed. Sleeves are used to protect any electric lines and water pipes that must pass through the concrete. Six-inch reinforcing wire mesh is embedded in the first layer of concrete above the granular fill (*Fig. 2–46*). Supported thus, the wire mesh rests 1½ inches below the surface of the completed poured slab.

After the concrete is poured and leveled or screeded, the surface is troweled to a smooth finish. Most troweling is now done with power trowels, since they do better finishing than hand trowels and are much faster. *Fig. 2–47* shows a section of a slab being screeded.

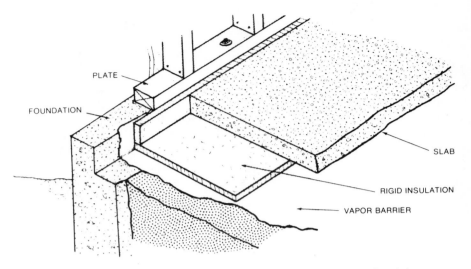

Figure 2–45 Vapor barrier and perimeter insulation are installed under slab.

Figure 2–46 Place protective sleeves over electric lines and water pipes before pouring slab.

Figure 2–47 Screeding a slab. Note reinforcing wire.

Concrete Stairs.   Porticos, platforms and stair walls should be an integral part of the foundation and poured with it if possible. However, if they are added later, they should be anchored to the main structure by means of keyways and reinforcing rods, which must be placed in the foundation walls during construction.

Fig. 2–48 shows details of a form for simple platform steps. The form risers are made to the same height as the finished stair riser. The bottom edge of the riser is beveled to enable the mason's trowel to reach the inside corner of the step (*Fig. 2–49*). For steps wider than 3 feet, riser stock should be 2 inches thick. Thinner boards will have a tendency to bulge.

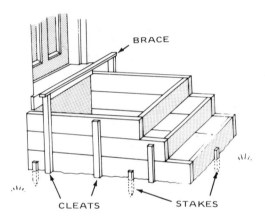

Figure 2–48 Form for concrete platform steps.

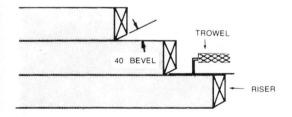

Figure 2–49 Clearance at bottom of riser provides space for trowel.

When concrete stairs are to be installed against one or two existing walls, simple forms can be used as shown in *Fig. 2–50*. The risers are cut wide enough to fit between the walls snugly. The riser forms are nailed to the 2 × 6 planks.

Curbs, Walks, and Driveways.   Forms for walks or driveways are made of 2 × 4 or 2 × 6 lumber. The longest pieces possible are used to minimize the number of joints. The walks are laid out directly on the soil unless there is a moisture problem, in which case a base of coarse granular fill should be used.

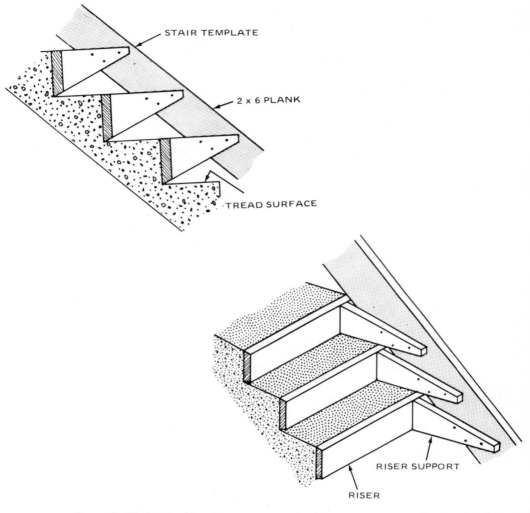

Figure 2–50 Simple form for concrete stairs between existing walls. *Top:* Risers are installed with aid of template. *Bottom:* Poured stairs before removal of form.

The forms are braced with stakes around the outer perimeter. They should be set to slope away from the building at about ¼ inch per foot. The top of the forms corresponds to the top surface of the walk. After the concrete is poured, a screed is used to level off the concrete (*Fig. 2–51*).

Control joints are tooled into the concrete at intervals of about 4 feet. This is usually done with a groover, though the joints can also be cut with a power saw fitted with a masonry blade. Tooling is done before the concrete sets, while sawing must be done after it sets.

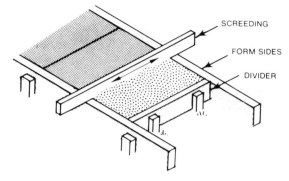

SCREEDING

FORM SIDES

DIVIDER

Figure 2–51 Screeding entails drawing a board back and forth over the concrete while moving it forward.

When curves are to be made in a walk, curved forms are needed. These can be made by laminating two pieces of ¼-inch plywood in a jig that will hold the boards in the desired shape (*Fig. 2–52*). While in the jig, the boards are nailed together so that they will retain their shape. The jig should be set up on a flat surface such as a subfloor. For small-radius curves, it may be necessary to steam or wet the wood.

Driveways are made in much the same manner as walks. However, they must include reinforcing wire mesh and they should be 4 to 6 inches thick.

Forms for curbs are made with 2-inch stock. They are installed in a trench with stakes, spreaders, and ties. *Fig. 2–53* shows a curb form made with wood spreaders and stirrups made from bent reinforcing rods.

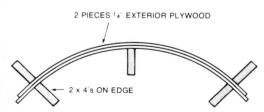

2 PIECES ¼" EXTERIOR PLYWOOD

2 x 4's ON EDGE

Figure 2–52 Detail of curved forms. The form boards are curved then joined together with shingle nails.

Figure 2–53 Install curb forms in a trench. Stirrups are made from reinforcing rods.

Floor Framing 3

After the foundation work is completed, the carpenter proceeds with the rough framing of the structure. The floor frame, which is assembled on the foundation, consists of sill plates, beams, joists, posts, and subfloor. Basically, the floor framing is the same whether the house is built on a slab or with a full basement. Proper construction and sound materials are important to ensure that the house will be rigidly built. Methods and materials may vary depending on availability, location, and other factors, but the basic construction principles are similar. *Fig. 3–1* shows a floor frame under construction.

Figure 3–1 A floor frame under construction.

Wood structures must be properly designed and built to withstand damage by moisture, fungi, and termites. Moisture can cause metals to rust, wood to decay, and concrete to crack. The fungi attack wood and cause decay only at certain temperatures and in excess of 20 percent moisture content. Good house construction must provide resistance against decay and the carpenter must see that all details are fully carried out. Good construction practices will provide (1) positive site and building drainage, (2) adequate control of termite and insect entry, and (3) ventilation and condensation control in enclosed spaces. Refer to Chapter 7, on insulation and moisture control, for further details.

Moisture Content of Wood

All wood retains a certain amount of moisture. For the beams and joists used in floor framing, the moisture content should be about 15 percent and must not exceed 19 percent. Moisture content and wood shrinkage are directly related. When the moisture content of the wood is reduced below 30 percent, shrinkage takes place. For every 1 percent loss of moisture below 30 percent, there is a $1/30$th shrinkage of the total possible shrinkage. At 15 percent, the wood will have undergone one-half of the total possible shrinkage. When moisture is added to wood, swelling takes place at a rate proportional to the percentage of moisture added. *Fig. 3–2* shows how shrinkage affects different parts of a log. The amount of shrinkage along the length of a piece of lumber is negligible. The greatest amount of shrinkage occurs tangent to the annular rings. (*Fig. 3–3*).

To minimize problems due to uneven shrinkage, the outer wall framing and the center beam should have equal moisture content. This is ac-

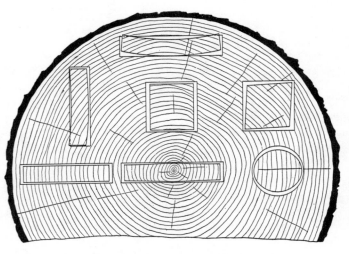

Figure 3–2 How shrinkage affects different parts of a log.

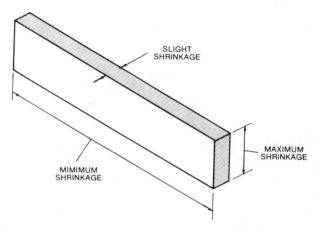

SLIGHT
SHRINKAGE

MAXIMUM
SHRINKAGE

MIMIMUM
SHRINKAGE

Figure 3–3 Shrinkage along the length of a board is negligible.

complished by using lumber not only of the same moisture content but also of the same total depth. If the total depth of wood at the foundation wall is 10 inches, then the center beam should also total 10 inches.

Types of Framing

There are three basic types of framing generally used in this country: platform, balloon, and post-and-beam. Platform framing is sometimes called Western framing, while post-and-beam framing is also known as plank-and-beam framing. Sometimes a combination of platform and balloon framing is used in a single structure. Although wall framing is covered in the next chapter, the various types of framing are mentioned here because they relate to the floor framing.

Platform Framing.   The platform frame (*Fig. 3–4*) consists of wall sections erected above the subfloor, one story at a time. Wall sections are constructed on the subfloor, thus affording the workers a safe platform to work on. The walls are assembled flat on the floor, then tilted into position, as in *Fig. 3–5*. The second floor is erected in a similar manner after the second-floor joists and rough flooring are in place. Firestops are automatically provided at each floor level.

To minimize the shrinkage problem in platform frame construction, all exterior walls and interior partitions are framed to equalize the shrinkage. To ensure equal shrinkage, the cross section of the lumber at the sills should equal that at the center of the structure. This also applies if a steel beam is used in place of wood girders. The lumber resting on the beam should have the same cross section as that used on the foundation (*Fig. 3–6*).

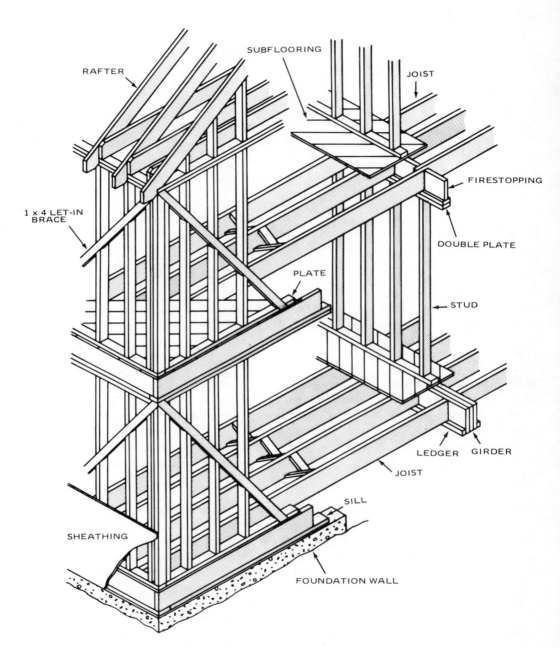

Figure 3-4 The platform frame.

Figure 3–5 Erecting wall frame of single-family dwelling.

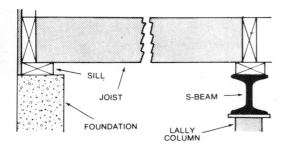

SILL

JOIST

FOUNDATION

S-BEAM

LALLY
COLUMN

Figure 3–6 To equalize shrinkage, framing members at sill and beam should have the same cross section.

Balloon Framing.   In balloon framing the studs extend in one piece from the sill to the top plate at the roof (*Fig. 3–7*). The second-floor joists rest on a ribbon board set into the studs. Firestops are used between the studs. These are needed to prevent air circulation between floor levels in case of fire (*Fig. 3–8*). Because of the unobstructed opening between floors, utilities are easily installed without the need for notching and cutting as in the case of platform construction. Because lumber shrinks very little along its length, the shrinkage problem is practically nonexistent. This factor makes balloon framing ideal for houses constructed with exterior walls of masonry, stucco, or brick veneer. It is also useful for trilevel houses. *Fig. 3–9* illustrates the difference in shrinkage in the two types of framing.

The third type of framing, post-and-beam, will be covered in Chapter 5. It differs from platform and balloon framing in that it utilizes exposed beams and plank ceilings.

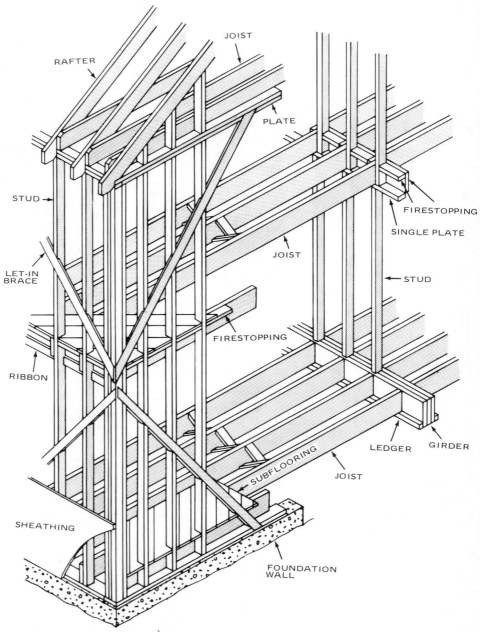

Figure 3–7 The balloon frame.

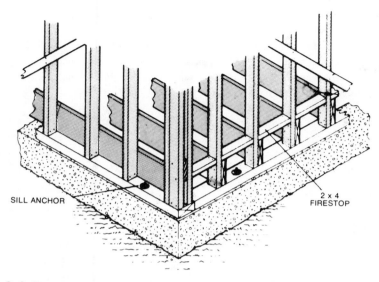

SILL ANCHOR

2 x 4
FIRESTOP

Figure 3–8 Firestops are placed between studs to slow down spread of fire and smoke.

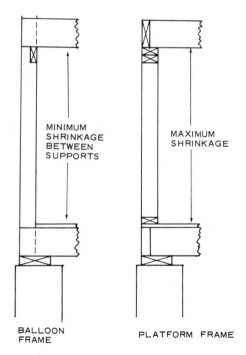

MINIMUM
SHRINKAGE
BETWEEN
SUPPORTS

MAXIMUM
SHRINKAGE

BALLOON
FRAME

PLATFORM FRAME

Figure 3–9 Comparison of shrinkage of members in balloon and platform framing systems.

Termite Protection

The damage caused by termites in the United States is quite extensive. At one time, termites were a problem only in southern states. However, today there are few areas in the country that are not infested by them. Because they eat wood fibers from within the timber, they can do considerable damage before they are discovered.

The termites responsible for most damage in lumber are the subterranean and dry-wood species. Of the two, the subterranean species does the most damage. *Fig. 3–10* shows the difference between a winged ant and a winged termite. The termite has a straight body compared to the hourglass shape of the ant.

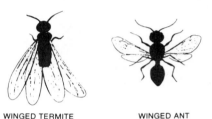

WINGED TERMITE   WINGED ANT

Figure 3–10 Termites are often confused with ants.

There are several ways to control termites and thus prevent damage to wood-frame members. One method is to poison the soil around the foundation. This, however, may pose ecological problems. It is better to use chemically treated lumber for construction instead. All of the lumber may be so treated if the area is highly infested. In less active areas, the lumber used below the first floor, including sills, joists, and subfloors, should be treated wood. These structural members are in the danger zone because they are in close proximity to the soil.

To minimize infestation, good drainage away from the foundation is essential. Termites must have moisture; they cannot live without it. They also thrive on wood scraps which are sometimes buried in the soil when the foundation is backfilled.

As a further precaution, termite shields may be used to thwart the ambitious termite. The shield consists of a metal strip, usually galvanized steel or aluminum, placed on the foundation wall under the sill and bent downward at an angle of 45 degrees. *Fig. 3–11* shows how these metal shields are used on foundation wall and pipes. While the shields do not prevent termite infestation, they form a barrier which forces the termites to build shelter tubes where they can easily be seen and destroyed. The shield must be continuous, without breaks or holes in the metal. All joints should be soldered or filled with tar.

Special attention should be paid to termite barriers for slab-on-ground construction, because termites may enter through joints between the slab

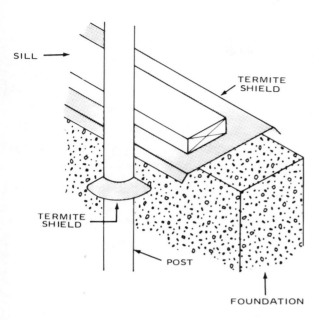

SILL →

TERMITE
SHIELD

TERMITE
SHIELD

POST

FOUNDATION

Figure 3–11 Method of applying termite shields at exterior foundation wall and around posts.

and wall or through openings made for plumbing or conduit. The use of termite shields, coal-tar pitch, or chemical soil treatment is recommended.

Sill Construction

The sill rests on the foundation wall and forms the bearing surface for walls and joists. It is used in frame and brick veneer structures but not when outside walls are solid brick or masonry.

Although the size may vary, 2 × 6s are commonly used for sills. Some workers prefer to set the sill in a bed of mortar; however, sill sealer consisting of 1-inch-thick fiberglass may be used as an alternative. The mortar or sealer will effectively fill in any voids to keep out insects and weather. The lumber used for sills should be either foundation-grade cedar or redwood or some other species that has been treated with preservative.

Choose flat, warp-free lumber for the sills. Cut the length of the first sill so that it equals the length of the foundation wall less two thicknesses of sheathing. Place this piece on the foundation wall and butt it against the anchor bolts. Prop the sill to prevent it from shifting. Mark the bolt locations by drawing a line at each side of the bolts. Next, measure the distance from the center of the bolts to the outer edge of the foundation. Deduct the sheathing thickness and transfer the dimension to the sill. Place the mark between the two lines previously drawn (*Fig. 3–12*). Since the bolts may be out of alignment, repeat this procedure for each bolt along the length of the wall. Bore the bolt holes ¼ inch larger than the bolt diameter.

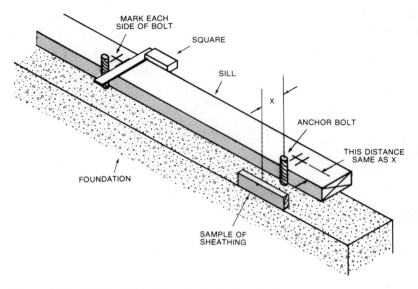

Figure 3–12 Method of laying out anchor-bolt holes.

Temporarily install the sills as each section is completed. This will facilitate cutting and fitting. When all of the sill pieces have been fitted, remove them from the wall and apply the sill sealer. If termite shields are required, place them over the sealer and then replace the sills. Use washers under the nuts and tighten them snugly (*Fig. 3–13*). The sill should remain straight and level. If the top of the foundation wall has deep low spots, these should be filled with mortar before the sills are bolted in place.

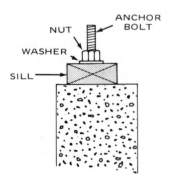

Figure 3–13 Anchor-bolt detail.

Beams and Girders

When the span to be crossed by joists is great, beams or girders are used to support them. Beams and girders serve the same purpose. When made of lumber they are called girders. If made of steel, they are called beams. (The two terms are often used interchangeably.) Girders and beams are load-bearing members and thus must be firmly supported. At the ends they rest on foundation walls or pilasters. At intermediate points they are supported by posts or columns.

Wood girders may be solid or built up. Steel beams usually have a cross section resembling the letter "I" and were formerly called I-beams. However, they are now called S-beams. Wide-flange beams, formerly called WF beams are now known as W-beams (*Fig. 3–14*). Steel beams are usually supported by Lally columns and girders with wood posts or Lally columns. *Fig. 3–15* shows a beam supported by a Lally column. The Lally column is a steel pipe flanged at each end and usually filled with concrete. Girder pockets in the foundation must be wide enough to provide for air circulation at the girder end. This will help prevent decay (*Fig. 3–16*).

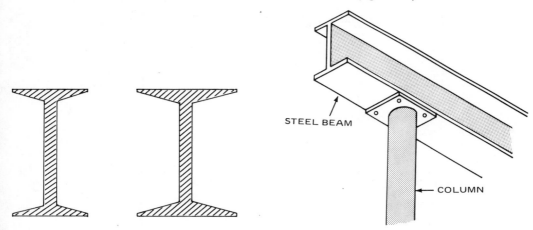

STEEL BEAM

COLUMN

Figure 3–14 Section through steel beams. *Left:* S-type. *Right:* W-type.

Figure 3–15 Bolt flange of column to steel beam. Use lag screws for wood girders.

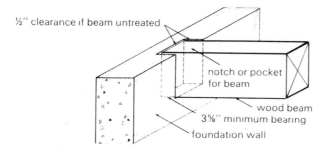

½" clearance if beam untreated

notch or pocket for beam

wood beam

3⅝" minimum bearing

foundation wall

Figure 3–16 Girder pockets require clearance at sides and end to provide air circulation.

The size of the girder is determined by the load it must carry. The cross section and length will be specified in the construction plans. Built-up girders of kiln-dried lumber are preferred over solid ones because they undergo less shrinkage. Built-up girders are usually made from two or more pieces of 2-inch stock. Splicing joints must be staggered and located over posts. For best results, the planks should be laminated by gluing. They may be fastened with either nails or bolts or a combination of these. Bolts should be of at least ⅝-inch diameter, staggered and spaced no more than 20 inches apart. Two-piece-girders are nailed from one side with 10d nails, staggered and 16 inches apart. Three-piece girders are nailed from both sides, using 20d nails, staggered and spaced 30 inches apart (*Fig. 3–17*).

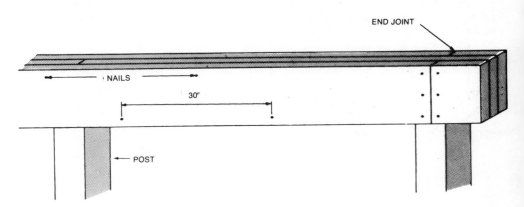

Figure 3–17 Arrangement and nailing details for built-up girders.

Floor Joists

Floor joists are main supporting members which carry floor loads to the sills and girders. They must be of sufficient strength and stiffness to withstand the intended load with a minimum of deflection. Joists lacking the proper stiffness will cause nails to pop in walls and ceilings. They also cause an uneasy feeling for occupants walking across floors supported by them, because the floor tends to "give" under the load. Generally the joists are 2 inches thick and either 8, 10, or 12 inches in depth. Joists are commonly spaced 16 inches on center, but other spacings may also be used. Their size and spacing must conform to local building codes.

Joist Layout. Joist assembly differs for platform and balloon framing. In platform framing the joists are nailed to the sill and header. In balloon framing the joists are nailed only to the sill. In balloon framing, the layout for the joists can be made directly on the sill, but for platform framing it is

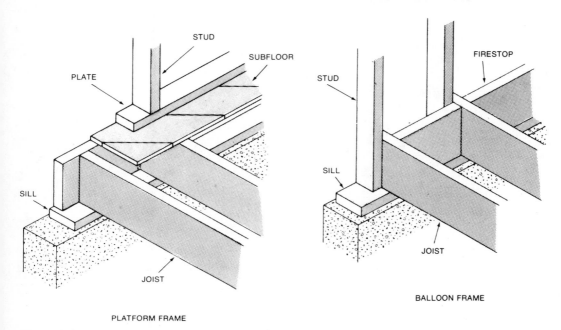

Figure 3–18 Joist assembly for platform and balloon framing.

more practical to make the layout on the header joist. The reason for this is clearly illustrated in *Fig. 3–18*. Joist layout is simplified if a joist rod is used. This is a long piece of wood with the joist spacings clearly marked on it. The markings are transferred to the header or sill (*Fig. 3–19*). The rod is also used to lay out the markings on the girder and sill (or header) on the opposite wall. To avoid errors, the joist location is indicated with a vertical line and the letter "X" as in *Fig. 3–20*. If the joists are butted or continuous, the X is placed on the same side of the vertical line at the opposite wall. If the joists are lapped, the X marking is placed on the opposite side of the line (*Fig. 3–21*). Make all the markings using the square. The spacings at the corners differ when joists are lapped. For 16-inch joist spacings, there is a 1½-inch difference between the first joist at opposite corners (*Fig. 3–22*). The locations for frame openings and doubled joists are also marked on the header or sill (*Fig. 3–23*).

Installing Joists. The nailing pattern used for joists is important. Use three 16d nails when nailing through a header into the joist. Use two 10d nails when toenailing into the sill. Check the building-code requirements for quantity and size of nails; these may vary from one locality to another.

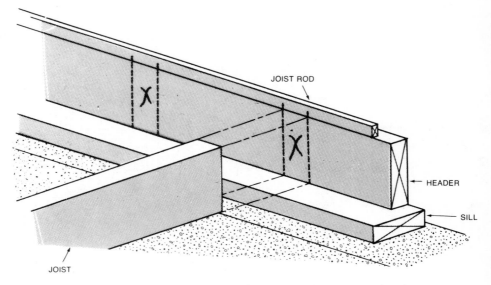

Figure 3–19 Use joist rod to lay out joist positions on header.

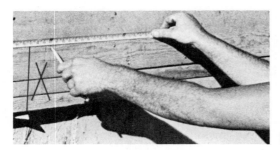

Figure 3–20 Header marked for joist position.

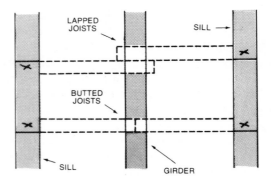

Figure 3–21 When joists are butted, place the X marking on the same side of the line. For lapped joists, place markings on opposite sides.

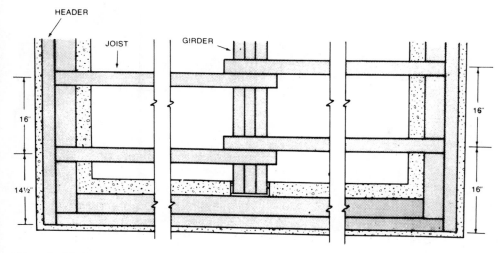

Figure 3–22 Spacing between header and first joist differs at opposite corners when the joists are lapped.

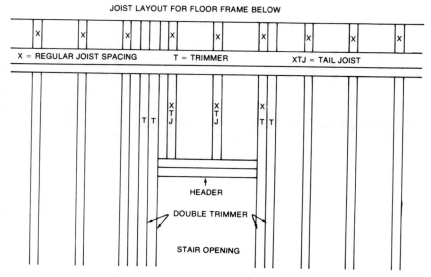

Figure 3–23 Header markings for joists. Regular joist spacing is not altered.

If the lumber used for the joist has a crown (curve), place it with the crown up. When loaded, the crown will tend to straighten (*Fig. 3–24*).

If the joists are butted, they are joined with overlay boards or metal plates made for the purpose (*Fig. 3–25*). Lapped joists should overlap at least 4 inches, and should be fastened with 10d nails.

Figure 3–24 Joist with crown should be installed with crown up.

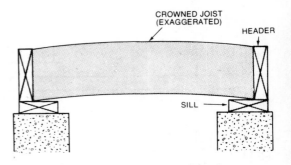

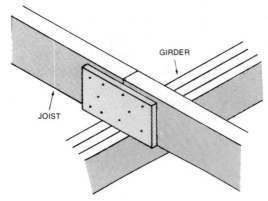

Figure 3–25 Overlay board is used to hold joists in line.

Joist Assembly at Girders. There are various methods of fastening the joists at the girders and beams. The simplest method is to rest the joist on top of the girder (*Fig. 3–26*). To gain more headroom in a basement, the joists can be notched and made to rest on a ledger nailed to the lower edge of the girder (*Fig. 3–27*). For flush ceilings under joists, stirrups or hangers may be used. These are shown in *Fig. 3–28*.

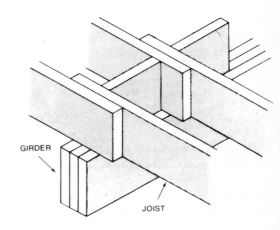

Figure 3–26 Lapped joists resting on girder.

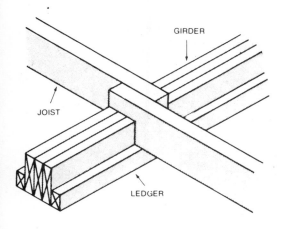

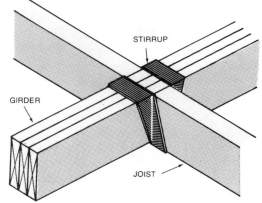

Figure 3–27 Notched joists resting on ledgers.

Figure 3–28 Joists supported by stirrups.

When joists are joined to steel beams, the method of fastening differs from that used for girders. In *Fig. 3–29*, a 2 × 4 nailer is bolted to the top of the beam. The joists rest on and are toenailed to this piece. If headroom is a problem, the joists may be made to rest on a ledger bolted to the lower part of the beam (*Fig. 3–30*). For flush headroom, a steel plate may be welded or bolted to the lower edge of the beam. The joist is then made to rest on this (*Fig. 3–31*). The notching must clear the beam to allow for shrinkage. Instead of a beam with a steel plate welded to it, a wide-flanged beam may be used. The flange of this beam is wide enough to support joists (*Fig. 3–32*).

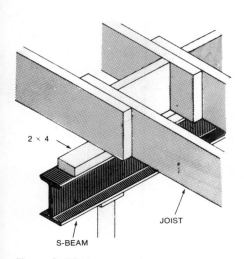

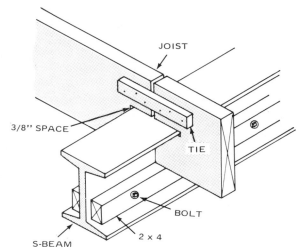

Figure 3–29 Joists resting on steel beam.

Figure 3–30 Joist fastened to steel beam. The space around beam allows for joist shrinkage.

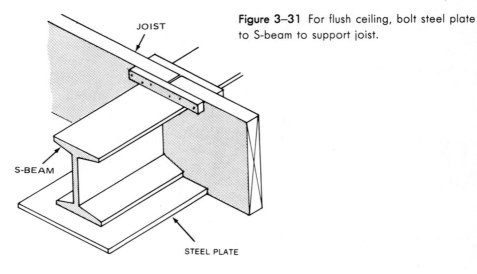

Figure 3–31 For flush ceiling, bolt steel plate to S-beam to support joist.

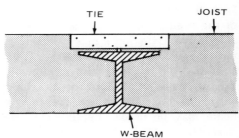

Figure 3–32 Wide-flanged beam (W-beam) supporting flush joists.

Floor-Frame Openings. When joists are cut through to make openings for stairwells, chimneys, and fireplaces, the floor frame is weakened considerably. To compensate for this, the shortened joists (called tail joists) are framed against doubled headers. These in turn are framed against double trimmers (*Fig. 3–33*).

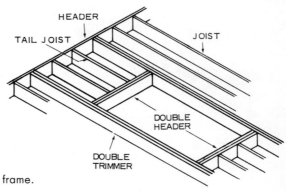

Figure 3–33 Tail joists in floor frame.

The nailing sequence for floor openings is important. The procedure described will permit end-nailing, thus eliminating the need for toenailing (*Fig. 3–34*). It is important to note that joist positions are never altered to accommodate an opening.

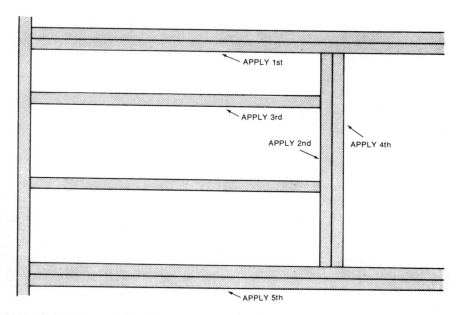

Figure 3–34 Floor opening framing procedure eliminates need for toenailing.

The opening sizes for stairwells are usually included in the architectural drawings. If they are not, the length and width of the opening must be determined according to the floor-to-floor height, stair width, angle of slope, and tread and riser sizes. (See the section on stairs for details.)

Doubled Joists. When a partition wall runs parallel to the joists, the joists supporting that wall must be doubled. The doubled joists may be spaced if the partition is to carry conduit, ducts, or pipes (*Fig. 3–35*). Use 2 × 4 spacer blocks between the joists.

Bridging. Studies made in recent years indicate that bridging is of doubtful value. This is especially so if the joists are fastened properly and the subflooring is properly nailed. However, most codes require bridging to be installed in mid-span at intervals not to exceed 8 feet. The bridging may be diagonal (also called cross-bridging) or solid. Diagonal bridging consists of 1 × 3 or 1 × 4 lumber placed crisscross between the joists (*Fig. 3–36*). Time can be saved by cutting the pieces on the radial-arm or miter saw. Set a gauge on the saw to ensure that all pieces will be uniform. To determine

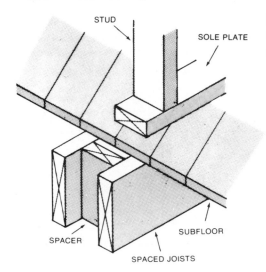

Figure 3–35 Method of spacing joists when partition above runs in same direction as joist.

STUD

SOLE PLATE

SUBFLOOR

SPACER

SPACED JOISTS

Figure 3–36 Diagonal bridging of floor joists.

the angles and length of each piece, place two joist cutoffs on a flat surface. The length is not important, but both should be equal. Space them carefully so that they are the same as the joist spacing. Now place a piece of 1-inch stock on edge diagonally, allowing the top edge to meet the corner of one block and the lower edge to meet the corner of the second block. Hold the stock firmly and mark the angular cuts from the underside

(*Fig. 3–37*). Cut several pieces and check them against the joists. If they fit, cut the required number.

Figure 3–37 *Left:* Method of laying out bridging template. *Right:* Bridging installed.

Snap a chalk line across the top of the joists, then fasten the bridges to either side of this line. Start the nails, two at each end, into the bridges and then toenail them flush with the top of the joist. Do not drive the bottom nails until the floor is in place and shrinkage is no longer a problem.

Solid bridging may be used instead of diagonal. It consists of solid lumber placed between each joist as shown in *Fig. 3–38.* To permit end-nailing into the blocks, the pieces can be staggered. There are disadvantages in using solid bridging. The pieces must be very accurately cut and shrinkage can be the cause of loose joints later.

Figure 3–38 Staggered solid backing permits end-nailing.

Metal bridging is often used. It is easily installed and does not require precise cutting and fitting. Prongs in the steel strut are driven into the joist, as shown in *Fig. 3–39*.

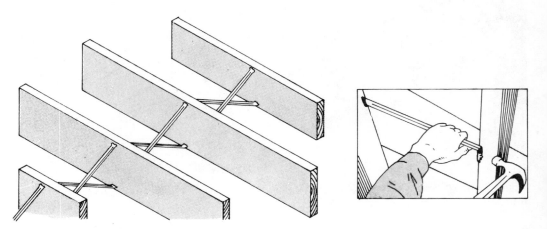

Figure 3–39 Compression-type metal bridging is easily hammered into place.

Overhangs. Often in residential construction the carpenter will be confronted with problems in floor framing brought about by wall projections. These are usually for bay windows and second-story overhangs. If the projection runs in the direction of the floor joists, the joists are simply extended, as in *Fig. 3–40*. If the extension exceeds 24 inches it may be necessary to use special anchorage for the joists at the opposite end. If the extension extends perpendicular to the joists, cantilever joists are used. These are fastened to a doubled joist (*Fig. 3–41*). The inboard ends of the joists are supported by a ledger or joist hangers. Since the thrust is upward, the ledger is fastened along the top edge of the doubled joist. Under normal conditions, the cantilever joist should extend inward twice the distance of the overhang.

Bathroom Floor Framing. Bathroom floors usually support heavier loads than other floors of the house, because of the weight of the fixtures and of ceramic tiles, if used. The joists supporting a tub are usually doubled (*Fig. 3–42*). If the floor frame is lowered to accommodate the extra thickness of tiles set in cement, the joists may have to be doubled and perhaps spaced closer together (*Fig. 3–43*). Another method of framing the floor for tiles is shown in *Fig. 3–44*. The joists are chamfered at the top to lessen the chance of the concrete fracturing at the corners.

Floor framing for entryways, kitchens, and other areas where slate, brick, or heavy flooring may be used is treated in a similar manner.

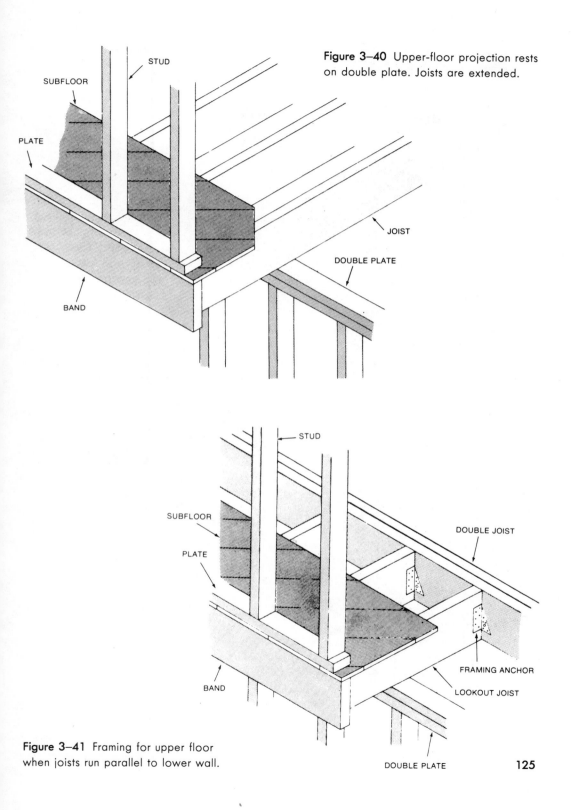

Figure 3–40 Upper-floor projection rests on double plate. Joists are extended.

STUD

SUBFLOOR

PLATE

JOIST

DOUBLE PLATE

BAND

STUD

SUBFLOOR

PLATE

DOUBLE JOIST

FRAMING ANCHOR

LOOKOUT JOIST

BAND

Figure 3–41 Framing for upper floor when joists run parallel to lower wall.

DOUBLE PLATE

125

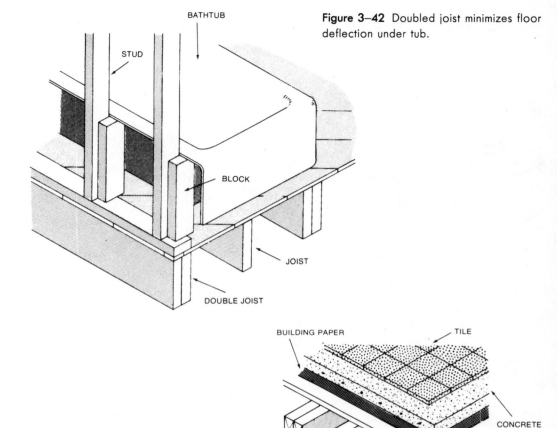

Figure 3–42 Doubled joist minimizes floor deflection under tub.

BATHTUB

STUD

BLOCK

JOIST

DOUBLE JOIST

BUILDING PAPER

TILE

CONCRETE

Figure 3–43 Heavy tile floor supported by doubled joists. Spacing between joists is reduced.

DOUBLE JOISTS

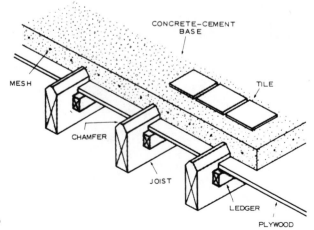

CONCRETE–CEMENT BASE

MESH

TILE

CHAMFER

JOIST

LEDGER

PLYWOOD

Figure 3–44 Floor framing for ceramic tiles.

Notching or Drilling Joists. It is often necessary to notch, drill, or cut joists to accommodate plumbing, heating, and electrical lines. The cutting or drilling must be done with care to prevent weakening of the structure.

A joist must never be notched beyond a point one-third the distance from its end. Also, the depth of the notch should not exceed one-sixth the joist depth (*Fig. 3–45*). Drilled holes must not exceed 2 inches in diameter, nor should they be closer than 2½ inches from the top or bottom edge (*Fig. 3–46*). If it becomes necessary to cut, drill, or notch beyond the limits stated above, the joist must be reinforced with scabs on both sides. In some cases it may be necessary to treat the area affected as a floor opening with tail beams and headers. In some situations it may be advisable to add another joist.

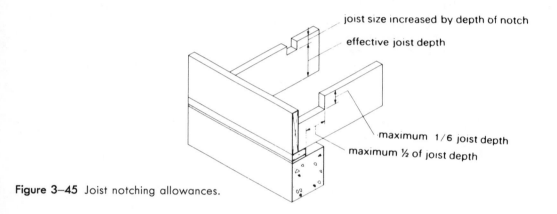

Figure 3–45 Joist notching allowances.

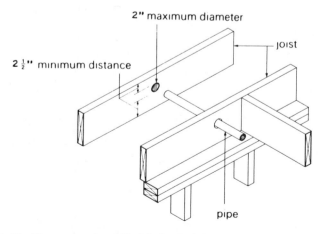

Figure 3–46 Allowances for drilled holes in joists.

Subflooring

The subfloor is placed over the joists and serves as a platform for the first floor. It supports the subsequent framing as well as the finish floor. The subfloor may consist of boards or plywood.

Board Subflooring. Board subfloors may be laid out straight or diagonally. If laid out straight, the boards must be placed perpendicular to the joists. The finish floor must then be placed perpendicular to the subfloor. However, if the boards are placed diagonally, the finish floor may be applied parallel to or across the joists. Diagonally placed boards also have a bracing effect on the framing. If the home is to be of two or more stories, the subfloor boards should run in opposite diagonal directions on alternate floors.

If strip flooring is to be used, the subfloor boards should be 1 inch thick and not more than 6 inches wide. Except for end-matched boards, all joints must be over a bearing member (*Fig. 3–47*). Boards should be kiln-dried or thoroughly air-dried, because the use of green lumber will frequently cause squeaks and cracks.

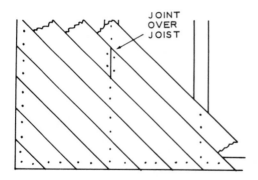

JOINT
OVER
JOIST

Figure 3–47 Unless end-matched boards are used, joints must fall over a joist.

Subfloor boards are fastened with 8d nails. Use two nails for boards less than 6 inches wide. For greater accuracy when laying boards diagonally, start the first board about 10 feet from the corner. Mark and measure a point 10 feet along the header and end joists. This will establish the 45-degree diagonal. Align the first board with these marks and nail securely (*Fig. 3–48*). Work outward away from the corner. The corner can be filled with the short cutoffs that accumulate as the work progresses. Leave a slight gap between boards to prevent rainwater from accumulating while the house is under construction.

Plywood Subflooring. Because of its ease in handling and economy, plywood subflooring is extensively used in this country. Large sheet sizes reduce installation time and provide a strong, smooth surface suitable for any

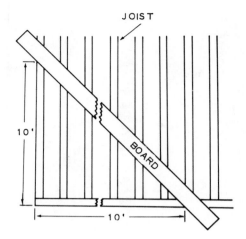

Figure 3–48 First board of diagonal subfloor should start at a distance of 10 feet from corner.

type of finish floor. Normally, interior-grade plywood is used for subflooring; however, if strip flooring is to be installed, plywood with an exterior glue line is recommended. Sheets 4 × 8 feet or larger are available in thicknesses of ½, ⅝, ¾, and ⅞ inch. Panels for subflooring are marked with rafter and joist spacings. For example, ³²/₁₆ is a typical marking on a sheet of ⅝-inch plywood. The 32 means that it is suitable for rafters spaced 32 inches on center. The 16 indicates the joist spacing permitted. The plywood thickness is determined by the floor-load requirements.

Plywood is installed with the long dimension of the panel, or outer plies, perpendicular to the joists. The joints in adjacent rows must be staggered. If the first row is stated with a 4 × 8-foot panel, the second row must be started with a 4 × 4-foot panel (*Fig. 3–49*). Lay the panels with the better face up and leave a ¹/₁₆-inch space at the ends and a ⅛-inch space at the sides. In humid or wet areas, double the amount of spacing at the joints. Use 6d nails for ½-inch plywood and 8d nails for ⅝- and ¾-inch thicknesses. Space the nails 6 inches apart around the perimeter, and use 10-inch spacing along intermediate members. Snap chalk lines to locate joist centers. To speed up installation, automatic nailers may be used to fasten subfloors (*Fig. 3–50*).

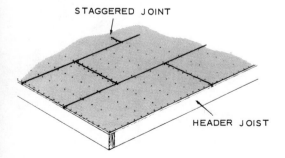

STAGGERED JOINT

HEADER JOIST

Figure 3–49 Plywood subflooring.

Figure 3–50 Automatic nailer in use.

2:4:1 Subfloor-Underlayment System. Tongue-and-groove plywood can be used to lay a combination subfloor and underlayment surface. The seven-ply panels are available in 1⅛- and 1¼-inch thickness and can be applied over 2-inch joists with maximum spacing of 32 inches center to center and over 4-inch joists with maximum spacing 48 inches center to center (*Fig. 3–51*). There must be a spacing of ¹/₁₆ inch at all end and edge joints. Nails should be 8d ring shank or 10d common smooth shank and spaced 6 inches apart. The 2:4:1 plywood provides a smooth, solid surface suitable for any type of finish floor covering—such as resilient tile, linoleum, carpeting, or hardwood flooring. It is one of the fastest, simplest floor-construction systems devised.

Figure 3–51 Thick panels serve as combination subfloor-underlayment.

Glued Floors

Newly developed adhesives permit the use of single-layer floors instead of the conventional two-layer system. The advantages of such a system include reduced labor and materials costs and elimination of squeaking and nail-popping. Floor stiffness is also increased by as much as 70 percent. When glued, the floor and joist become fused into an integrated T-beam unit.

Use underlayment grade tongue-and-groove plywood with a thickness of ½, ⅝, or ¾ inch. The outer plies (face grain) must run perpendicular to the joists. Use ¹/₁₆-inch spacing at all joints. Apply a bead of glue to each joist to be covered by a panel. Repeat the procedure for each panel. For added stiffness, glue may also be applied to the edge grooves. Secure the panels with 6d annular-threaded nails, 12 inches on center at all supports. *Fig. 3–52* shows adhesive being applied with a caulking gun.

Figure 3–52 Gluing underlayment with caulking gun.

Wall and Ceiling Framing 4

Wall framing includes the vertical studs and horizontal members which make up the exterior and interior walls of a structure. The framing also supports ceilings, upper floors, and the roof. The walls must be capable of supporting dead loads (the weight of the wall and other members which it supports) as well as live loads such as earthquakes and winds. The properly designed wall will transmit all of these forces to the foundation.

Conventional wall framing consists of nominal 2 × 4-inch studs spaced 16 inches on center (*Fig. 4–1*). However, 24-inch spacing is not uncommon, especially when thicker covering materials are used. The lumber used for wall construction must be sound, warp-free, and reasonably dry. A moisture content of 15 percent is ideal.

Figure 4–1 Residential frame house under construction using 16-inch stud spacing.

A typical wall frame is shown in *Fig. 4–2*. It consists of sole plates, studs, top plates, headers, cripples, trimmers, and sills. A brief description of the various wall members follows.

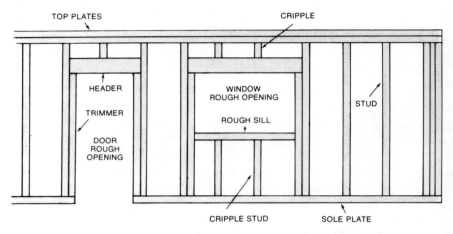

Figure 4–2 Typical wall-frame section. Stud spacing is not altered for openings.

Sole Plate.   This is the lowest horizontal member of the wall. It supports the studs and is nailed through the subfloor into the floor joists and the headers.

Studs.   Studs comprise the vertical members of a wall. They are made of 2 × 4s and are spaced 16 or 24 inches on center. The center-to-center spacing allows the use of standard-size wall panels (*Fig. 4–3*).

Studs that are shortened because of door or window openings are called *cripple studs*. They are spaced with the same center-to-center spacing as regular studs.

Trimmer studs are used adjacent to regular studs at door and window openings. They serve to stiffen the sides of the openings. They also bear the weight of the headers resting upon them (*Fig. 4–4*).

Headers carry the load above door and window openings to the trimmers. They are usually made by nailing two pieces of 2-inch stock with a spacer in between (*Fig. 4–5*). The spacer is used to make the header as thick as the stud on which it rests. The size of a header depends upon the load. Typical sizes are shown opposite. These are for outside walls in one-story buildings. Headers are placed on edge for maximum efficiency.

When the load is extra heavy or the span especially wide, trussed headers may be used. They shrink less than wide solid wood headers—a common cause of plaster cracks on interior walls. *Fig. 4–6* shows two types of trussed headers commonly used.

Figure 4–3 Modular stud spacing permits use of standard-size wall covering.

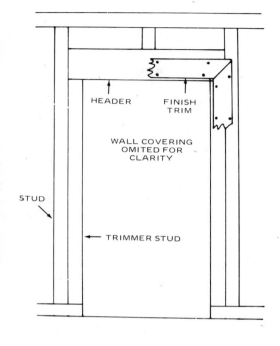

Figure 4–4 Trimmer studs support the load of headers. They also provide nailing surface for trim.

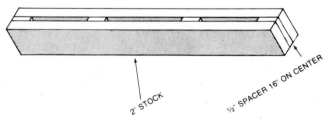

Figure 4–5 Header construction for door and window openings. Spacers are used to increase thickness.

OPENING SIZE	HEADER (2 pcs. on edge)
3'6''	2 × 4
6'0''	2 × 6
8'0''	2 × 8
10'0''	2 × 10
12'0''	2 × 12

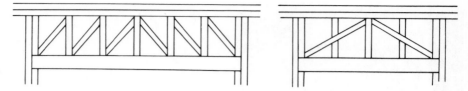

Figure 4–6 Trussed headers are often used for wide openings.

Another type of header for wide spans is the box type. As its name implies, it is built like a box with a ½-inch plywood skin glued to one or both sides of the frame (*Fig. 4–7*). Because of the lightweight construction, shrinkage is kept to a minimum.

As in all phases of construction, local building codes must be consulted for specific requirements.

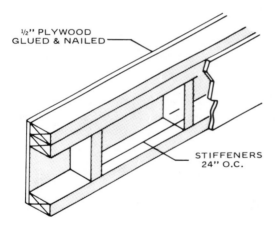

½" PLYWOOD
GLUED & NAILED

STIFFENERS
24" O.C.

Figure 4–7 A box header for wide spans. Design keeps shrinkage to minimum.

Corner Construction

There are various methods of forming corners for a wall frame. Three methods are shown in *Fig. 4–8*. In each method three studs are used, but in different configurations. In one, a filler strip is used to form the nailing surface for interior finish.

In platform construction, where wall sections are usually preassembled, the corners are formed when the wall sections are raised into position. In this method the second stud is placed in the end wall spaced with 2 × 4 blocking.

Another method is to build and erect the corner posts first. They are carefully plumbed, then used as a guide for plumbing the preassembled

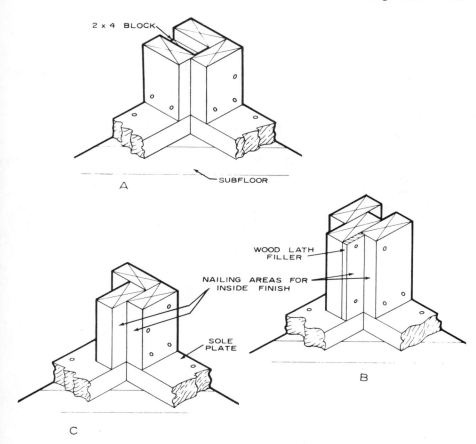

2 x 4 BLOCK

SUBFLOOR

A

WOOD LATH
FILLER

NAILING AREAS FOR
INSIDE FINISH

SOLE
PLATE

B

C

Figure 4–8 Three types of corner stud assembly: A. Standard outside corner. B. Special corner with lath filler. C. Special corner without lath filler.

wall sections. With this method, however, the preassembled wall sections cannot be sheathed beforehand.

Intersecting Walls

When a partition wall intersects an outside wall, some means must be provided to tie the two walls together firmly. Also, corners must be formed to provide a nailing surface for the interior wall-covering material. *Fig. 4–9* shows two methods of construction. In one, extra studs are placed on the outside wall. The partition is attached to these. In the second method, the stud spacing of the outside wall is maintained. Blocking and a backing board of 1 × 6 or 2 × 6 are installed to form the intersection. Anchorage must be firm in order to prevent air spaces and cracked plaster joints.

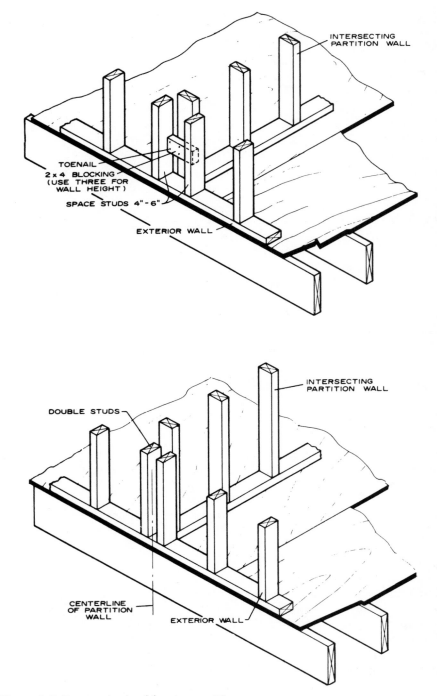

Figure 4-9 Two methods of forming partition corners.

Exterior Wall Construction

As mentioned previously, two types of wall framing are commonly used—platform and balloon. Of the two, the platform method is the more popular. Balloon framing is more practical for two-story homes when stucco or other masonry exterior wall covering is used.

Laying Out the Sole Plate. The first step involved in wall framing is to mark the sole and top plates. The marking will show the location of the studs, rough openings, trimmers, and cripple studs. Choose straight 2 × 4s and place them around the perimeter of the structure on the subfloor. Let the side wall pieces align with the ends of the outside edge of the sill. If it is necessary to add 2 × 4s to reach the end of the building, cut the plate at the center of a stud location. The end wall plates are placed between those of the side walls (*Fig. 4–10*).

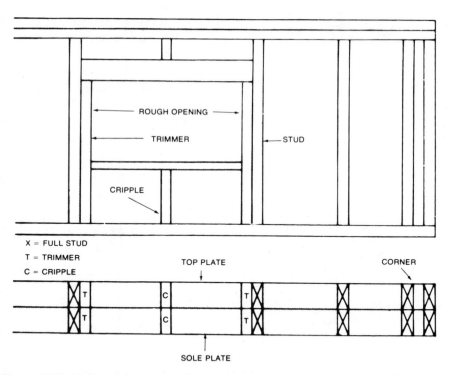

Figure 4–10 Method of marking sole and top plates to construct wall shown above.

To prevent errors due to movement, tack the 2 × 4s to the subfloor. Use working drawings to locate the studs, trimmers, and cripples. Draw two lines 1½ inches apart to represent each stud space. Mark the regular studs with an "X." Use the letter "C" for cripples and "T" for trimmers (*Fig. 4–11*).

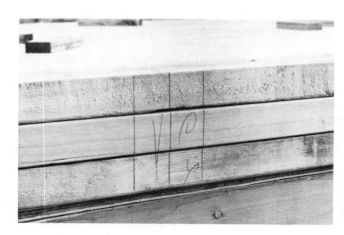

Figure 4–11 Mark sole plate to indicate location of studs.

If 16-inch stud spacing is used, the edge of the first stud will measure 15¼ inches from the outside corner of the frame. Thereafter, all intermediate studs will be 16 inches on center or 16 inches from edge to edge (*Fig. 4–12*).

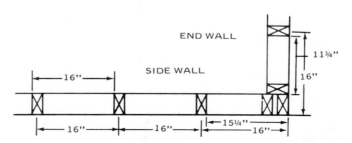

Figure 4–12 Stud spacing at end and side walls.

To minimize errors, lay out the center lines for doors and windows. Working from the center line, measure to each side one-half the opening width. This will indicate the inside edge of the trimmer stud. Next to this, lay out the regular stud (*Fig. 4–13*). Do not vary stud spacings because of openings. If necessary, add extra studs. Transfer the markings to another set of 2 × 4s which will be used for the top plate. Repeat the procedure for the opposite and end walls.

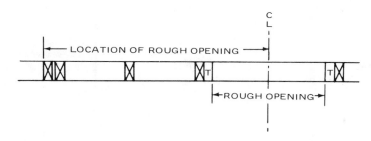

Figure 4–13 Measure rough openings equally from center of opening.

Stud Pattern.   The stud pattern is used by the carpenter to increase efficiency and to prevent errors. It is usually made from a straight piece of 1 × 4 or 2 × 4 stock. It is marked off with the floor levels, ceiling heights, and door and window heights. Cross sections of the various members are shown full size. (*Fig. 4–14*).

Markings should include the sole, rough sill, header, and top plates. For platform construction, a 10-foot story pole is sufficient. For balloon construction, you will need one to extend past the top plate of the second story.

Figure 4–14 Mark stud pattern with the various wall members such as floor levels, doors, and windows.

Building the Wall Sections.   Use the stud pattern to mark off the length of the various stud members. Cut them to size and place them in approximate position on the subfloor. The subfloor must be clean and free of any obstruction. Place the sole and top plates on edge about 8 feet apart. Nail the sole and top plates to the regular studs as shown in *Fig. 4–15*. Use two 16d nails per stud. Before installing the door and window studs, nail the trimmers to them. Be sure to align the bottom edges carefully.

After the regular studs and trimmers are in place, install the door headers, window headers, and window rough sills. Position them tightly in place, then fasten them with 16d nails. Next, add the cripple studs above

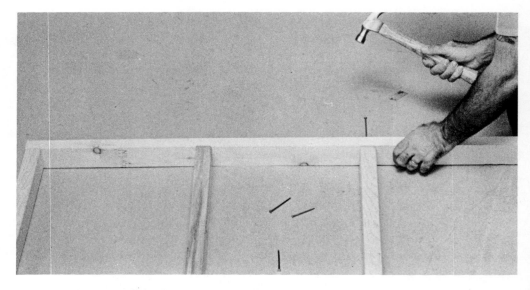

Figure 4–15 Nailing sole plate to studs.

the doors and windows. Toenail the lower ends of the cripples to the headers with 8d box nails. The box nails have a thin shank and are less likely to split the wood. Install studs or blocking if needed for intersecting partitions.

If exterior wall bracing is required, it should be installed as shown in *Fig. 4–16*. Use 1 × 4 lumber and notch the studs to allow the brace to lie flush with the studs. Temporarily tack the brace to the studs, then mark each stud. Remove the brace and cut the sides of the notch with a saw.

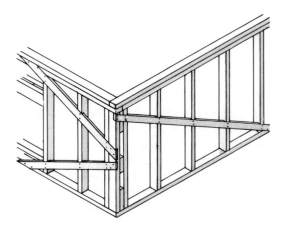

Figure 4–16 Set let-in braces flush and nail to each stud.

Use a chisel to remove the waste between the saw cuts. The brace should be nailed at each stud. This is called "let-in" bracing. If 2-inch lumber is specified for the bracing, it is not let in. Instead, the 2-inch stock is cut diagonally to fit between each pair of studs. Each block is toenailed into place (*Fig. 4–17*).

DIAGONAL BRACING

Figure 4–17 Cut block bracing diagonally and toenail it between studs.

Some workers prefer to install the sheathing after the wall frame is in place. Others find it more expedient to sheath the frame while it is flat on the subfloor. If the sheathing is installed before the wall frame is erected, be sure the frame is square. Check the diagonals. If they are equal, the unit is square (*Fig. 4–18*). To insure that the frame remains square, tack it to the subfloor.

Erecting the Wall. Except for very large sections, the preassembled wall sections can usually be erected without the use of a crane (*Fig. 4–19*). Before the section is raised, install temporary braces at the ends. These will be used to plumb and support the wall. If the wall frame is not sheathed before it is raised, diagonal bracing should be applied to the outside to prevent the frame from swaying.

When the frame is in proper position and plumbed, fasten the temporary braces, then proceed to fasten the sole plate to the subfloor. Use 16d nails driven through the subfloor and into the joists. Follow the same procedure for each wall section.

Figure 4–18 Checking diagonals of wall frame.

Figure 4–19 Erecting wall sections.

Partitions

Partitions are installed after the exterior walls are in place. Common practice is to install only load-bearing partitions at this point. Other partitions can be installed after the roof is in place and the structure has been made watertight. Load-bearing partitions are those running perpendicular to the ceiling joists. Non–load-bearing partitions run parallel to the joists.

Locate and mark the partitions on the subfloor by snapping chalk lines. Construct and install partitions as was done for the exterior walls (*Fig. 4–20*). Erect the long partition first, followed by the shorter cross-partitions and closets. It is not necessary to use headers in non–load-bearing partitions. Also, lighter construction may be utilized to keep costs down.

Figure 4–20 Erecting a partition wall.

When non–load-bearing partitions occur between ceiling joists, they should be secured to blocking nailed between joists, as shown in *Fig. 4–21*. A 1 × 6 serves as a nailer for the ceiling covering material. Other methods of providing nailers for these partitions are shown in *Fig. 4–22*.

After partitions are erected, cut away the sole plate between doors. Use a handsaw and cut flush along the trimmer (*Fig. 4–23*).

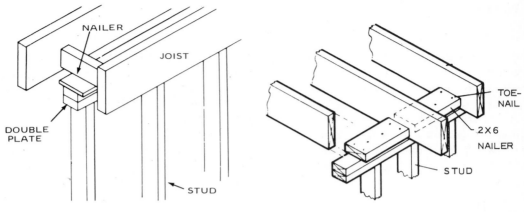

Figure 4–21 Method of attaching non–load–bearing partition to frame wall.

Figure 4–22 Application of nailer when stud wall is at right angle to the joist.

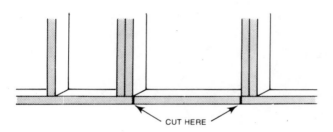

Figure 4–23 Cut sole plate between trimmers for doorway.

Special Framing.   Walls containing plumbing will require wider-than-normal studding and plates. For a wall containing a 4-inch cast-iron soil stack, use 2 × 6 studding and plates. As an alternative to using heavier than 2 × 4 studs, some carpenters add wood strips to the edge of the regular studs to increase their thickness (*Fig. 4–24*). *Fig. 4–25* shows another method of constructing the wall. Here, the 2 × 4 studs are turned sideways. Blocking is used between the two studs to add rigidity.

Figure 4–24 Add furring to studs to increase wall thickness.

Figure 4–25 Framing for soil stack.

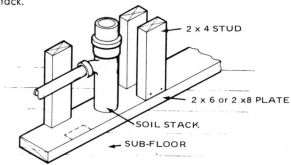

There are occasions when floor joists must be notched or drilled. Care must always be taken that these procedures do not affect the strength of the member.

Heating and air conditioning also pose special problems for the carpenter. Framing must allow for ducts and registers. If the ducts run from one floor to the next, the carpenter must align the studs for both floors. If a register does not come within a stud space, the stud will have to be cut and a header installed. The same applies for openings made for convectors, medicine chests, and similar items (*Fig. 4–26*).

Blocking is used to support tubs, as shown in *Fig. 4–27*. Backing is used to support various fixtures such as washbasins, toilet tanks, towel

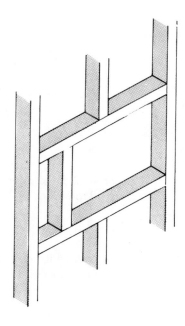

Figure 4–26 Framing for medicine chest.

bars, and shelves. Wall backing may be installed by notching, as in *Fig. 4–28*, or with nailing strips (*Fig. 4–29*).

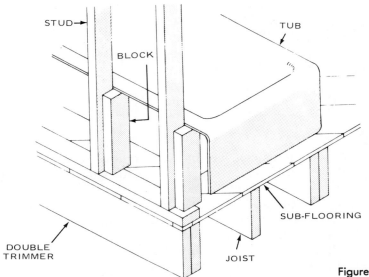

Figure 4–27 Blocking for bathtub.

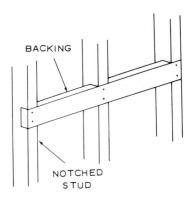

Figure 4–28 Notched backing for fixture.

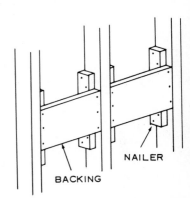

Figure 4–29 Backing for fixtures attached to nailers.

Walls combining masonry and framing must be built with care. Never let a wall frame rest on masonry. Use a header, as shown in *Fig. 4–30*. The size of the header will depend on the span.

Figure 4–30 A header is used to support wall frame above masonry.

Wall Sheathing

Wall sheathing is used to cover the framework of a structure. It may consist of solid lumber, plywood, fiberboard, or gypsum board. In addition to serving as a flat base for the outer covering material, it also minimizes air infiltration when it is properly installed. Some sheathing materials have insulating qualities.

Wood sheathing consists of nominal 1-inch boards in 6-, 8-, 10-, or 12-inch widths. The boards may have a square edge, shiplap, or tongue-and-groove pattern, and they may be applied horizontally or diagonally. The boards are cut so the joints occur at the center of a stud. If end-matched boards are used, the joints need not be placed over a stud (*Fig. 4–31*). Use two 8d nails at each stud for 6- or 8-inch boards and three 8d nails for 10- and 12-inch material. When sheathing is installed diagonally, corner bracing is not required. The sheathing is installed diagonally at a 45-degree angle with a direction change at each corner.

Plywood sheathing is widely used today. It is available in 4 × 8-foot and longer sheets. Installation time is much less than in the case of board

Figure 4-31 Application of board sheathing.

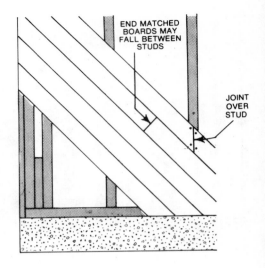

END MATCHED
BOARDS MAY
FALL BETWEEN
STUDS

JOINT
OVER
STUD

sheathing. Plywood is the strongest of the sheathing materials and is available with interior or exterior glue lines. Thicknesses range from $5/16$ inch to $5/8$ inch, but the most commonly used thicknesses are $5/16$-inch and the $3/8$-inch panels. A minimum $3/8$-inch thickness is recommended when exterior finish is to be nailed to the sheathing. It may be applied vertically or horizontally (*Fig. 4-32*). Greater stiffness is achieved when it is applied horizontally. Use 6d nails spaced 6 inches apart at the edges and 12 inches along intermediate studs.

Often plywood is used in combination with fiberboard or gypsum. In this application, the plywood is used vertically at the corners, and corner bracing is not required.

Several types of fiberboard, of various densities, are available for sheathing. Some of the materials used in the manufacture of fiberboards include wood pulp, sugar cane, and cornstalks. They are pressed into flat boards either $1/2$ or $25/32$ inch thick; these are usually asphalt-impregnated so they will be water-resistant. Fiberboard is available in 2 × 8, 4 × 8, and 4 × 9-foot sizes. The 2 × 8 boards are installed horizontally; the 4 × 8 and 4 × 9-foot boards are installed vertically. Use $1\frac{1}{2}$-inch roofing nails for the $1/2$-inch boards and $1\frac{3}{4}$-inch roofing nails for the $25/32$-inch boards. Space the nails 3 inches on center around the edges and 6 inches on center on intermediate studs.

Corner bracing is needed for the horizontal application and for $1/2$-inch boards used vertically. Bracing is not needed for $25/32$-inch material used vertically.

Gypsum sheathing consists of fiberglass-impregnated gypsum sandwiched between water-repellent paper. Standard-size sheets are 2 × 8, 4 ×

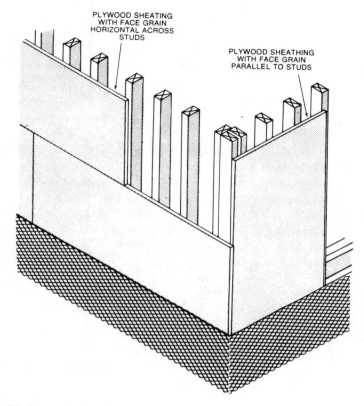

PLYWOOD SHEATING
WITH FACE GRAIN
HORIZONTAL ACROSS
STUDS

PLYWOOD SHEATHING
WITH FACE GRAIN
PARALLEL TO STUDS

Figure 4–32 Plywood sheathing may be applied vertically or horizontally.

8, and 4 × 9 feet by ½ inch thick. Gypsum sheathing is used like fiberboard and requires corner bracing. It is not suitable as a nailing base.

Nonsheathed Walls. It is possible to eliminate wall sheathing entirely, provided that a proper, structurally sound sheet siding is used. Generally, vertical applications do not require bracing, but horizontal sidings usually require let-in corner bracing.

Ceiling Frame

The ceiling joists are installed after the wall framing is completed. In addition to providing a nailing surface for the ceiling finish material, they tie opposite walls together and resist rafter thrust. Normally, they span the width of the house (*Fig. 4–33*). However, there are occasions when the joists may run in two directions to reduce the length of the span. When

Figure 4–33 Joists running the width of the house.

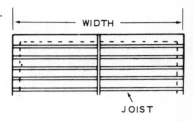

this is done, a nailer must be installed above the double top plate which runs parallel to the joist (*Fig. 4–34*).

Joist size depends on various factors such as span, spacing, load, and wood species used. If an attic is to be used for living quarters, the joist requirements would be heavier than for an unoccupied attic. Whenever possible, the joist spacing should be the same as for the rafters. This will make for stronger construction, since the joists can be nailed to the rafters as well as the top plate (*Fig. 4–35*). If a double top plate is used, it is not necessary to align the joists over studs.

Joists are spliced or lapped over load-bearing partitions (*Fig. 4–36*). If the joists are lapped, use a spacer block between the two as shown in *Fig. 4–37*. This will permit straddling of the rafters.

Flush beams are used to support joists when wide ceilings are not supported by partitions. A ledger strip along the lower edge of the joist may be used to maintain the flush ceiling (*Fig. 4–38*). Metal hangers or steel straps may also be used (*Fig. 4–39*).

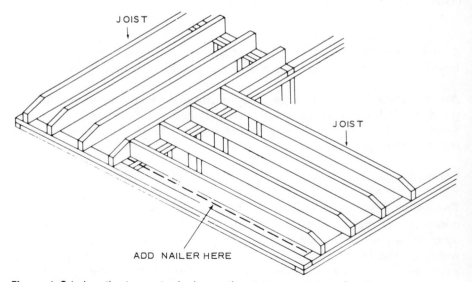

Figure 4–34 A nailer is required when ceiling joists run in two directions.

Figure 4–35 Fasten ceiling joists to top plate and rafters.

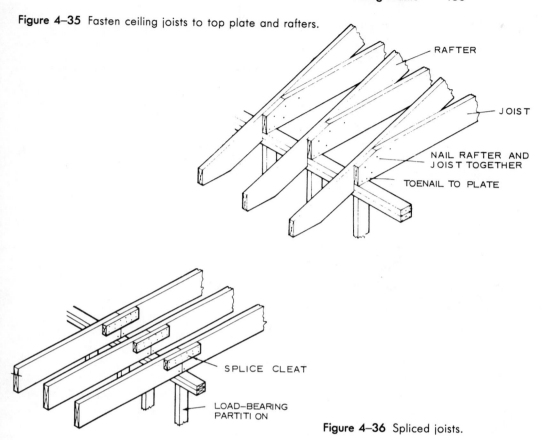

RAFTER

JOIST

NAIL RAFTER AND
JOIST TOGETHER

TOENAIL TO PLATE

SPLICE CLEAT

LOAD–BEARING
PARTITION

Figure 4–36 Spliced joists.

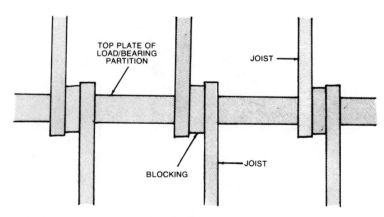

TOP PLATE OF
LOAD/BEARING
PARTITION

JOIST

BLOCKING

JOIST

Figure 4–37 Use blocks to space lapped joists.

Figure 4–38 Ledger strip supports joists for flush beam.

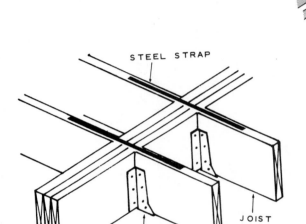

Figure 4–39 Framing for flush beam using a metal hanger.

In areas where high winds may occur, it is advisable to anchor the ceiling and roof frame. *Fig. 4–40* shows a strap anchor used for this purpose.

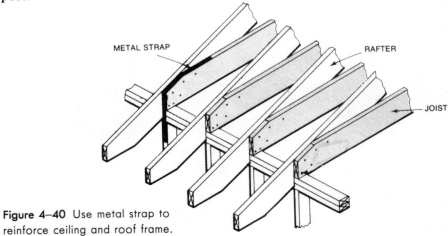

Figure 4–40 Use metal strap to reinforce ceiling and roof frame.

Often in low-sloping hip roofs, the joists running parallel to the edge of the roof may project beyond the rafters. To overcome this problem, stub joists are used (*Fig. 4–41*). They should be installed perpendicular to the regular joists.

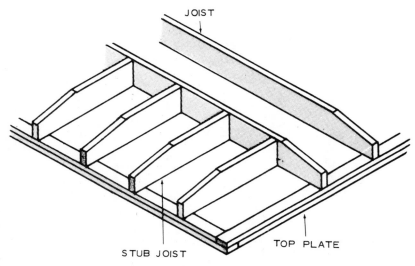

JOIST

STUB JOIST

TOP PLATE

Figure 4–41 Stub joists used for low-sloping hip roof.

Openings in the ceiling frame are treated the same as they are in floor framing. However, small openings for hatchways do not require doubled joists or headers. When making openings for disappearing stairs, be sure to check rough opening dimensions with the manufacturer of the unit, because sizes vary. *Fig. 4–42* shows a typical disappearing stairway.

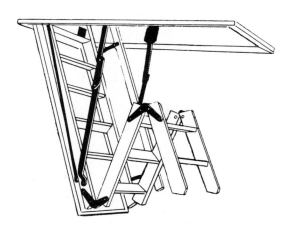

Figure 4–42 Folding stairs.

Roof Framing 5

The roof covers and protects the structure and all members within. It must be strongly built to withstand the pressures of high winds, heavy rains, and severe snowstorms, as well as to carry the weight of its own structural members. All framing parts must be calculated, measured, and cut precisely for a snug, sound roof. Learning to use and apply the framing square will greatly simplify and speed up roof construction.

Roof Styles

The most common types of roof styles are shown in *Fig. 5–1*; they include flat, shed, gable, hip, gambrel, and mansard roofs. They vary from the simplest flat roof to the intersecting roof with unequal pitch, which is perhaps the most complicated to build. Often a combination of several styles is incorporated in one roof.

Flat Roof. In this type of construction, the ceiling joists support the roof-covering materials. The roof is usually made with a slight slope so it will shed water. Since the frame of a flat roof carries the ceiling and roof loads, construction must be sturdy.

Shed Roof. This is similar to the flat roof. It is pitched in one plane and is often referred to as a lean-to. It may be freestanding or it may lean against another part of a structure, as in the case of a porch.

Gable Roof. The gable roof is a simple roof consisting of two straight slopes rising and meeting at the ridge. The triangular wall thus formed at each end is called a gable.

Hip Roof. This consists of four sloping sides which rise to meet at the ridge. If the plan of the structure is square, all four sides meet at the ridge to form a *pyramid roof*.

Gambrel Roof. The gambrel roof is a variation of the gable roof. It has two slopes on each side; the lower slope is steeper than the upper.

157

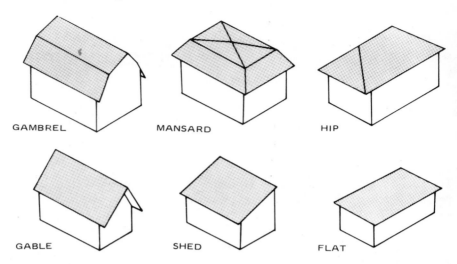

Figure 5–1 Six basic roof styles.

Mansard Roof. This is a variation of the hip roof. It has two slopes on each of the four sides. As in the gambrel roof, the lower slope is steeper than the upper.

Roof Frame Members

The basic roof frame consists of the plate, common rafters, and the ridgeboard. These are the members which make up the simple gable roof. The more complex hip roof also contains hip, valley, and jack rafters. *Fig. 5–2* shows the various members of the roof frame. The parts are described as follows:

Ridgeboard (also called *ridgepole*): The horizontal member at the top of the roof (the ridge) to which the sloping members (rafters) are fastened.

Plates: The top horizontal members of the walls, on which the sloping members (rafters) rest.

Common Rafters: Those members extending from the plate to the ridge, at right angles to both.

Hip Rafters: The outside corner rafters which extend diagonally from the plate to the ridge.

Valley Rafters: The inside corner rafters between two intersecting roof surfaces. They extend diagonally from the plate to the ridge.

Jack Rafters: Similar to common rafters with one end cut off at an angle to fit against hip or valley rafters. When cut at the upper end to fit the hip rafters, they are called hip jacks. When cut at the lower end to fit the valley rafters, they are called valley jacks.

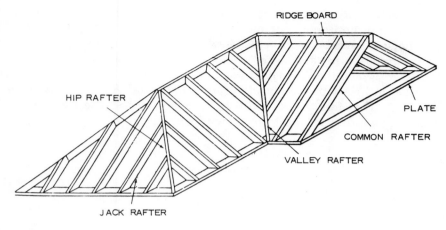

Figure 5–2 Roof frame members.

Cripple Jacks: When both ends of a rafter are cut to fit against hip and valley rafters, the shortened rafter is designated a cripple jack.

The various cuts made in a common rafter are shown in *Fig 5–3*. These consist of the following:

Ridge Cut: The top or vertical cut, which rests against the ridgeboard.

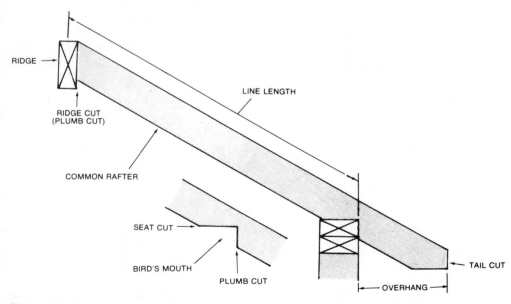

Figure 5–3 Cuts of the common rafter.

Bird's Mouth: A notch cut at the lower end of a rafter. It consists of a seat (level or horizontal) cut and a plumb (vertical) cut when the rafter has a tail and extends beyond the plate. Reminder: In carpentry, "plumb" refers to the vertical plane and "level" refers to the horizontal plane.

Heel: When the rafter does not extend past the top plate, a plumb cut is made in line with the outside edge of the plate. This is called the heel (*Fig. 5–4*). Therefore, a rafter without an overhang has a seat and heel cut. However, a rafter can also be made without a heel.

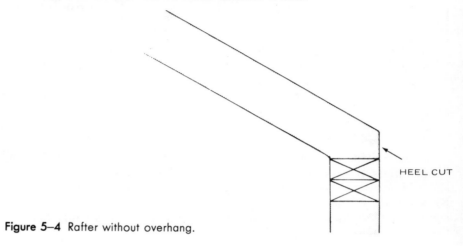

HEEL CUT

Figure 5–4 Rafter without overhang.

Tail: That portion of a rafter extending beyond the building line. The measurement of the rafter does not include the tail when the rafter is being laid out. Additional cuts made in hip and valley rafters are *side cuts*, also referrred to as *cheek cuts* (*Fig. 5–5*).

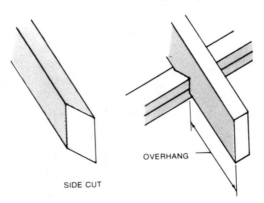

SIDE CUT

OVERHANG

Figure 5–5 *Left:* Side or cheek cut. *Right:* Tail of rafter.

In roof frame layout, the terms commonly used are span, run, rise, length of rafter, and pitch. As shown in *Fig. 5–6*, the *span* is the distance be-

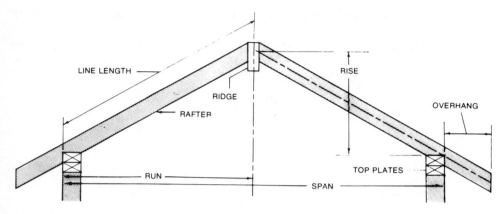

Figure 5–6 Illustrations of the basic terms used in rafter layout.

tween two opposite walls, measured from the outside of the plates. The *run* is the horizontal distance from the outside of one plate to the center of the ridge. Generally it is equal to one-half the span. If the ridge is not centered between two walls, the run will be less or more than half the span. When compared to the right triangle (*Fig. 5–7*), the run is the same as the base of the triangle.

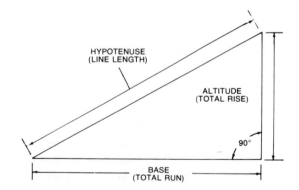

Figure 5–7 Relation of right triangle to the common rafter.

The *unit rise* is expressed as the number of inches a roof rises for every foot of run. The *total rise* is the total distance the rafter extends above the plate. It is measured vertically and is comparable to the altitude of the triangle.

The *unit length* of a rafter is the line length of the rafter for every foot of run and unit of rise. The *length* of a rafter is the shortest distance between the outside edge of the plate and the center of the ridge line. Multiplying the unit length by the number of feet of run will give the rafter length.

The *pitch* of a roof is the ratio of the rise to the span, and is always indicated as a fraction. For example, a roof with a 6-foot rise and a 24-foot span will have a ¼ pitch. This is arrived at by the formula

$$\frac{Rise}{Span} = Pitch$$

$$\frac{6'}{24'} = \frac{1}{4} \text{ pitch}$$

A 4-foot rise and the same span would result in a ¹⁄₆ pitch.

$$\frac{4'}{24'} = \frac{1}{6} \text{ pitch}$$

Slope is the ratio of rise to run. It is expressed in architectural drawings with a symbol, shown in *Fig. 5–8*. You will note that the rise in inches is indicated for every foot of run. Thus, the roof illustrated has a 6 in 12 slope, which means that it rises 6 inches for every 12 inches of run. Some common roof pitches and slopes are shown in *Fig. 5–9*.

Fundamentals of Roof Framing

Roof framing involves the understanding and application of the principles of the right triangle. The carpenter with a knowledge of geometry will have little difficulty in visualizing the various rafters in roof framing. Geometry deals with the measurement of points, lines, planes, and angles as they relate in space. In essence, the carpenter is making practical use of this branch of mathematics. However, while it is possible to lay out a roof mathematically, it would not be practical to do so. Instead, the carpenter makes use of the framing square. It eliminates much of the mathematical calculations otherwise involved in framing layout and is used to determine rafter lengths as well as top, bottom, and side cuts. (To refresh your memory about the framing square, refer to the section on this tool in Chapter 1.)

You will recall that the framing square is made in the form of a right angle with two legs. The wider and longer of the two legs is called the body or blade and the narrower leg is the tongue. The body represents the base of a right triangle, and the tongue is the altitude. If the ends of the legs, A and C, are connected, this will form the hypotenuse of the triangle (*Fig. 5–10*). In rafter layout, the base is used as the unit run and the altitude is used as the unit rise. The hypotenuse is equivalent to the line length of the rafter.

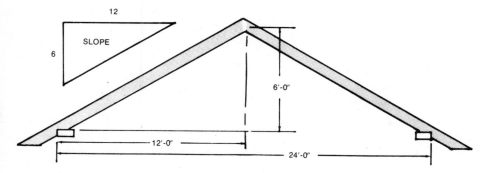

Figure 5–8 Triangular symbol is used to denote slope.

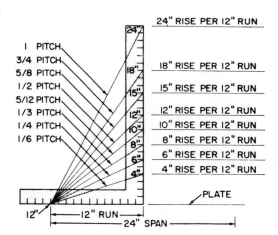

Figure 5–9 Common roof pitches.

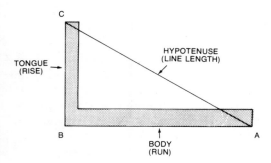

Figure 5–10 The framing square serves as a right triangle.

Top and Bottom Cuts.   The top or plumb cut (also called ridge cut) is the cut at the upper end of the rafter where it rests against the ridgeboard (or the opposite rafter if there is no ridgeboard). The bottom or heel cut is at the lower end of the rafter which rests on the plate. The top cut is parallel to the center line of the roof and the bottom cut is parallel to the plane of the plates. The top and bottom cuts are therefore at right angles to each other. To obtain the top and bottom cuts of the common rafter, use 12 inches on the body (unit run) and the unit rise (rise per foot of run) on the tongue of the framing square. (Note that the unit run for any common rafter or jack rafter is always 12 inches or one foot.) Place the square on the rafter and align the 12-inch mark on the body and the unit rise on the tongue with the edge of the rafter. A line drawn along the tongue edge will give the top or vertical cut (ridge cut) and a line drawn along the body will give the horizontal or bottom (seat) cut, as illustrated in *Fig. 5–11.*

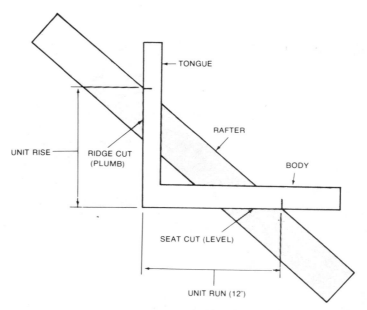

Figure 5–11 Lay out seat and ridge cuts with the framing square.

Rafter Table.   The table stamped on the face side of the square gives the line length of various rafters (*Fig. 5–12*). To find the unit line length of a common rafter, refer to the inch marks along the outer edge of the square body. Then refer to the tables, remembering that these marks represent the rise in inches of a rafter per foot of run (unit rise). For example, to find the line length of a common rafter where the run is 10 feet and the rise is 6 inches per foot of run, look at the 6-inch mark along the outer edge of the

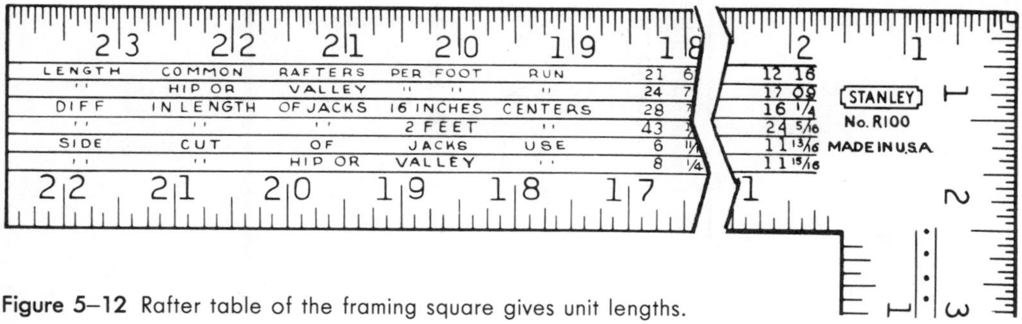

LENGTH	COMMON	RAFTERS	PER FOOT	RUN	21 6
'' ''	HIP OR	VALLEY	'' ''	''	24 7
DIFF	IN LENGTH	OF JACKS	16 INCHES	CENTERS	28 7
'' ''	'' ''	'' ''	2 FEET	'' ''	43 3
SIDE	CUT	OF	JACKS	USE	6 11/
'' ''	'' ''	HIP OR	VALLEY	'' ''	8 1/4

12 18
17 09
16 1/4
24 5/16
11 13/16
11 15/16

STANLEY No. R100 MADE IN U.S.A

Figure 5–12 Rafter table of the framing square gives unit lengths.

body of the square. Directly under the 6 in the first row, marked "length of common rafters per foot of run," you will note the figure 13.42. This means that for each foot of run, the line length of the rafter is 13.42 inches. Therefore, since the run is 10 feet, you can obtain the length of the common rafter by multiplying 13.42 by 10 as follows:

$$13.42 \times 10 = 134.20 \text{ inches}$$

$$\frac{134.20}{12} = 11.183 \text{ feet or } 11'2\frac{3}{16}''$$

The figures in the rafter table represent the hypotenuse of a right triangle. You can check this by placing a straightedge across the framing square so that it crosses the 12-inch mark on the body and the 6-inch mark on the tongue. If you measure across these points you will find that the hypotenuse is 13.42, the same as the figure printed under the 6-inch mark on the rafter table.

Measure Line.   The measure line obtained from the rafter table is the length of the common rafter from the outside corner of the plate to the center of the ridge (*Fig. 5–13*). It is an imaginary line and its position on the rafter is determined by the length of the seat cut. *Fig. 5–14* shows how the length of the seat cut can alter the position of the measuring line on the rafter. Note that this line is not the true length of the rafter and is used only in layout. You can obtain the true length of the rafter by subtracting one-half the thickness of the ridgeboard (*Fig 5–15*). If the rafters are installed without a ridgeboard, then the measure line can be considered the actual length.

Rafter Sizes.   The cross-section size of rafter stock depends on several factors. The length, spacing, and load must be considered. Usually this will be determined by the architect, and the carpenter needs only to check the specifications. Sizes may vary from 2 × 4 up to 2 × 10.

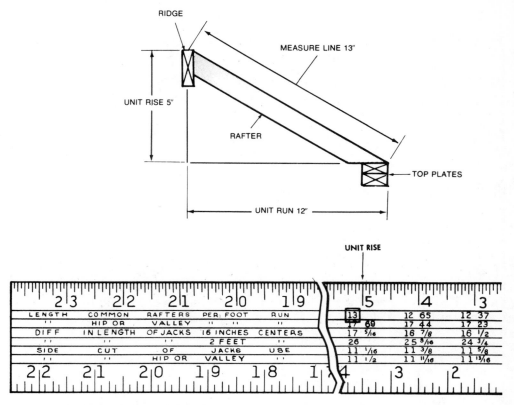

	2 3	2 2	2 1	2 0	1 9		5	4	3
LENGTH	COMMON	RAFTERS	PER FOOT	RUN		13	12 65	12 37
''	HIP OR	VALLEY	'' ''	''		17 69	17 44	17 23
DIFF	IN LENGTH	OF JACKS	16 INCHES	CENTERS		17 5/16	16 7/8	16 1/2
''	''	''	2 FEET	''		26	25 5/16	24 3/4
SIDE	CUT	OF	JACKS	USE		11 1/16	11 3/8	11 5/8
''	''	HIP OR	VALLEY	''		11 1/2	11 11/16	11 13/16

Figure 5–13 Figure from table gives measure line of rafter.

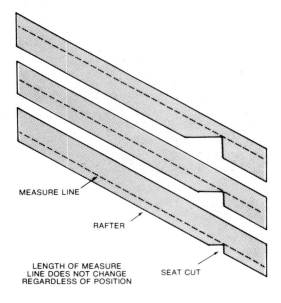

MEASURE LINE

RAFTER

LENGTH OF MEASURE LINE DOES NOT CHANGE REGARDLESS OF POSITION

SEAT CUT

Figure 5–14 Position of the measure line on the rafter depends on the size of the seat cut.

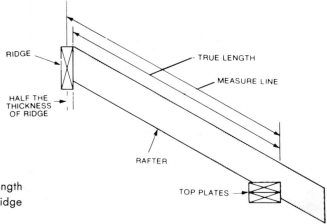

Figure 5–15 To obtain the true length of the rafter, deduct one-half the ridge thickness.

Rafter Layout. The standard practice in laying out common rafters is to make a master pattern. This is then used to mark the required number of rafters. Always be sure to choose a straight, flat piece of stock for the rafter pattern.

Using the figures of the previous example, where the line length of the common rafter is 11'2³/₁₆'' long and the unit rise is 6 inches, proceed as follows: Locate points A and B on a measure line to indicate the length of the rafter. Position this line according to the length of the seat cut as explained previously. Apply the square with the 12-inch mark of the body coinciding with point A. Set the 6-inch mark of the tongue on the measure line and draw a line along the body to form the seat cut (*Fig. 5–16*). For the top cut, slide the square, still positioned to the cut of the roof, so the tongue coincides with point B. Draw a line along the tongue to form the

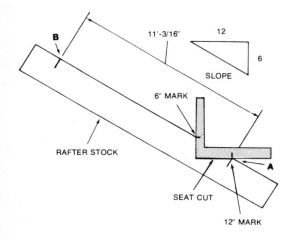

Figure 5–16 Laying out seat with square set to cut of roof.

top or plumb cut. The term "cut of the roof" means that the square is placed on the edge of the stock with the tongue and body marks corresponding to the rise and run of the roof. (The addition of the tail and deduction for the ridge thickness will be discussed later.) This method makes use of the rafter tables. Another method of laying out rafters is called the step-off or step-and-repeat method.

Step-Off Rafter Layout.   If the framing square were large enough, it could be placed on the rafter and both top and bottom cuts could be marked off quite simply (*Fig. 5–17*). Of course this is not practical, so the conventional framing square is used in step-off fashion. The number of steps or positions used depends on the number of feet in the run. For example, if the run were 4 feet, the square would be stepped 4 times, as in *Fig. 5–18*. For a run of 12 feet, the square would be repositioned 12 times.

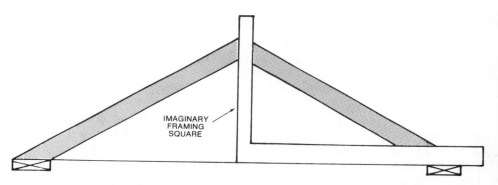

IMAGINARY
FRAMING
SQUARE

Figure 5–17 If the framing square were large enough, it would not be necessary to use the step-off method to lay out top and bottom cut.

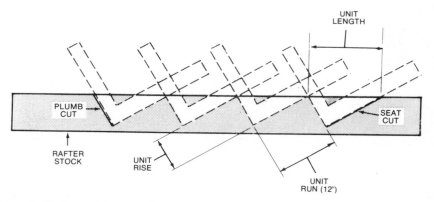

UNIT
LENGTH

PLUMB
CUT

SEAT
CUT

RAFTER
STOCK

UNIT
RISE

UNIT
RUN (12")

Figure 5–18 For a 4-foot run, step the square four times.

Assume that the roof has a span of 13 feet and a unit rise of 9 inches. It also has a 14-inch overhang and a 2 × 6 ridgeboard (*Fig. 5–19*). The total run (which is one-half the span) will be 6 feet 6 inches.

Support the stock on a pair of sawhorses and position the square to the cut of the roof. In our example, set the 9-inch mark on the tongue and the 12-inch mark on the body. Place the square toward the left end of the stock. Special framing clamps are available to insure that the square is kept in this position firmly (*Fig. 5–20*). The clamps must be carefully set so that

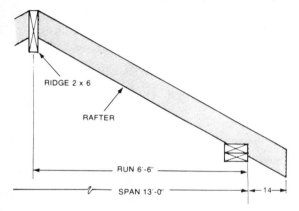

RIDGE 2 x 6

RAFTER

RUN 6'-6"

SPAN 13'-0"

14

Figure 5–19 Rafter detail for step-off layout.

Figure 5–20 Fasten adjustable gauge clamps to square with knurled screws.

the rise and run marks on the outer edge of the square will be aligned with the edge of the stock.

With the square in position, use a sharp pencil to draw the top cut (plumb) line along the tongue (*Fig. 5–21*). This line represents the center line of the ridge. Since the run contains 6 odd inches (6 feet 6 inches total), mark off the odd unit next. Do this by measuring 6 inches along the body, and then place a mark on the stock. This measurement is made perpendicular to the ridge or plumb line (*Fig. 5–22*). (If the odd unit were 4½ inches, you would simply place the mark 4½ inches from the plumb line.)

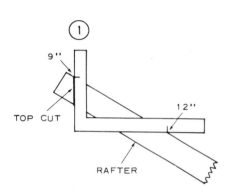

Figure 5–21 With square set to the cut of the roof, mark the top cut along the tongue. This is position 1.

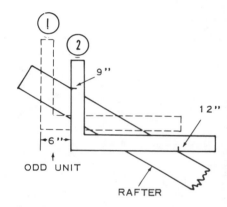

Figure 5–22 Step off the odd unit (6 inches) by moving the square to position 2.

Next, mark off the full units. Move the square (still set to the cut of the roof) along the stock until the tongue is even with the 6-inch odd unit mark. Draw a line along the body at the edge of the stock. This is the first of the full units. Repeat this procedure as many times as there are feet of run. In our example this would be six times (*Fig. 5–23*). The 12-inch mark on the body of the last position (#7 in figure) represents the building line. Proceed to complete the rafter layout as explained in the diagram.

The bird's mouth is required when the rafter has an overhang or tail that extends beyond the plate. It consists of a plumb and a level cut. The level cut forms the seat and rests on top of the plate; the plumb cut aligns with the building line; and together the two form the bird's mouth. The plumb cut, when extended upwards, becomes the heel. If the rafter lacks an overhang, the heel cut becomes the end of the rafter (*Fig. 5–24*).

As the heel dimension is increased, the seat measurement decreases. The size of the heel dimension is usually increased if greater strength is required because of an oversized tail or overhang. Since the ridge plumb cut

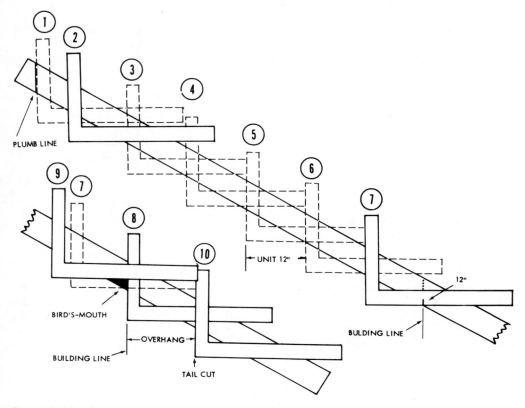

PLUMB LINE

BIRD'S-MOUTH

UNIT 12"

BULDING LINE

12"

OVERHANG

BUILDING LINE

TAIL CUT

Figure 5–23 After odd unit is laid out, move the square six full units (12 inches each) to position 7. The 12-inch mark indicates the building line (heel). Move the square another full unit to position 8. Lay out the plumb line for the bird's mouth. Determine the height of the heel (2 inches in this case), then move the square back to position 9. Lay out the seat cut. This will form the bird's mouth. From the building line, move the square 14 inches to position 10, then lay out the tail cut.

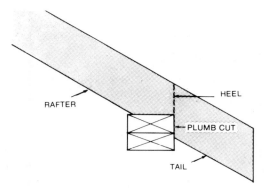

RAFTER

HEEL

PLUMB CUT

TAIL

Figure 5–24 If the rafter lacks an overhang, extend the plumb cut upward to form the heel.

and heel cut are parallel to each other, the heel dimension can be altered without affecting the pitch or shape of the roof. This is shown in *Fig. 5–25*.

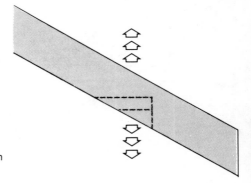

Figure 5–25 Altering the heel dimension will not affect the slope of the rafter.

Overhang Layout. The tail or overhang of the rafter is the part that extends beyond the building line. It is treated as an addition to the rafter length. The amount of overhang is always given as a projection from the building line. Therefore, the projection is considered the run of the overhang (*Fig. 5–26*).

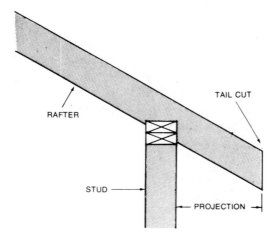

Figure 5–26 Measure the overhang from the building line to the tail cut.

Tail ends can be made in a variety of profiles. The most common are the square cut, plumb cut, and a combination of both (*Fig. 5–27*). Lay out the overhang by taking the measurement from the plumb or heel cut. Start with the full unit if the overhang is 12 inches or more, then add the odd unit. In our example, the overhang is 1 foot 2 inches (*Fig. 5–28*).

Figure 5–27 Three variations of tail cuts.
Left: Square.
Center: Plumb.
Right: Combination of level and plumb.

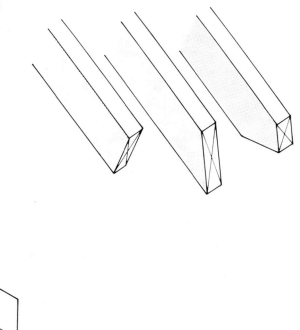

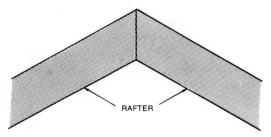

1'-2"

OVERHANG

Figure 5–28 Overhang layout.

Shortening the Rafters at the Ridge. In a roof without a ridgeboard, the two opposing rafters would meet at the center, as shown in *Fig. 5–29*. With a ridgeboard, the ends of the rafters must be cut an amount equal to one-half the thickness of the ridgeboard. This cut is made parallel to the first line drawn on the rafter. It is measured at right angles (perpendicular) to the plumb line (*Fig. 5–30*). Since the ridgeboard in our example is 2 × 6 (which is 1½" × 5½" actual size), the amount of shortening is ¾ inch.

RAFTER

Figure 5–29 Shortening allowance for ridge thickness is not required when rafters meet without a ridgeboard.

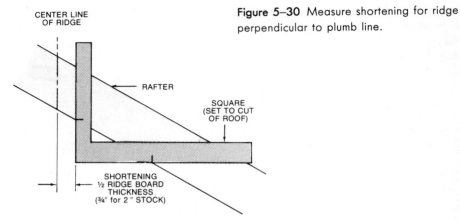

CENTER LINE
OF RIDGE

RAFTER

SQUARE
(SET TO CUT
OF ROOF)

SHORTENING
½ RIDGE BOARD
THICKNESS
(¾" for 2 " STOCK)

Figure 5–30 Measure shortening for ridge perpendicular to plumb line.

After the layout is completed, carefully cut the pattern to size. Nail short blocks near each end of the pattern to serve as guides when you are making the rafters (*Fig. 5–31*).

Before marking and cutting all of the required rafters, cut two only and check them for accuracy. Install them temporarily and make sure they meet at the ridgeboard when they are fully seated on the plate. Also check the total rise, which should equal that on the specifications.

Now mark and cut the remaining rafters. Use only the master template for marking. The use of a power saw to cut the rafters speeds the work and makes for greater accuracy. Various tools may be used for cutting. The radial-arm saw is excellent for this purpose. Work may be ganged

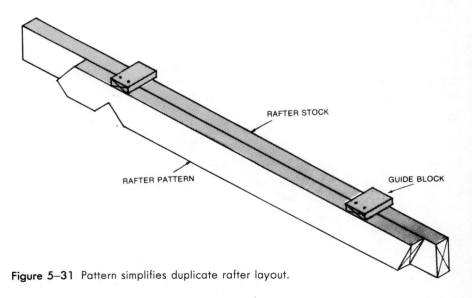

RAFTER STOCK

RAFTER PATTERN

GUIDE BLOCK

Figure 5–31 Pattern simplifies duplicate rafter layout.

and special cutter heads may be used to cut the bird's mouth in one pass. Ridge and tail cuts are also efficiently cut on this machine.

When rafters are cut on the job, most carpenters prefer to use the portable saw, as shown in *Fig. 5–32*. After they have been cut, the rafters should be placed against the frame of the structure so they will be readily available.

Figure 5–32 Cut rafters with portable saw and stack them against frame.

Installing the Common Rafter

The gable roof is the easiest of the pitched roofs to install. After you have cut all the common rafters, choose the stock for the ridgeboard. This should be straight, true, and free of defects. If more than one length is required, locate the joints so they occur at the center of a rafter. Except for roofs with extended rakes, the total length of the ridgeboard in a gable roof is equal to the length of the roof measured from plate to plate.

Mark the rafter locations on the top plate and ridgeboard. If the rafter spacing is the same as that for the ceiling joists, the rafters will be installed adjacent to each joist. Usually, the ceiling joists are spaced 12 inches on center and rafters 16 inches on center. In such cases, the rafters and joists will coincide at 4-foot intervals (*Fig. 5–33*).

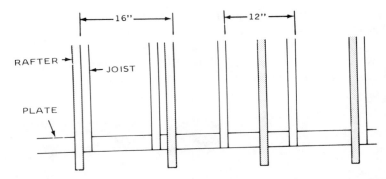

Figure 5-33 When ceiling joists are spaced 12 inches on center and rafters are spaced 16 inches on center, the joists and rafters will coincide at 4-foot intervals.

Select flat, straight rafters for the gable ends. Nail a rafter to the plate while an assistant supports the free end. Do the same for the opposite rafter, then nail the top ends of the rafters to the ridgeboard. If the ridgeboard is to be pieced because of its length, do not join the sections beforehand. It will be easier to work with the shorter lengths. Install another pair of rafters about 12 feet from the end. Plumb the rafters and install temporary bracing. Continue to install the intermediate rafters in pairs. Face-nail one rafter in each pair through the ridgeboard (*Fig. 5-34*); toenail the opposite rafter to the ridgeboard. Use 16d nails for end nailing and 8d box nails for toenailing. The box nails, being thinner than common nails, will have less tendency to split the lumber. When a rafter falls adjacent to a joist, it must be nailed to both the joist and the plate (*Fig. 5-35*). For added strength, framing anchors may be used (*Fig. 5-36*). In high-wind areas, framing straps are also utilized.

In hurricane-prone areas, the roof frame may be braced, as in *Fig. 5-37*. Here, the braces act as trusses, thus reducing the span of the rafters. Collar-beam ties are also used to reinforce the roof frame (*Fig. 5-38*). In uninhabited attics, the collar beams are made of 1 × 6 stock and nailed to every fourth rafter. If the attic may eventually be used as a dwelling room, the beams should be 2 × 4s or 2 × 6s and placed at each rafter, where they can serve as ceiling joists.

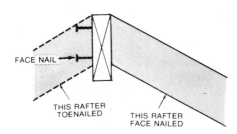

Figure 5-34 In each pair of rafters, face-nail the first rafter; the second must be toenailed.

Figure 5–35 Nail each rafter to adjacent joist and top plate.

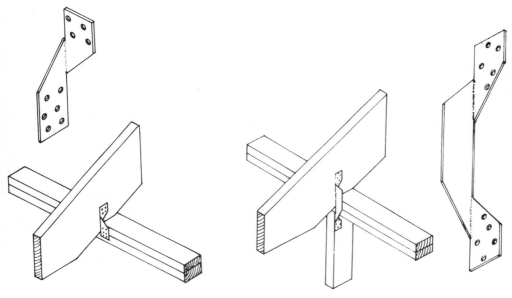

Figure 5–36 Special anchors strengthen roof framing.

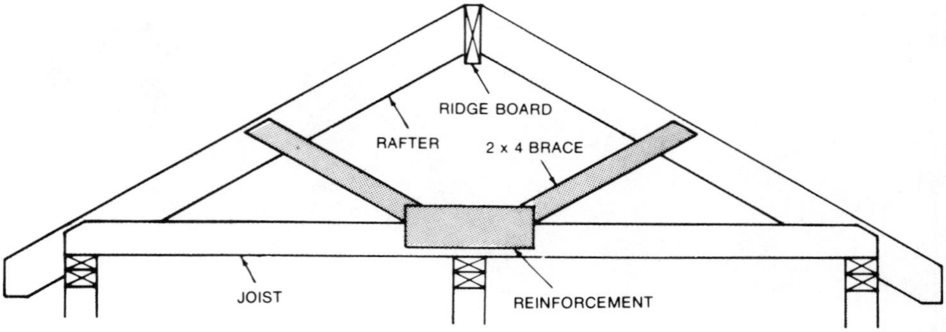

RIDGE BOARD

RAFTER   2 x 4 BRACE

JOIST   REINFORCEMENT

Figure 5–37 Truss-type bracing is recommended for hurricane-prone areas.

Figure 5–38 Installing collar-beam ties.

Framing the Gable End.   The gable end studs should be installed flush with the end wall plate. To do this, notch the top of each stud to fit around the rafter, as shown in *Fig. 5–39*. Align the studs with those in the wall below. If a window is to be installed in the gable end, measure one-half the opening width at each side of the gable center. Start the first studs outside these marks. Space the balance of the studs at the regular interval. Mark the diagonal line by tracing along the underside of the rafter. Be sure to hold the stud plumb when you are marking. If the end rafter is bowed or crowned, install temporary braces to straighten it while measuring and installing the gable studs.

Figure 5–39 Framing at gable end wall.

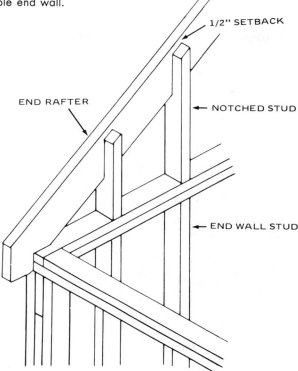

The studs of the gable end wall have a common difference as they decrease in length. The difference between two adjacent studs will be the common difference for all the other studs. Lay out the remaining studs by increasing or decreasing each succeeding stud by this amount. The tops of the notched studs should be set back about ½ inch to allow for rafter shrinkage. Nail the stud flush with the end plate and rafter. Use 8d box nails at the plate and 16d nails at the top.

When the roof plans call for a gable overhang (extended rake), the end rafter must be supported by short lookouts. These are constructed as shown in *Fig. 5–40*.

The Hip Roof. The hip roof may be described as a gable roof sliced diagonally at each end in a plane lying downward and outward from the ridge. The line formed where the two sloping sides meet is called the hip. The corner rafters of such a roof are called hip rafters. The hip rafter is shown in the plan view of *Fig. 5–41*. The hip roof may be of equal or unequal pitch. In the equal-pitch roof, both slopes are equal. In the unequal-pitch roof, the slopes differ. In the plan view of an equal-pitch roof, the diagonal of the square represents the hip rafter (*Fig. 5–42*). The sides of this square

are 12 inches and represent the unit run of a common rafter. The diagonal measures 16.97 inches. Thus, the run of the hip rafter is 16.97 inches for every foot of run of the common rafter. Because the number 16.97 is so close to 17, the run of the hip rafter is said to be 17 inches. The relative position of the hip rafter is shown in the square prism (*Fig. 5–43*).

Figure 5–40 Framing for overhang at gable end.

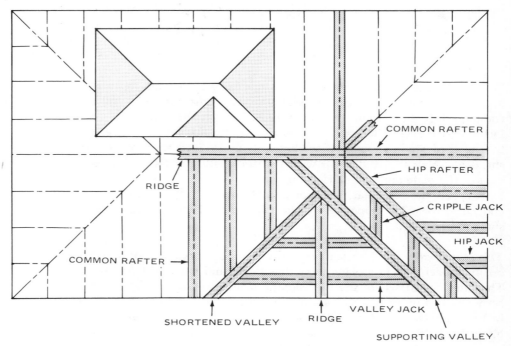

Figure 5–41 Plan view of hip roof with parts listed.

Figure 5–42 The diagonal of a 12-inch square represents the unit run of the hip rafter.

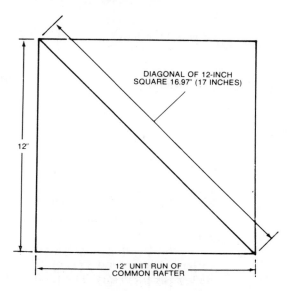

DIAGONAL OF 12-INCH
SQUARE 16.97" (17 INCHES)

12"

12" UNIT RUN OF
COMMON RAFTER

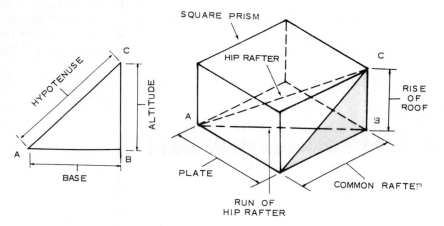

SQUARE PRISM

HIP RAFTER

HYPOTENUSE

ALTITUDE

C

A

B

BASE

C

A

B

PLATE

RUN OF
HIP RAFTER

RISE
OF
ROOF

COMMON RAFTER

Figure 5–43 Rafter parts as related to the square prism.

When you are laying out the hip rafter, follow the procedure as for the common rafter, with one exception: Use the 17-inch mark instead of the 12-inch mark on the body of the square (*Fig. 5–44*). Because of the added load on hip and valley rafters, they should be at least 2 inches wider than common rafters. The extra width also provides a full bearing surface for the side cuts of the jack rafters.

You can find the odd unit for a hip rafter by letting the extra inches in the run of the common rafter equal the sides of a square. The diagonal

Figure 5–44 The unit run for the hip rafter is 17 inches.

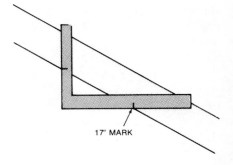

17" MARK

measurement of that square will be equal to the actual odd-unit dimension of the hip rafter. Note that the measurement is made perpendicular to the tongue of the framing square when you are laying out the odd-unit dimension of the hip rafter (*Fig. 5–45*).

The framing square's tables may also be used for laying out the hip rafters. The second line of the rafter table is stamped "Length of hip or valley per foot run." The figures under the inch mark indicate the length of the hip or valley rafter per foot run of the common rafter. To find the length of the hip rafter, multiply the figure given in the table by the number of feet of run in the common rafter. For example, find the length of a hip rafter where the span is 10 feet and the unit rise is 8 inches. Locate the inch mark on the outer edge of the body which corresponds to the unit rise

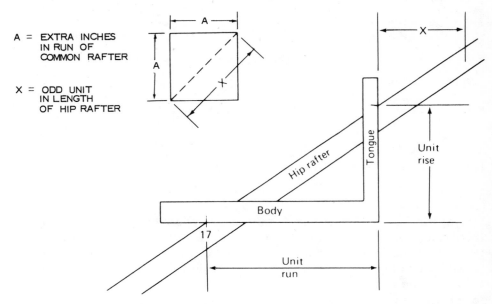

A = EXTRA INCHES
IN RUN OF
COMMON RAFTER

X = ODD UNIT
IN LENGTH
OF HIP RAFTER

Figure 5–45 Find the odd unit for the hip rafter by measuring the diagonal of a square whose sides equal the odd inches in the run of the common rafter.

of the roof (*Fig. 5–46*). Under the 8-inch mark, on the second row, you will find the number 18.76. Multiply this number by 5 (the run of a 10-foot span) to obtain the hip rafter length. Thus 18.76 × 5 = 93.80 inches. This is the length of the hip rafter. To obtain the answer in feet, divide 93.80 by 12, as follows:

$$\frac{93.80}{12} = 7.81 \text{ feet or } 7'9\frac{3}{4}''$$

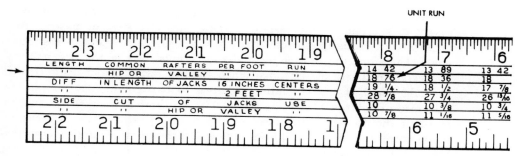

Figure 5–46 Rafter table gives hip rafter length per foot of run.

Ridge of Hip Roof.   The true length of the ridge of a hip roof is equal to the center-to-center distance between the first and the last common rafters along the side of the roof. This is the same as taking the length of the roof and subtracting the span. However, for practicability, the ridge is made to extend to the ends of these rafters, as shown in the plan view in *Fig. 5–47*. This may be expressed as follows:

The length of a hip roof ridgeboard is equal to the length of the

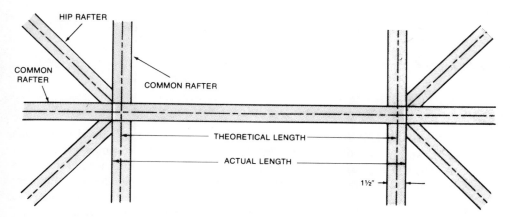

Figure 5–47 When the ridge of a hip roof is framed to the common rafters, add the thickness of one rafter to the theoretical length to find the actual length of the ridge.

structure minus the span plus the thickness of the common rafter. If the length of the structure (from outside of plate to outside of opposite plate) is 22 feet, the span 10 feet, and the common rafters are 2 × 6, then the ridgeboard length will be 12 feet 1½ inches (thus 22' − 10' + 1½"). The 1½ inches represents the thickness of the common rafter, which in our example is a 2 × 6.

Shortening the Hip Rafter

All measurements are taken from the true ridge length. It is therefore necessary to shorten the hip rafters, because the ridge extends past its true length. In the equal-pitch roof, the hip intersects the ridge at a 45-degree angle. Therefore, the shortening at the ridge must be equal to one-half the 45-degree thickness of the ridge when the hip rafter is joined directly to the ridgeboard. If the hip rafter is joined to two common rafters at the ridge as in the "tripod" method (*Fig. 5–48*), the shortening must be equal to one-half the 45-degree thickness of the common rafter. Always measure perpendicular to the plumb line and draw the shortening line parallel to the plumb cut (*Fig. 5–49*).

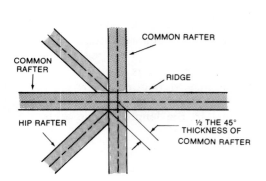

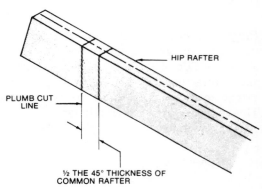

Figure 5–48 Method of shortening the hip rafter.

Figure 5–49 Shortening must be measured parallel to the plumb cut.

Side Cut of Hip Rafter. In addition to the shortening, side cuts are also necessary to permit the hip rafter to fit between the common rafters. Although the hip rafter is at 45 degrees relative to the common rafters, the side-cut angles are not 45 degrees. Theoretically, they would be 45 degrees only if the hip has zero rise. However, as the rise of the hip changes, so does the angle of the side cut. To lay out the side cut, lay out a line equal to one-half the thickness of the hip rafter on both sides of the rafter, parallel to the shortening lines (*Fig. 5–50*). Join these at the top of the rafter.

Figure 5–50 Measure side cuts of hip rafter parallel to shortened line.

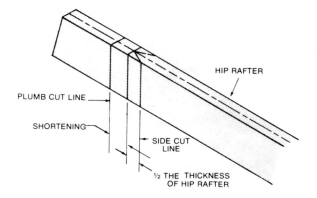

PLUMB CUT LINE

SHORTENING

SIDE CUT LINE

½ THE THICKNESS OF HIP RAFTER

HIP RAFTER

When they are cut on these lines, the hip rafter will fit snugly between the common rafters.

The sixth line of the framing-square table may also be used to lay out the side cut. Assuming a roof with an 8-inch unit rise, locate the 8-inch mark on the body of the square. Below this mark on the sixth line (marked "side cuts for hip and valley rafters") you will find the figure 10⅞. Apply the square to the rafter with the 10⅞-inch mark on the body and the 12-inch mark on the tongue. Position the square to the back edge of the rafter and let the tongue intersect the shortening mark. Draw a line from the center out. Flop the square over and repeat for the second cut. Continue this line at the side of the rafter, keeping it parallel to the shortened line along the side of the rafter (*Fig. 5–51*).

The tail cuts at the ends of the rafters are laid out in a manner similar to that employed for the ridge cut above. The same angles are used for both.

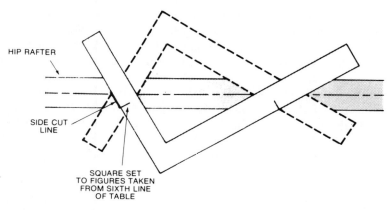

HIP RAFTER

SIDE CUT LINE

SQUARE SET TO FIGURES TAKEN FROM SIXTH LINE OF TABLE

Figure 5–51 Using the framing square to lay out the side cuts.

Backing Off the Hip Rafter.   The edges of the hip rafter lie in a plane above that of the jack rafters (*Fig. 5–52*). But if they were installed in this manner, the roof sheathing would not lie flat at the hip. There are two ways to eliminate this problem. The top of the hip rafter can be beveled (backed) or the rafter can be dropped at the seat. When dropped, the top edges of the rafters will be flush with the tops of the jack rafters (*Fig. 5–53*).

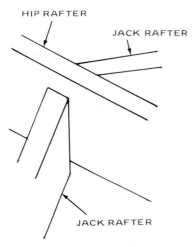

Figure 5–52 Edge of hip rafter lies in a plane above the jack rafter.

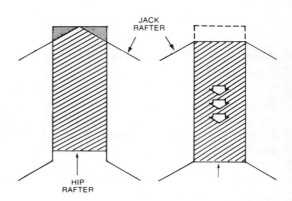

Figure 5–53 The hip rafter can be beveled (backed) as at *left* or dropped, as at *right*.

To determine the amount of backing needed, place the square on the stock with the unit rise on the tongue and the unit run (17) on the body. Draw a diagonal line near the edge, as in *Fig. 5–54*. Measure along this line a distance equal to one-half the thickness of the rafter. For 2-inch nominal stock, the distance would be ¾ inch. Mark this point on the rafter, then draw a line through this point parallel to the edge of the rafter. Draw a center line along the top of the rafter, then use a plane to bevel the rafter between the two lines.

To drop the rafter, you can use the measurement obtained for the backing described above to lower the seat cut. Again, set the square on the stock with the body set at the 17-inch mark and the tongue on the unit rise. Draw a level line equal to one-half the rafter thickness. Measure the plumb distance and cut away a portion of the seat equal to this amount (*Fig. 5–55*).

Hip Rafter Overhang.   The overhang length of the hip rafter is greater than that of the common rafter. This is shown in the plan view (*Fig. 5–56*).

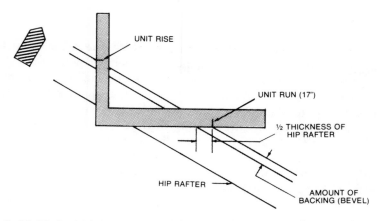

Figure 5–54 Method of determining the amount of backing needed for a hip rafter.

Figure 5–55 Method of determining drop of hip rafter.

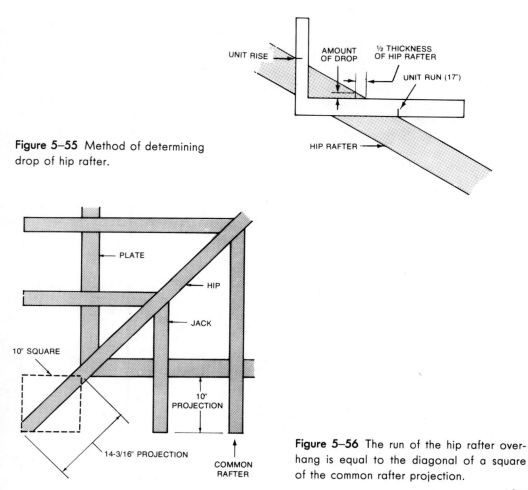

Figure 5–56 The run of the hip rafter overhang is equal to the diagonal of a square of the common rafter projection.

To determine the projection, draw a square with the sides equal to the common rafter projections. The diagonal measurement will be the amount of the hip projection. For a 10-inch overhang, the diagonal will measure $14^{3}/_{16}$. To lay out the projections on the rafter, place the square on the stock to the cut of the roof. Align the tongue of the square with the building line on the rafter, then mark off the distance obtained by measuring the diagonal of the square ($14^{3}/_{16}$ inches). Draw a line through this mark, keeping it parallel to the building line. You will note that this procedure is similar to that used when you are marking off the odd unit for the hip rafters. If fascia boards (facing across the ends of rafters) are to be used, add the side cuts.

Valley Rafters.   Valley rafters are similar to hip rafters. They are placed where two sloping surfaces of intersecting roofs meet to form an inside angle. The intersecting roofs may be of equal or unequal pitch. The equal pitch exists when the span and slopes of the two roof sections are equal. In such a roof, the common rafters of both sections will be equal and the valley rafters will butt against both ridges (*Fig. 5–57*). If the span of the intersecting roof is less than the main roof, the situation changes. Here, only one valley rafter extends to the ridge of the main roof, as shown in *Fig. 5–58*. This is called the supporting valley rafter. Since it extends from the plate to the ridge, it is treated like a hip rafter when it is being laid out. Layout may be by the step-off method or by use of the second line of the

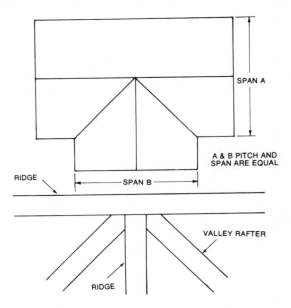

Figure 5–57 In intersecting roofs with equal pitch and spans, the valley rafters butt against both ridges.

Figure 5–58 When the pitch and spans of intersecting roofs are unequal, only one valley rafter butts against the main roof ridge.

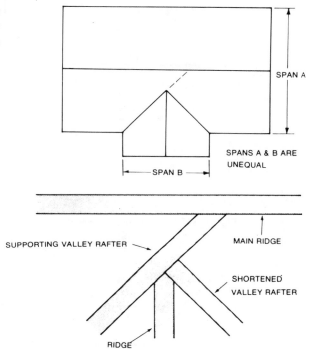

SPAN A

SPANS A & B ARE UNEQUAL

SPAN B

SUPPORTING VALLEY RAFTER

MAIN RIDGE

SHORTENED VALLEY RAFTER

RIDGE

framing square. When you are employing the step-off method, use a unit run of 17 inches on the body of the square, just as with the hip rafter.

The valley rafter must be shortened at the ridge, using one-half the 45-degree thickness of the ridge. The side cuts are equal to one-half the valley rafter thickness. Where the spans of the main and intersecting roofs are equal, double side cuts are required at the top of each valley rafter. If the spans differ, a single side cut is made on the supporting rafter. No side cuts are used on the shortened valley rafter. Instead, the secondary ridge will require double side cuts. Shorten the ridge one-half the 45-degree thickness of the supporting valley rafter (*Fig. 5–59*). Note that all shortening and side-cut measurements are made perpendicular to the plumb cut.

The side cuts at the plate should be angled toward the overhang. If the rafters are not exposed, the side cuts can be eliminated. In this case, make a straight cut across the side lines (*Fig. 5–60*). If there is no overhang, make the side cuts at the building line to provide a nailing surface for the fascia boards. Otherwise, the side cuts should be made at the end of the tail to serve the same purpose.

Jack Rafters. Jack rafters are like shortened common rafters. They lie in the same plane as common rafters and therefore have the same run as the

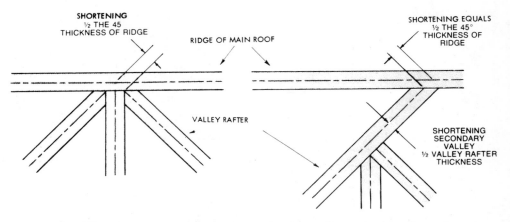

Figure 5–59 Procedure for shortening and making side cuts in the valley rafters.

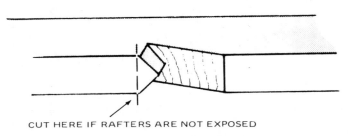

CUT HERE IF RAFTERS ARE NOT EXPOSED

Figure 5–60 If rafters are concealed, tail of hip rafter can be cut square; otherwise, it should be made with side cuts.

common rafters. There are three types of jack rafters. Those that run from the plate to the hip rafter are called *hip jacks*. Those between the ridge and valley rafters are called *valley jacks*, and those that lie between hip and valley rafters are *cripple jacks*. The various jacks are shown in *Fig. 5–61*. Hip and valley jacks have one straight and one side cut, while cripple jacks have side cuts at each end.

Common Difference of Jacks. Generally, jack rafters are spaced the same as common rafters, either 16 or 24 inches on center. When equally spaced, they have the same common difference in length (*Fig. 5–62*). The second rafter is twice the length of the first, the third is three times the length of the first, and so on. Lines three and four of the rafter table give the difference in length of jack rafters. Line three is for 16-inch spacing, and line four is for 24-inch spacings (*Fig. 5–63*).

Figure 5–61 Various jack rafters.

4X
3X
2X
X

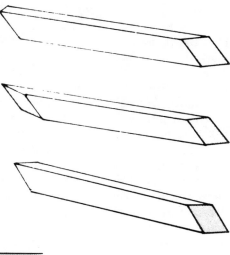

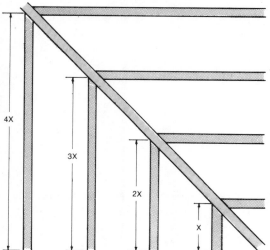

Figure 5–62 Hip (and valley) jacks have a common difference.

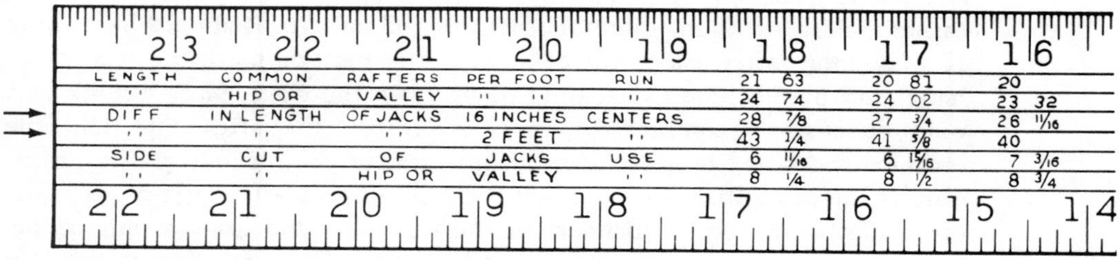

23 22 21 20 19 18 17 16
LENGTH COMMON RAFTERS PER FOOT RUN 21 63 20 81 20
" HIP OR VALLEY " " " 24 74 24 02 23 32
DIFF IN LENGTH OF JACKS 16 INCHES CENTERS 28 7/8 27 3/4 26 11/16
" " " 2 FEET " 43 1/4 41 5/8 40
SIDE CUT OF JACKS USE 6 11/16 6 15/16 7 3/16
" " HIP OR VALLEY " 8 1/4 8 1/2 8 3/4
22 21 20 19 18 17 16 15 14

Figure 5–63 Third and fourth lines of the framing square give the differences in jack-rafter lengths.

Finding Jack-Rafter Length.   To find the jack-rafter length of a roof with a rise of 8 inches per foot and a spacing of 16 inches, proceed as follows: Find the figure 8 on the body of the square. On the third line under the 8, marked "difference in length of jacks," you will find the number 19¼. This means that the first or shortest jack rafter is 19¼ inches. To find the length of the third jack, multiply by 3. In this case, the length of the third jack rafter will be 19¼ × 3, which equals 57¾ inches. From each of the measurements, you must deduct one-half the 45-degree thickness of the hip or valley rafter for the shortening.

To find the jack-rafter lengths by layout, place the square on the side of the stock and to the cut of the common rafter—that is, the 12-inch mark on the body and the 8-inch mark on the tongue. Place a mark on the stock along the blade. Keeping the square in the same position, slide it along to a distance equal to the rafter spacing, 16 inches in this example. Place a mark along the tongue. You will find that the two marks are 19¼ inches apart, the same as the figure given on the rafter table (*Fig. 5–64*).

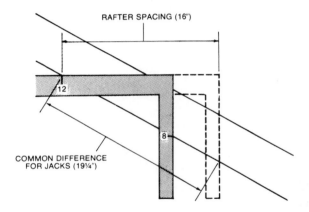

RAFTER SPACING (16")

12

8

COMMON DIFFERENCE
FOR JACKS (19¼")

Figure 5–64 Layout method of finding common difference of jack rafters.

Jack-Rafter Pattern.   The layout of jack rafters is simplified by the use of a jack-rafter pattern. This is similar to the common rafter pattern described earlier. Select a straight piece of stock for the pattern. It should be as long as a common rafter. Using the common rafter pattern, mark the seat, overhang, and plumb lines. Measuring down from the ridge, place a mark at a point equal to the common difference taken from the rafter table. This line must be shortened to allow for the thickness of the hip rafter. The procedure is the same that is used when the hip rafters are being laid out. Take one-half the 45-degree thickness of the hip rafter and draw a plumb line. Be sure to take the measurement on a level line (*Fig. 5–65*). Square this line across the top of the pattern and mark its center point. Now, measuring from the plumb line, take one-half the jack-rafter thickness and draw another plumb line. Draw a diagonal line from this mark at the edge of the stock through the center mark. This line represents the side cut.

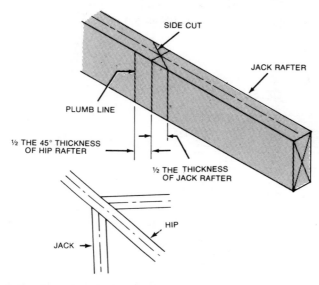

Figure 5–65 Jack-rafter shortening and side cut.

Repeat for each of the common difference lengths along the pattern (*Fig. 5–66*). To save time, use a T-bevel or a jack template to mark off the remaining side cuts.

The framing square table can also be used for laying out jack-rafter side cuts. The fifth line of the table is marked "side cut of jacks use." To obtain the side-cut angle, take the figure shown in the table on the body of the square and 12 inches on the tongue. Mark along the tongue side.

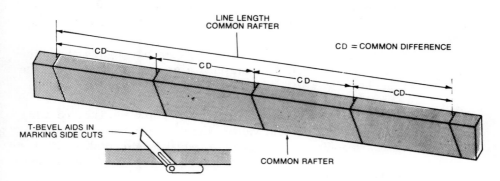

Figure 5–66 Jack-rafter pattern.

After the pattern is completed, use it to lay out the required number of jacks. You will need a pair of left and right jacks for each position on the hip rafters. The lengths of each pair will be the same, but the angle cuts will be in opposite directions (*Fig. 5–67*).

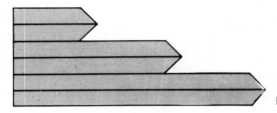

Figure 5–67 Lay jack rafters out in pairs.

Care must be taken when you are cutting jacks, because the cuts consist of compound angles. Many carpenters use portable radial-arm saws on the jobsite. Such saws can be set to make the plumb and side cut in one pass.

Valley Jacks. The valley jack is similar to the hip jack. It is spaced like the hip jack and likewise has a common difference. The basic difference between the two is that the valley jack extends from the ridge to the valley rafter with the side cuts made at the bottom. Also, since the valley jacks do not extend to the plate, they do not have overhangs.

Lay out the pattern for the valley jacks, following the same procedure used for the hip jacks. Use the same allowances for shortening and for side cuts. Cut the valley jacks in pairs, as was done with the hip jacks.

Installing Jacks. Mark the spacing for the jacks along the plate and ridge. The spacing will be the same as that of the common rafters. Install the jacks in pairs, starting midway along the hip or valley. This will prevent movement of the hip or valley. If necessary, you can use temporary braces. To further insure that the hip or valley rafters remain straight, do not seat the nails until all the jacks are in place, then drive them home securely. Use 10d nails.

Valley jacks are nailed flush at the ridgeboard but slightly above the top edge of the valley rafter. This prevents a gap at the joint when the sheathing is applied (*Fig. 5–68*).

Cripple Jacks. The rafter extending from a valley to a hip is called a hip valley cripple jack. Since hip and valley rafters are parallel to each other, the cripple jacks spanning them will be of equal length. Both ends of a cripple jack have diagonal side cuts. For the hip valley cripple jacks these angles are in the same direction (*Fig. 5–69*).

Figure 5–68 Proper position of valley jacks when they are being nailed to the valley rafter.

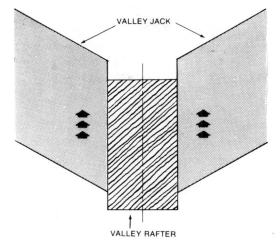

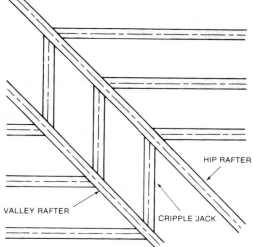

Figure 5–69 Cripple jacks are equal in length.

The run of a hip valley cripple jack is equal to the distance from the center of the hip to the center of the valley measured along the plate line (*Fig. 5–70*). To lay out the hip valley cripple jack, use the same procedure as for the common rafter. Shorten both ends by one-half the 45-degree thickness of the hip and valley rafters. For the side cuts, take one-half the thickness of the cripple jack.

When a rafter spans two valleys, as would occur between a short and supporting valley, it is called a valley cripple jack rafter (*Fig. 5–71*). You will note that unlike the hip valley cripple, this jack has side cuts with opposing angles at the opposite ends.

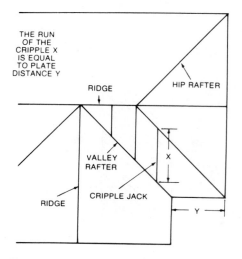

Figure 5–70 The run of the cripple jack is equal to the length of the plate between the hip and valley.

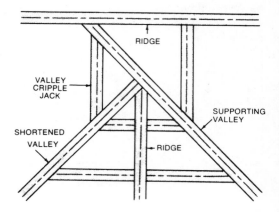

Figure 5–71 The valley cripple jack is framed between two valley rafters.

The run of the valley cripple jack is twice the run of the valley jack that intersects the valley cripple jack (*Fig. 5–72*). Measurements are taken from the center lines. Shortening and side cuts are treated as for the hip valley cripple jacks.

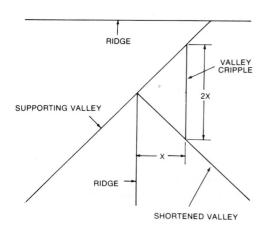

Figure 5–72 The run of the valley cripple jack is twice its distance from the valley rafter intersections.

Construction of Various Roof Types

Shed Roof. The shed roof has a single slope and is often used for porches, shed dormers, and sheds (*Fig. 5–73*). In porch construction, the rafters are similar to common rafters of the gable roof. They contain a seat and plumb top cut (*Fig. 5–74*). In a shed roof, the rafters contain two overhangs and two seat cuts, as shown in *Fig. 5–75*. The rafter length is laid out as a common rafter. The run of the rafter is taken to the inside plate edge of the higher wall.

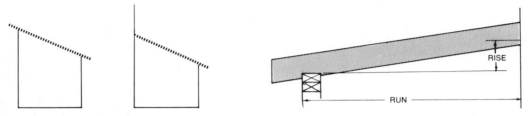

Figure 5–73 Two simple shed roofs. *Left:* Freestanding. *Right:* Porch.

Figure 5–74 Detail of porch roof.

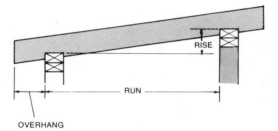

Figure 5–75 Detail of shed roof.

Dormer Roofs. Dormers are projections built into a sloping roof. They are usually installed to provide light and air or additonal interior space. The most common types are the shed and gable. The *shed dormer* provides greater use of floor space in upstairs rooms. The width of the shed dormer is not limited; it can be made to extend to almost the full length of the roof (*Fig. 5–76*).

Figure 5–76 Typical shed dormer.

When the shed roof extends to the ridge, a top cut should be made at the top of each rafter, as shown in *Fig. 5–77*. If the shed roof stops short of the ridge, a double header must be installed across the top of the opening.

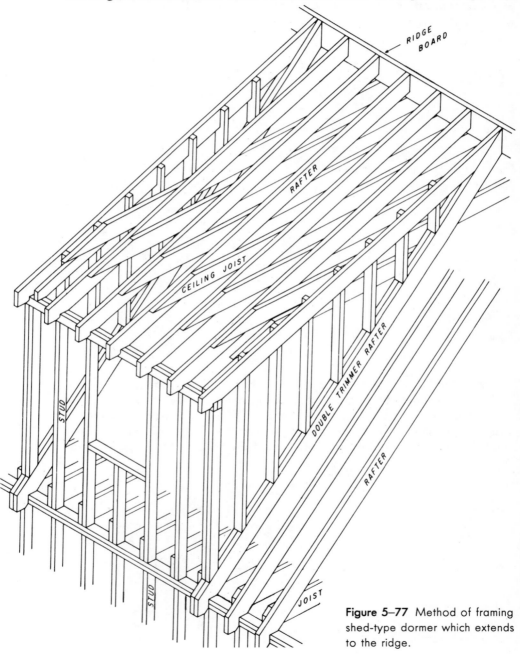

Figure 5–77 Method of framing shed-type dormer which extends to the ridge.

The front wall is usually extended from the main wall plate. Double trim-mer rafters are installed to support the side wall studs.

The *gable dormer* is like a miniature roof and may contain a window or ventilator. Like the shed dormer, it may extend from the plate but is usual-ly set back from the plate of the main roof and stops short of the ridge. Construct double headers and trimmers in the usual manner (*Fig. 5–78*).

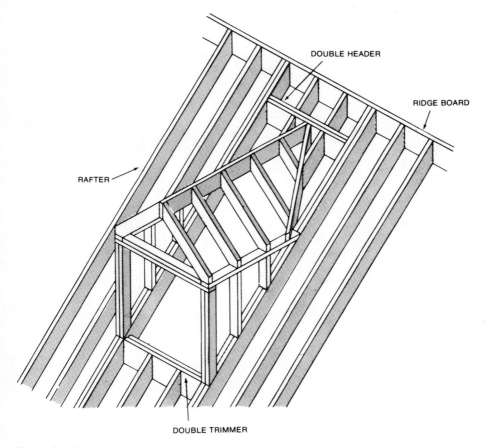

Figure 5–78 Double headers are used to frame dormer.

Place the headers between the main roof rafters at a point where the dor-mer ridge and valley rafters meet. The slope of the dormer roof should match that of the main roof.

Gambrel Roof. This roof has a single ridge but two sets of common raf-ters (*Fig. 5–79*). The intersecting rafters are supported by a purlin plate,

Figure 5–79 The gambrel roof consists of two slopes.

which in turn is supported by partitions, or it may be tied to the opposite purlin plate by means of collar beams (*Fig. 5–80*). The gambrel roof is laid out as two separate roofs. The purlin serves as the plate for the upper rafters and the ridge for the lower ones (*Fig. 5–81*). The plumb and level cuts of the rafters are similar to those of the common rafter.

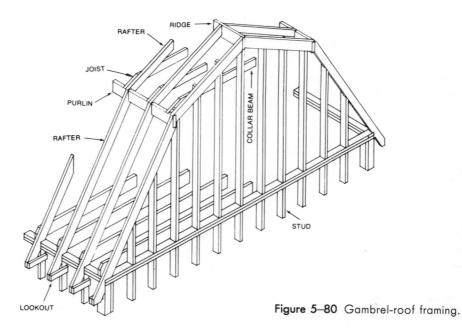

Figure 5–80 Gambrel-roof framing.

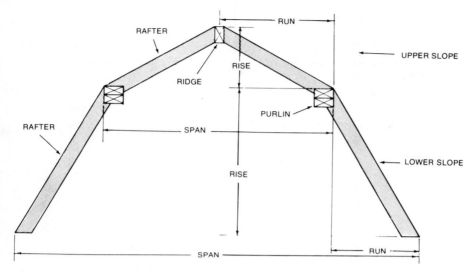

Figure 5–81 In a gambrel roof, the purlin serves as the ridgeboard for the lower slope rafters and as the sole plate for the upper slope rafters.

Flat Roof.   The construction detail of a flat roof is shown in *Fig. 5–82*. It is perhaps the easiest roof to construct. Since the roof joists carry the roof load and ceiling below, they are made of heavier stock such as 2 × 10s or 2 × 12s. The roof joists overhang the top plates of the outside wall. If the overhang is to extend over the side walls, lookout rafters are employed. These are tied to doubled roof joists. To permit drainage, a slight slope is required.

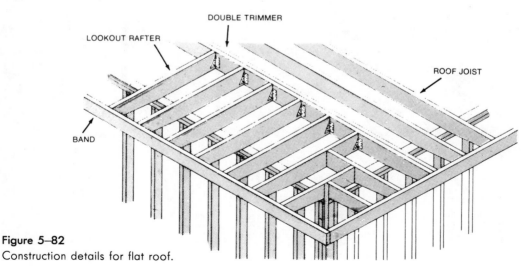

Figure 5–82
Construction details for flat roof.

Other Roof Elements

Collar Beams.   The rafters of a roof are reinforced and stiffened by the use of collar beams. Generally made of 1 × 6 stock, they span the rafters on opposite sides of the roof (*Fig. 5–83*). They are installed at every third or fourth rafter. If the attic space is used as living quarters, the collar beams should be made of 2 × 4 or 2 × 6 stock and placed at every rafter. They then serve as ceiling joists.

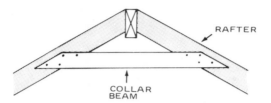

RAFTER

COLLAR
BEAM

Figure 5–83 Reinforce rafters with collar-beam ties.

The length of a collar beam depends upon its height. As its height increases, its length decreases. The length of the beam can be determined in several ways. Most carpenters prefer this method. Take a straight board that spans and extends past the rafters. Place it at the desired height and check to make sure it is perfectly level. Tack it in place, then mark along the outer edges of the rafters. Remove the board, cut it to size, and then use it as a template to mark and cut the remaining beams (*Fig. 5–84*).

Figure 5–84 A collar beam being installed.

Roof Openings.   Roof openings are required for such things as chimneys and skylights. Large openings require the interruption of the roof framing and are treated the same as floor openings. Headers are placed between rafters with their faces plumb (*Fig. 5–85*). For small openings, the rafters are installed in the usual manner. The openings are then cut and framed afterward. Large openings are constructed during the framing and rafter spacing is not altered. Headers are doubled and trimmers are added to each side.

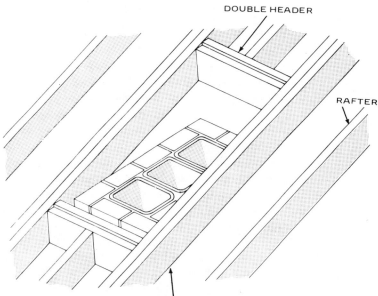

Figure 5–85
Framing for roof opening.

For chimney openings the headers in the roof must line up with those in the floor below. Plumb the lines carefully; then install the headers, as in *Fig. 5–86*. Allow 2-inch clearance all around so the framing will clear the masonry.

Chimney Saddles.   The saddle is a gabled structure placed behind the chimney to shed water. It is not built in as part of the roof frame. Instead, it is constructed as a separate unit to be added to the roof after the roof is sheathed.

The span of the saddle should be equal to the width of the chimney. When you are laying out the common rafters, they should be dropped sufficiently to allow for the thickness of the saddle sheathing as well as the roof covering. If the rafter is not dropped, the roof boards of the saddle will project beyond the chimney.

Because the size is small, most carpenters make a full-size layout of

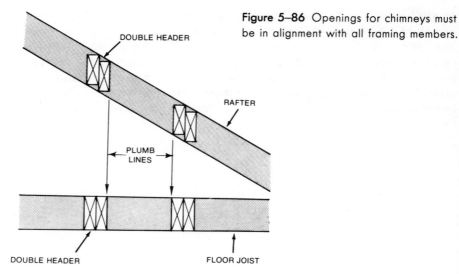

Figure 5–86 Openings for chimneys must be in alignment with all framing members.

the end view of the saddle. The procedure is as follows: On a sheet of plywood or other suitable surface, lay out a horizontal base line AB equal to the width of the chimney. Also, lay out a vertical center line that intersects line AB at point C. Place the square with the body on line AB and the tongue on the center line. Mark off the unit run and unit rise at points E and F. Draw a line through points E and F; this represents the unit line length. Now draw line BD parallel to line EF. This will represent the line length of the saddle common rafter. Draw line AD. The resulting triangle will be a full-size front view of the saddle (*Fig. 5–87*).

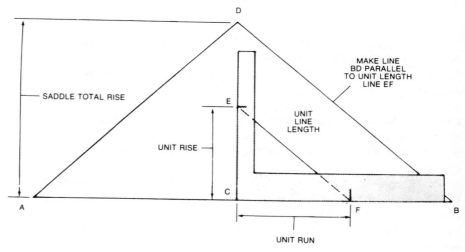

Figure 5–87 Layout procedure to find saddle total rise.

When using line AD as the top of the roof covering, draw a line below and parallel to it to represent the thickness of the roof covering. Below
this line draw another to represent the saddle sheathing. Now draw still
another line to represent the width of the common rafter. Next, parallel to
the base line, lay out the thickness of the valley strip (*Fig. 5–88*). At the top
of line CD, draw the ridgeboard with line CD as its center line. Now lay
out the dropped common rafter with plumb and seat cut full size. The level
line at the lower end of the common rafter requires a cheek cut so it will
match the slope of the roof.

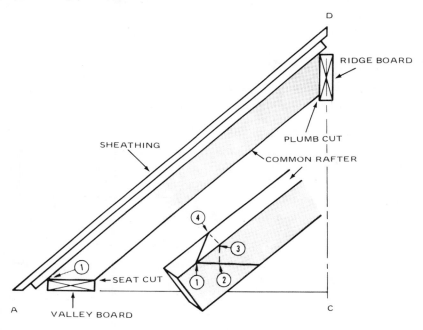

Figure 5–88 Full-size layout of saddle with allowances for ridge and valley strip.

Transfer the layout to the rafter stock. The shortening for the ridge
and the allowance for the valley strip have been taken care of in the full-
size layout. The level cheek cut must be laid out on the rafter stock. Measuring along the level line from the outer edge (1), place a mark equal to
the thickness of the rafter stock (2). Plumb this point to the top edge of
the rafter (3). Square this point across the top edge of the stock (4), then
connect points 1 and 4. This represents the level cheek cut. When cut on
this line, the rafter will be plumb on the roof with the level cheek cut
matching the slope of the main roof.

The *valley strip* should be laid out next. Use a piece of 1 × 4 and find
the length by stepping off with the square set to the unit rise on the
tongue and 17 on the blade. Step off the square for as many feet as there

are in the run of the common rafter. To lay out the top and bottom cuts, reset the square with the unit length of the common rafter set on the tongue and 12 on the blade. Obtain the unit length by measuring line EF in the full-size layout previously made. Draw the top cut along the tongue, then move the square to the opposite end and draw the bottom cut along the blade (*Fig. 5–89*). Make two valley strips, one right and one left.

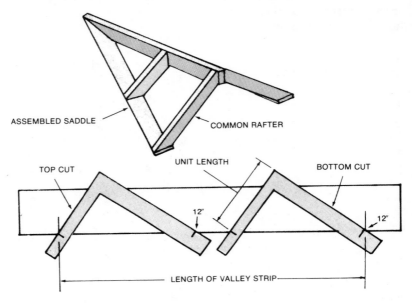

Figure 5–89 Layout of valley strip.

The *ridgeboard* rests on the valley strips. The valley cut is made in the ridge piece and is the same angle as was used for the seat cut for the main roof common rafter. The length of the ridge is equal to the run of the saddle common rafter. Deduct the drop of the ridge to allow for the amount the rafters were dropped.

The *jack rafters* are laid out with the sizes varying according to the common difference. Cut all pieces to size, then assemble.

Intersecting Roofs with Unequal Pitch

Unequal-Pitch Roofs. Various problems arise in building the unequal-pitch roof because the rise per foot of run for each span is different. Layout procedures differ for roofs with and without overhangs. Backing and dropping of the rafters and layout of cheek cuts bring on additional problems. To simplify the layout and to lessen the chance of error, it is advisable to make a full-size plan layout of the hip or valley section of the roof in-volved. A ½ or ¼ or ¹/₁₂ scale may be used, instead, but the rafter sections should be drawn full size so that the side-cut angles can be taken directly from the layout. The layout will also show the total runs for the various jack rafters.

When you are building an unequal-pitch roof, construct both sections separately, using standard procedures. Install ridge pieces and all of the common rafters in the normal manner. The work remaining will include the layout, cutting, and erecting of the hip, valley, and jack rafters.

Large-Scale Layout.   The large-scale layout can be made on the subfloor. Choose a clear area that will enable you to work unhindered. In our discus-sion, we shall assume that a full-size layout is made of a roof without overhang. (The sizes shown are chosen for illustration purposes only. Rare-ly would such sizes exist in an actual roof.) The roof consists of a gabled main section with a gabled projection of narrower span at the side, result-ing in an unequal-pitch roof. To differentiate between the two, we shall re-fer to them as the major and minor spans.

The major roof has a span of 8 feet and the minor roof a span of 5 feet 4 inches. Both have a total rise of 4 feet (*Fig. 5–90*). In this case the to-tal rise is the same, but it need not be. The rafter spacing is assumed to be 16 inches on center.

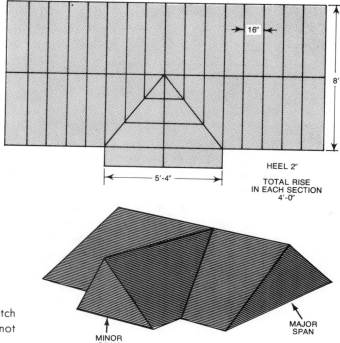

Figure 5–90 In the unequal-pitch roof, the rise per foot of run is not the same in each section.

Lay out the plates and ridges of the intersecting areas of both roofs in full size, as shown in *Fig. 5–91*. Note that the jacks do not meet at the valley. They would meet only if the valley were at a 45-degree angle to the ridge lines.

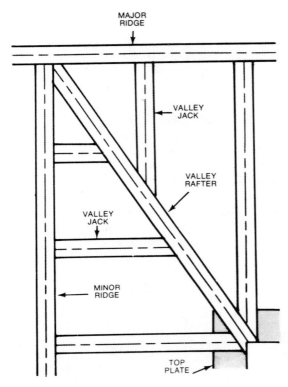

MAJOR RIDGE

VALLEY JACK

VALLEY RAFTER

VALLEY JACK

MINOR RIDGE

TOP PLATE

Figure 5–91 A full-size layout of intersecting roofs will clearly show the various cuts.

Unequal Valley Layout. The unit run of a hip or valley rafter in an equal-pitch roof is equal to the diagonal of a 12-inch square and therefore equals 17 (actually 16.97). In an unequal-pitch roof, the unit run is equal to the diagonal of a rectangle whose sides are equal to the unit runs of the major and minor roof sections. (*Fig. 5–92*). To obtain the total run of the valley, lay out a full-size rectangle with one side equal to half the major span and the adjacent side equal to half the minor span. The diagonal will equal the total run of the valley (*Fig. 5–93*).

To lay out the full-size valley rafter, draw a right triangle using the total run of the valley as its base and the total rise as the altitude. The hypotenuse will be the line length of the valley rafters (*Fig. 5–94*).

Using the line length, lay out the rafter outline, positioning it on the hypotenuse of the triangle so that it has the required seat and heel cuts,

Figure 5–92 The unit run of hip and valley rafters in an unequal-pitch roof is equal to the diagonal of a rectangle whose sides are equal to the runs of the major and minor spans.

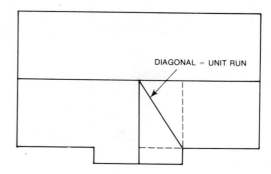

DIAGONAL = UNIT RUN

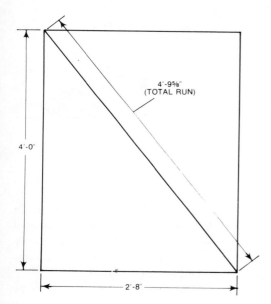

4'-9⅝"
(TOTAL RUN)

4'-0"

2'-8"

Figure 5–93 Determining the total run by full-size layout of major and minor spans.

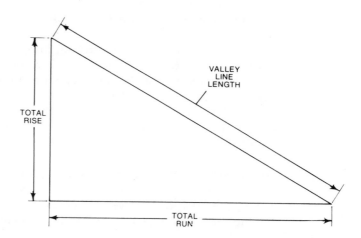

VALLEY
LINE
LENGTH

TOTAL
RISE

TOTAL
RUN

Figure 5–94 Layout to obtain line length of valley rafter.

which should be the same as those for the common rafters (*Fig. 5–95*). Transfer the outline to the rafter stock.

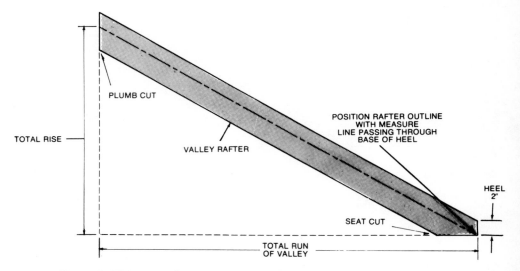

Figure 5–95 Layout of the valley rafter using the line length.

Ridge and Tail Cuts. The ridge and tail cuts of the valley rafter are in the form of a double cheek or double side cuts. Because of the unequal pitch, the run of the cheek cuts will also be unequal (see detail in *Fig. 5–96*). Since the allowances and side cuts on the drawing are full size, the dimensions can be transferred to the rafter stock for the rafter layout.

The plumb-cut line previously drawn on the rafter will have to be adjusted to allow for the ridge shortening. Take the measurement for the shortening directly from the full-size layout. Be sure to mark the allowance perpendicular to the original plumb line, and note that the cut at the ridge is slightly off-center.

After you draw the shortening line on the stock, draw the side-cut lines. Note how the angles vary greatly from one side to the other. Connect the lines at the top of the rafter to form the pointed end, which will butt against the intersecting ridges.

Follow the same procedure for the side cuts of the tail at the lower end of the rafter. When properly laid out and cut, the side cuts will be flush with the outer edges of the plates when installed.

Allowance for Hip Rafters. Since valley rafters are set below the jack rafters, it is not necessary to drop or back them. Hip rafters must be dropped or backed, however. When the difference in pitch is not too great, the hip may be dropped or backed in the manner described for equal-pitch roofs. When the differences are excessive, it may be necessary to drop the hip, then back the high side (*Fig. 5–97*).

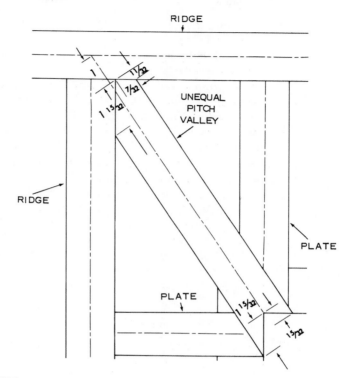

Figure 5–96 Ridge and tail cuts for unequal-pitch valley rafter.

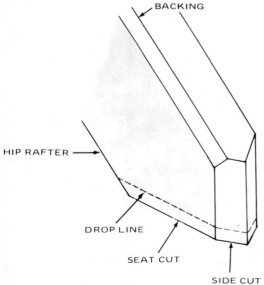

Figure 5–97 Hip rafters must be backed or dropped.

Valley Jack Layout.   The valley jack rafter of an unequal-pitch roof has ridge and cheek cuts similar to those of the equal-pitch roof. However, the side cuts and common differences of the major roof will differ from those in the minor roof section. The jack rafters in the major section will have plumb cuts similar to those of the common rafters in that section. Likewise, the plumb cuts of the jack rafters in the minor section will match those of its common rafters.

A simple and practical method of laying out the jack rafters is to make use of the common rafter. The largest jack rafter and the common rafter are equal in length. They also have the same plumb and seat cut. The only difference is the side or cheek cut at the lower end of the jack rafter where it butts the valley rafter. Since the common difference between jacks is always the same, the common rafter is simply divided into as many equal spaces as there are jacks (*Fig. 5–98*). Take all measurements from the top ridge cut, marking the length along a center line drawn on the top edge of the rafter. The cheek-cut angle and the deduction for one-half the thickness of the valley rafter are taken from the full-size layout. A T-bevel set to the proper angle will simplify the marking of the various jacks.

The same procedure is used for the second roof section, except that the cheek cuts are made in the opposite direction. Follow the procedure for laying out hip jacks.

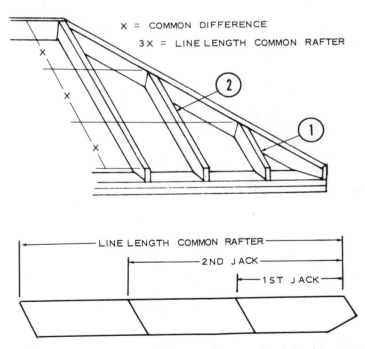

Figure 5–98 In laying out jacks, the common rafter must be divided into equal parts.

Unequal-Pitch Roof with Cornice.   The addition of a continuous cornice introduces added problems to the framing layout. *Fig. 5–99* shows the rafters of an unequal-pitch roof without overhang resting on the plate. Compare this with *Fig. 5–100*, which has an overhang. It is obvious that it would be impossible to install a continuous fascia board or plancier (which is the underside of a cornice) on such a roof. The problem is solved by laying out rafters from the fascia line instead of the plate. It may also become necessary to raise the wall height for the section with the steeper pitch, so that the rafters will have sufficient bearing on the plate (*Fig. 5–101*). The plan view shown in *Fig. 5–102* is the same as the one used earlier in *Fig. 5–90*, except that a 1-foot cornice projection has been added. The span is no longer measured from wall plate to wall plate. Instead, it is measured from plumb cut to plumb cut of the rafter tails. A full-size layout will provide the necessary information for runs, cuts, and allowances.

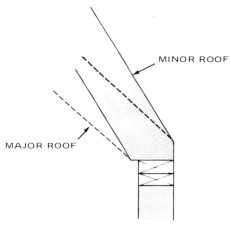

Figure 5–99 Heel cuts are in alignment on the unequal-pitch roof without overhang.

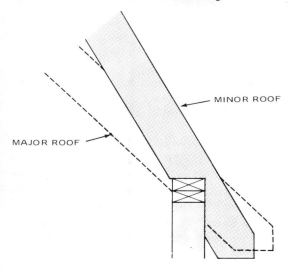

Figure 5–100 Heel cuts do not align when the roof has a cornice.

Figure 5–101 Wall height may have to be raised to bring heel cuts into alignment.

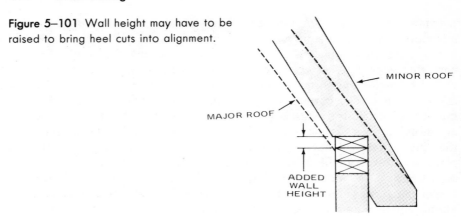

Figure 5–102 Unequal-pitch roof with overhang.

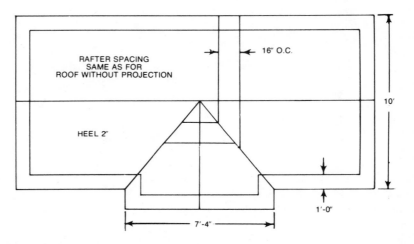

Note that the valley does not pass over the intersection of the wall plates because of the unequal runs. It therefore becomes necessary to make a single side cut at the plumb cut of the bird's mouth where it bears against the plate (*Fig. 5–103*).

A full-size or scale layout of the side view, including both slopes and plate lines, should be made (*Fig. 5–104*). It will be helpful in leveling the fascia lines and in determining the required plate buildup. The top of the fascia line is used as a reference point when the span is being laid out. Since the slope of the steeper roof must be changed, the pitch of that roof is also altered. This is clearly shown in the side-view layout. The hips, valleys, and jacks are laid out in the manner described for unequal-pitch roofs without overhang.

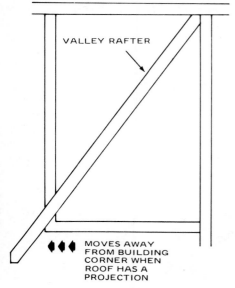

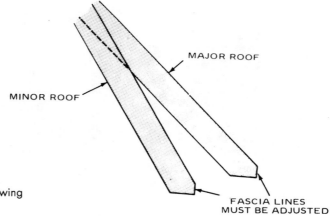

Figure 5–103 Position of the valley rafter changes when a projection is introduced.

Figure 5–104 Full-size or scale drawing will help in leveling fascia lines.

Roof Trusses

Roof trusses are widely used in residential and commercial construction. Their use permits wide spans without intermediate supports. This is a great advantage, especially in architectural designs calling for spacious rooms without supporting partitions or posts. Also, their use permits rapid assembly of the roof frame, and thus the structure can be enclosed and protected in less time. The truss or trussed rafter consists of upper and lower chords and diagonals. The upper chords represent the rafters and the lower chords correspond to the ceiling joists. The diagonals serve as braces. A typical truss is shown in *Fig. 5–105.*

Figure 5–105 A typical roof truss.

Trusses are generally mill-made, because controlled conditions and manufacturing methods assure low cost and high quality. However, they can also be made on the jobsite by the carpenter.

Over the years, truss design has been highly developed for use in both residential and commercial construction. Special hardware and connectors are made in various sizes and styles. Some require special equipment for assembly, while others can be installed by the carpenter on location.

Although many truss designs are available, three basic types are generally used in residential construction. These are the W-type, king-post, and scissor trusses (*Fig. 5–106*).

The W-truss (also called Fink) is economical and easy to assemble. When it is made with split-ring connectors, it can be folded flat to simplify

Figure 5-106 Typical trusses.
Top: W-type;
Center: King-post;
Bottom: Scissors.

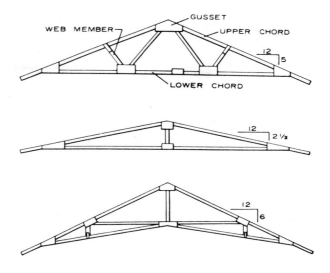

shipping. This truss design is widely used in house construction. The simplest of trusses is the king-post. It has upper and lower chords and a center vertical post. It is more economical than the W-type, but the allowable spans are less when the same size members are used. The scissor truss is ideally suited for houses with slanting "cathedral" ceilings.

Truss design must take into account live and dead loads and roof slope. Tables listing lumber grades and construction details should be consulted.

Trusses are usually made with a master template used as a guide. Some carpenters prefer to make a chalk-line layout on the floorboards. Various members are then laid out and cut to match the layout. When this procedure is followed, it is important to make the layout with great precision.

If wood gussets are used, they should be cut from ⅜- or ½-inch exterior-grade plywood. Assembly should be with nails and glue. In high-moisture areas, waterproof glue is recommended; otherwise, a casein-type glue will suffice. Nailing patterns and surface grain direction of the gussets are important (*Fig. 5-107*). Use 4d nails for ⅜-inch plywood, 6d for ½-inch and ¾-inch plywood.

To check the accuracy of the assembled truss, flop it on the pattern layout. Any discrepancies will show up clearly.

When large quantities of trusses are required, jigs may be used to advantage. Stops and guides permit the placement of each member quickly and accurately. Drilling templates are also useful in speeding up production. If you assemble trusses with bolts and connectors, be sure to use washers under the head and nut ends of each bolt.

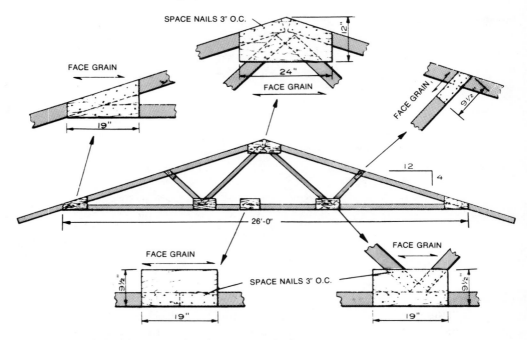

Figure 5–107 Details for W-truss with 26-foot span.

The lower chords of fully loaded trusses will settle. To allow for this, a slight camber is introduced into the lower chord (*Fig. 5–108*). The amount of camber depends on various factors, including the span, load, and slope of the roof. The flatter the roof, the greater the stresses upon it.

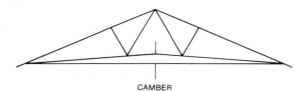

Figure 5–108 Introduce camber into lower chord to allow for sag.

Truss Assembly. In residential construction, trusses are usually erected by a crew of three. The procedure followed is to mark the location of each truss on the top plate. The truss is then placed in position with the peak down. One worker is stationed at each truss end, while a third rotates the truss into position using a rope or pole (*Fig. 5–109*). The end workers align the truss with the marks on the plate and fasten them by toenailing. Metal connectors are often used to supplement the toenailing (*Fig. 5–110*). Temporary braces are used to support the trusses until the sheathing is applied.

Figure 5–109 Rotate trusses into position by means of a truss pole.

Figure 5–110 Metal reinforcing plate for trusses.

Roof Sheathing

The type of roof sheathing used depends to some degree on the roofing to be applied. For asphalt and other composition materials, solid sheathing is required. Wood shingles and shakes used in damp climates may be installed on spaced sheathing (*Fig. 5–111*). The materials used for sheathing include shiplap and common boards, plywood, and nonwood materials. Regardless of the type used, all sheathing should be smooth, be securely fastened, and should provide a suitable base for roofing nails and fasteners.

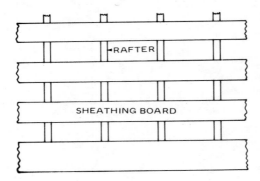

Figure 5–111 Spaced sheathing boards provide ventilation for wood shingles in warm areas.

When roofs are flat or of low pitch, sheathing boards may be installed diagonally to resist racking. Although all roof-sheathing lumber should be seasoned, it is especially important that the boards are well seasoned when asphalt shingles are used. The use of poorly seasoned lumber may result in buckling of the asphalt. The maximum moisture content should not exceed 12 percent.

Spaced Sheathing.   The use of spaced sheathing for shingles and shakes is highly recommended for damp areas. The spacing permits air to circulate, thus preventing damage to the roof. Wood nailers, consisting of 1 × 3s or 1 × 4s, are spaced to the weather. That is, the center-to-center spacing of the nailers should be the same as the shingle exposure. For example, if the shingle exposure is 5 inches and 1 × 3 nailers are used, the strips should be installed 2½ inches apart.

Plywood Sheathing.   Plywood is ideal for sheathing. The large-size panels add rigidity to the roof, they are easily installed, and they provide a sound base for the roof covering. The plywood is installed with the face grain perpendicular to the rafters (*Fig. 5–112*). In areas in which damp conditions exist, use standard sheathing with an exterior glueline. For all roofs with exposed open soffits (the underside of a cornice), the perimeter panels should have an exterior glueline (*Fig. 5–113*). Be sure that all end joints occur over the center of a rafter.

Figure 5–112 Application of plywood roof sheathing.

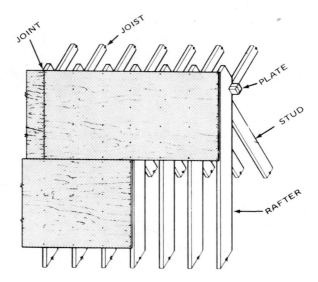

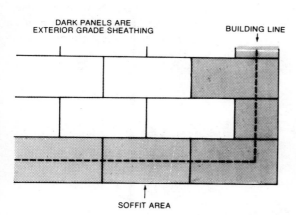

Figure 5–113 Shaded area indicates exterior-grade sheathing, required with open soffits.

To reduce costs, some carpenters use two thicknesses of sheathing when the roof has exposed soffits. For example, ⅜-inch panels are used throughout the roof except at the edges, where ½-inch panels are used. To provide a flush surface at the joint area, shims are used, as shown in *Fig. 5–114.*

The rafter spacing and roof covering must be considered when the carpenter is determining the thickness of the plywood sheathing. For wood and asphalt shingles on a roof with 16-inch rafter spacings, use ⁵⁄₁₆-inch plywood. Use ⅜-inch plywood for 24-inch rafter spacings. Heavy roofing, such as tile, slate, and asbestos cement, requires ½-inch plywood for 16-inch rafter spacings and ⅝-inch thickness for 24-inch rafter spacings.

Figure 5–114 Detail for thickness change in sheathing.

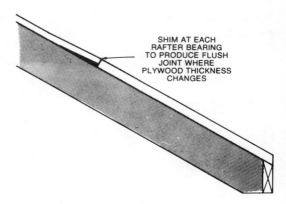

SHIM AT EACH
RAFTER BEARING
TO PRODUCE FLUSH
JOINT WHERE
PLYWOOD THICKNESS
CHANGES

Nail plywood sheathing at each bearing, 6 inches apart along the edges and 12 inches elsewhere. Use 6d nails for plywood up to ⅜-inch thickness and 8d nails for ½-inch or thicker plywood.

Blocking must be used under sheathing edges which run perpendicular to the rafters. As an alternative, sheathing clips may be used. These clips are made of aluminum and are available in thicknesses from ⁵⁄₁₆ to ¹³⁄₁₆ inch. The clips are slightly tapered and are installed by being slipped over the plywood edges (*Fig. 5–115*). A ½- to ¾-inch sheathing clearance is required around chimney openings. Framing clearance around chimneys is 2 inches (*Fig. 5–116*).

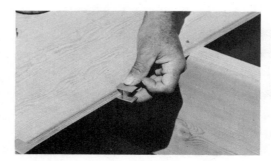

Figure 5–115 Use sheathing clips to stiffen plywood edges.

Lumber Sheathing.   Common or shiplap boards are used to sheath roofs for asphalt and other roof-covering materials where continuous support is needed. When the joints are end-matched, the joint need not occur over a bearing. Roof boards must be flat and between 6 and 8 inches wide. Wider boards, if used, may cause shrinkage problems. Start installing the roof boards at the eave, using two 8d nails at each bearing.

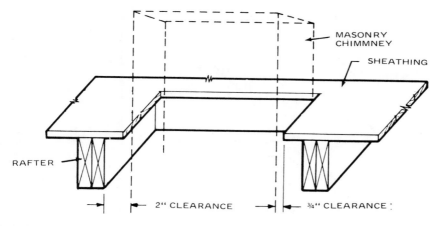

Figure 5-116 Sheathing and framing clearance around chimney.

Attics

Attic space in a house can be made accessible for storage by the addition of folding stairs. A small gable or shed dormer will increase the amount of air and light available.

If the attic has sufficient headroom—a minimum of 7 feet 6 inches over one-half the room and minimum of 5 feet at the outer edges—it is usable for living space (*Fig. 5-117*). The walls of the room can be constructed by nailing kneewall studs (at least 5 feet long) to each rafter at one end and to a sole plate at the opposite end. Blocking can be nailed between the studs and rafters to provide nailing surface for the finish wall (*Fig. 5-118*).

For a cathedral ceiling, apply the ceiling finish directly to the rafters (*Fig. 5-119*). Otherwise, nail collar beams between the rafters at least 7½ feet above floor level. Then nail blocking between the collar beams and rafters to provide a nailing surface for the finish wall and ceiling materials.

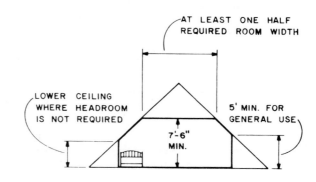

Figure 5-117 Headroom requirements for attic rooms.

Figure 5-118 Installation of kneewalls and blocking.

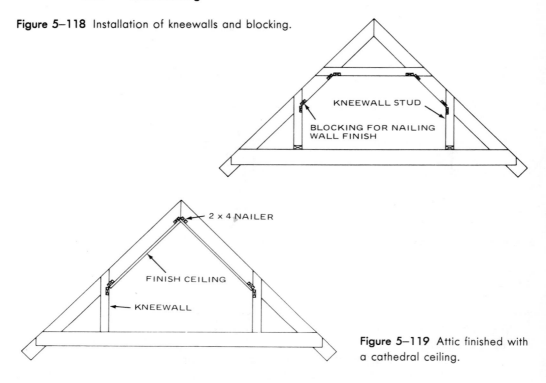

KNEEWALL STUD

BLOCKING FOR NAILING
WALL FINISH

2 x 4 NAILER

FINISH CEILING

KNEEWALL

Figure 5-119 Attic finished with
a cathedral ceiling.

For maximum use of the attic space in a gable roof, consider the construction of a shed dormer, which may be made any width and may even extend across the entire length of the house (*Fig. 5–120*). (For construction details, see p. 197.)

Proper flashing is essential where the dormer walls intersect the roof of the house, to prevent water leakage into the interior. When roofing felt is used under shingles, it should be turned up the wall at least 2 inches. Shingle flashing, in the form of tin or galvanized metal shingles, must be used at this junction. The metal shingles should be bent at 90 degrees so they will extend up the wall over the sheathing a minimum of 4 inches. One piece of flashing is applied with each shingle course (*Fig. 5–121*). Sid-

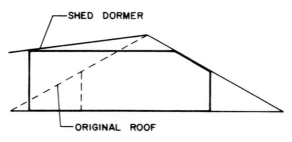

SHED DORMER

ORIGINAL ROOF

Figure 5-120 Shed dormer for
additional attic space.

Figure 5–121 Flashing at dormer walls.

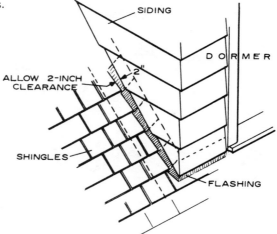

ing is applied over the flashing with a 2-inch space between the bottom of the siding and the roof. The cut ends of the siding should be treated with a water-repellent preservative.

Before deciding to convert an attic to living quarters, consider whether there is sufficient room for the addition of a stairway. A straight-run stairway usually requires a space about 3 feet wide and 11 feet long with headroom clearance of about 6 feet 8 inches minimum. Also, there must be room for a small landing at the top and bottom of the stairway.

Like the rest of the house, the attic walls, ceiling, and floor must be properly insulated and vapor barriers installed to help control the flow of moisture and heat to the outside. Particular attention must be paid to attic ventilation, since some moisture will always manage to escape from the house to collect upstairs in the attic. The installation of inlet vents at the soffits and outlet vents near the ridge will provide natural air circulation which will reduce moisture and help cool the attic. Gable vents will be helpful when of adequate size (*Fig. 5–122*). (Consult relevant chapters for details on insulation, vapor barriers, and interior finish; these details also apply to attics.)

Plank-and-Beam Framing

Plank-and-beam (also called post-and-beam) framing is a simplified method of framing. Posts, beams, and planks are combined in a system that uses fewer and larger members than conventional framing (*Fig. 5–123*). Because the pieces are spaced farther apart, construction time and costs are reduced somewhat. *Fig. 5–124* illustrates plank-and-beam framing. Non–load-bearing walls are placed between posts that support the beams. The posts are usually of 4 × 4 lumber and are generally spaced 4 feet to 8 feet on center.

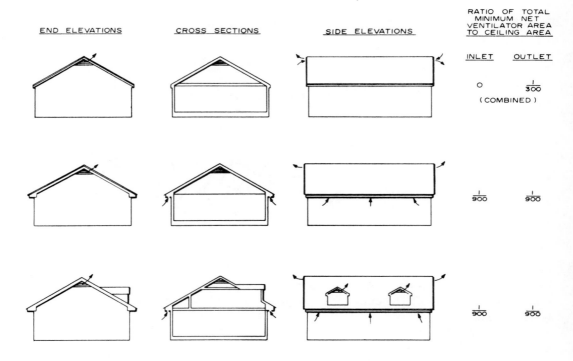

END ELEVATIONS	CROSS SECTIONS	SIDE ELEVATIONS	RATIO OF TOTAL MINIMUM NET VENTILATOR AREA TO CEILING AREA	
			INLET	OUTLET
			O	$\frac{1}{300}$
			(COMBINED)	
			$\frac{1}{900}$	$\frac{1}{900}$
			$\frac{1}{900}$	$\frac{1}{900}$

Figure 5–122 Method of ventilating gable roofs and the amount of ventilation required for various types. Ratios should be multiplied by total ceiling area to find the size of opening required for the vent. The open area required should be completely unobstructed. Where 16-mesh screen is used to cover the area, vent area should be doubled.

Figure 5–123 Plank-and-beam framing utilizes fewer but heavier members.

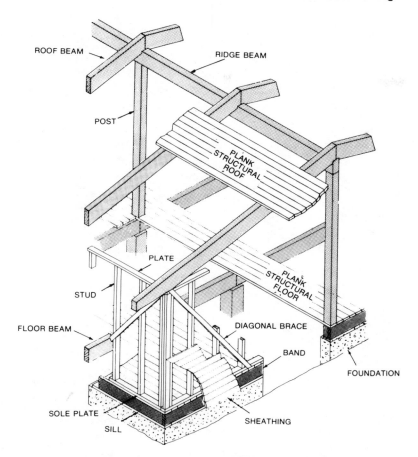

ROOF BEAM

RIDGE BEAM

POST

PLANK STRUCTURAL ROOF

PLATE

PLANK STRUCTURAL FLOOR

STUD

FLOOR BEAM

DIAGONAL BRACE

BAND

FOUNDATION

SOLE PLATE

SHEATHING

SILL

Figure 5–124 Typical framing for a plank-and-beam structure.

The wall sections between the posts should be framed and finished the same as conventional walls. They serve to enclose the structure and to provide lateral support to the posts. Lacking vertical loads, the wall-frame members may be installed horizontally to accommodate exterior vertical siding. Because of the spacing of the posts, large doors, windows, and window walls can easily be accommodated. Glass panels can be set directly into the posts, further simplifying the construction. Wide overhangs and exposed ceiling beams are characteristic of plank-and-beam construction. The underside of the roof deck is often used as the ceiling surface, thus further reducing costs.

The difference between plank-and-beam framing compared to conventional framing is shown in *Fig. 5–125*. In both illustrations the wall area covered is the same. Note the quantity of framing members in each. It

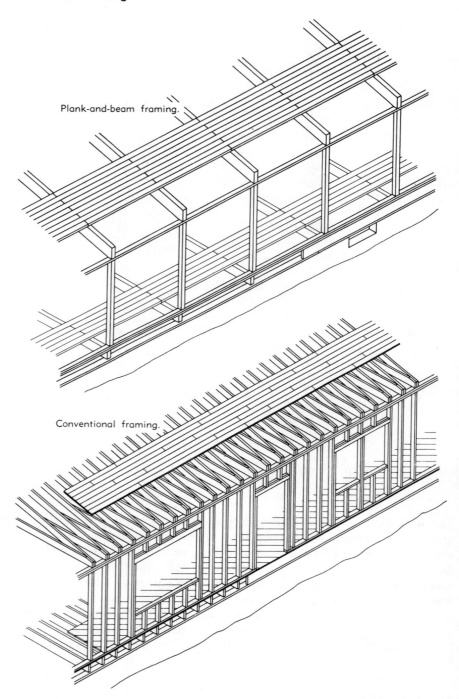

Plank-and-beam framing.

Conventional framing.

Figure 5–125 Comparison of plank-and-beam system with conventional framing.

must be pointed out, however, that the roof and floor decks used in the plank-and-beam system are made of 2-inch nominal stock compared to the 1-inch nominal material used in the conventional framing.

Due to the wide spacing of beams and the lack of floor joists, heavy concentrated loads pose some problems for this type of construction. Extra framing is needed to support heavy items such as refrigerators and bathtubs.

Posts. Since the walls in plank-and-beam construction are non–load-bearing, the roof and ceiling loads are transmitted from the decking to the beams and from the beams to the posts. The posts may be supported by foundation walls or on piers placed under each post (*Fig. 5–126*). Both types of foundations must rest on footings. The posts must be of sufficient size to support the loads placed on them. In residential construction 4 × 4 stock is common. The posts must also be large enough to provide sufficient bearing surface for the beams resting on them. When supporting beams join over a post, the bearing surface can be increased by the use of bearing blocks installed as shown in *Fig. 5–127*. The connections of the post at the floor plate and beam must be sound in order to provide uplift resistance.

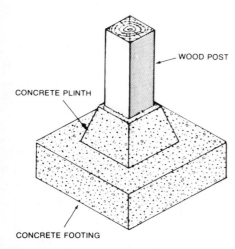

WOOD POST

CONCRETE PLINTH

CONCRETE FOOTING

Figure 5–126 Pier foundation for wood post.

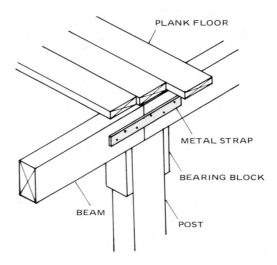

PLANK FLOOR

METAL STRAP

BEARING BLOCK

BEAM

POST

Figure 5–127 Bearing blocks and metal strap provide support for beams that join over a post.

One approved method, using metal angles, is shown in *Fig. 5–128*. Other types of fasteners are shown in *Fig. 5–129*.

Most fasteners used in plank-and-beam construction are visible and should be installed with care. If at all possible, the connectors should be hidden or made less conspicuous. Special concealed fasteners are available for some applications. Bolts or lag screws should be used for fastening.

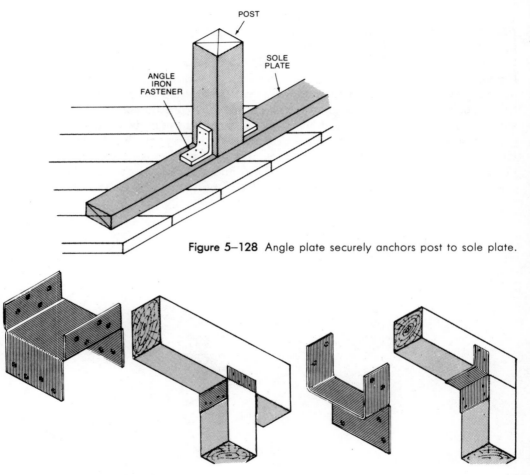

Figure 5–128 Angle plate securely anchors post to sole plate.

Figure 5–129 Two commonly used post-and-beam anchors.

Beams. The size of the beams used in plank-and-beam construction depends on the span, spacing, load, and other factors. Floor beams may be of solid lumber, laminated, or of built-up construction. Several types are shown in *Fig. 5–130*. The beams may rest on sills, as in conventional framing (*Fig. 5–131*), or they may be set in pockets in the foundation wall (*Fig. 5–132*). When they are set in pockets, the floor-to-grade height is reduced. This also results in lower building height.

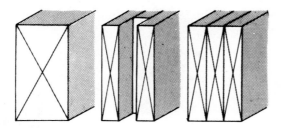

Figure 5–130 Various beam types. *Left:* Solid; *Center:* Built up; *Right:* Laminated.

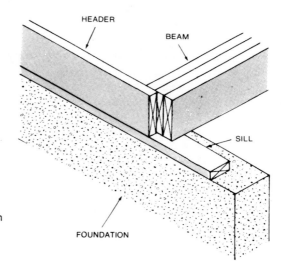

Figure 5–131 Framing at exterior wall with beam bearing on sill.

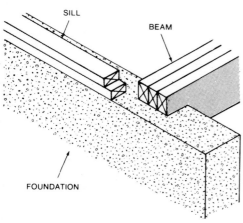

Figure 5–132 Framing at exterior wall with beam set in foundation wall.

Since roof beams are usually exposed, appearance is a factor and the choice of lumber is important. When the cross section is small enough, solid lumber is generally used. It can be rough-sawn or planed, depending on the appearance desired. Larger beams are usually laminated or built up. Either type can be given a finished appearance by any of the methods shown in *Fig. 5–133.*

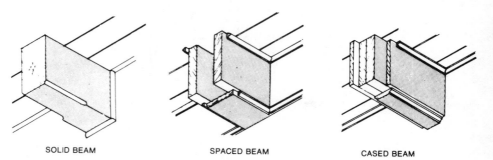

SOLID BEAM   SPACED BEAM   CASED BEAM

Figure 5–133 Methods of finishing exposed beams.

Because of the limited amount of concealed ceiling space, the placement of electrical wiring can be a problem. Spaced beams with adequate blocking can be used for wire runs. The bottom of the beam can then be covered with a board and molding as shown in *Fig. 5–134.* Also shown in this figure are methods of dressing up exposed beams. Wiring can also be concealed by grooving the top of the beam to serve as a raceway for the wire.

The roof beams in plank-and-beam construction are of two types, longitudinal and transverse. Longitudinal beams run parallel to the ridge

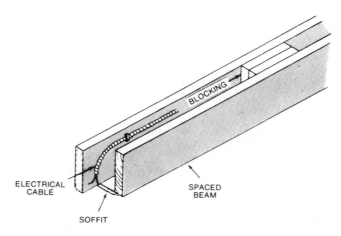

ELECTRICAL
CABLE

BLOCKING

SPACED
BEAM

SOFFIT

Figure 5–134 Wire runs can be concealed between spaced beams.

(*Fig. 5–135*) and are supported by the end walls or columns. Transverse beams run perpendicular to the length of the building, extending from the ridge to the side walls (*Fig. 5–136*). They are generally spaced 6 feet to 8 feet on center and are supported by partition walls or posts.

The roof beams may rest on top of the ridge beam or at the side, as shown in *Fig. 5–137*. Metal reinforcements provide resistance against horizontal thrust.

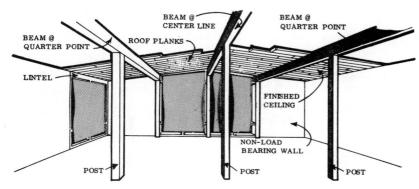

Figure 5–135 Longitudinal beams in plank-and-beam construction.

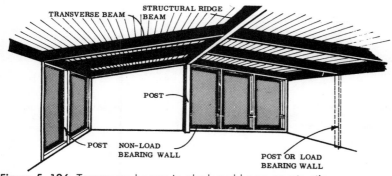

Figure 5–136 Transverse beams in plank-and-beam construction.

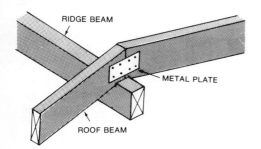

Figure 5–137 Roof beam supported by ridge beam.

Plank-Floor Deck. The flooring used in plank-and-beam construction usually consists of 2 × 6 or 2 × 8 tongue-and-grooved or splined planks (*Fig. 5–138*). These span the beams which may be placed on 4-foot or 8-foot centers. They serve as a subfloor and transmit the floor load to the beams.

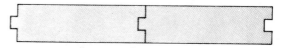

Figure 5–138 Cross section of 2-inch flooring used in plank-and-beam construction.

Strip wood and similar flooring may be applied directly over the planks. Sheet materials and tiles require an underlayment. Planks should be installed so that they are continuous over more than one span; this will result in added strength and stiffness (*Fig. 5–139*).

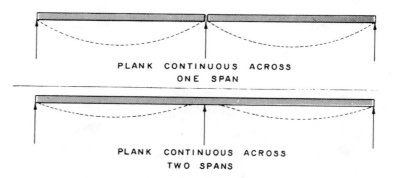

PLANK CONTINUOUS ACROSS ONE SPAN

PLANK CONTINUOUS ACROSS TWO SPANS

Figure 5–139 Comparison of stiffness of plank on single span with plank over two spans.

Plank-Roof Deck. Wood or fiberboard decking is widely used in plank-and-beam construction. Wood decking consists of wood planking tongued and grooved in various sizes, the most common being 2 × 6, 3 × 6, and 4 × 6. The 2-inch stock is used for spans up to 8 feet. The thicker planking is used for spans of up to 10 or 12 feet. They may be used to span longitudinal or transverse beams. Since the underside of roof decking is usually exposed, use care when choosing and applying the decking.

Face-nailing and blind-nailing through each tongue is sufficient for 2-inch decking. The thicker material should be facenailed, toenailed, and edge-nailed, as shown in *Fig. 5–140*.

In areas where cold weather is a factor, rigid insulation over the wood decking will cut down heat loss. A vapor barrier must be used between the decking and the insulation (*Fig. 5–141*).

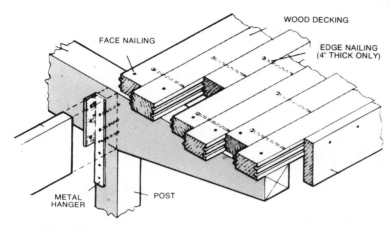

Figure 5–140 Roof decking details in plank-and-beam construction.

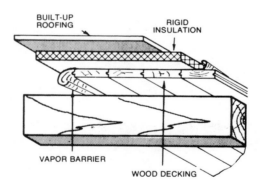

Figure 5–141 Method of installing rigid insulation on wood roof deck.

Plank-and-Beam Walls. Although the framing walls in plank-and-beam construction are non–load-bearing walls, they do give lateral support to the posts. To resist racking, the wall sections between open glazed areas must be fully framed and sheathed as is done in conventional framing. Let-in bracing is also used if necessary.

Load-bearing partition walls must be placed over beams heavy enough to carry the load. Non–load-bearing partition walls which run parallel to the plank floor require additional support. Two methods of support are shown in *Fig. 5–142*. For a partition with openings, a small beam is placed on the underside of the floor. Otherwise a small beam is used to replace the conventional floor plate. The beam can be made of two pieces of 2 × 4 on edge or a single piece of 4 × 4.

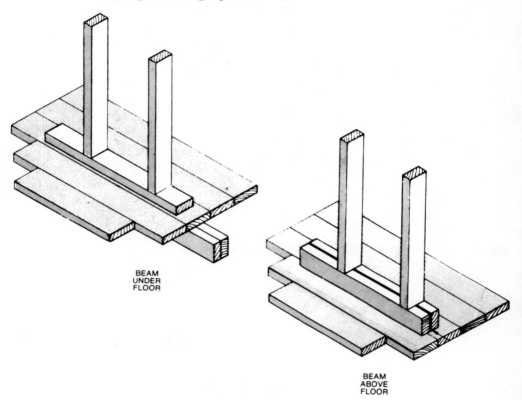

BEAM
UNDER
FLOOR

BEAM
ABOVE
FLOOR

Figure 5–142 Support for non–load-bearing partition parallel to floor planks.

6 Exterior Finish

The exterior finish includes roof and wall coverings, window and door frames, and exterior trim. Careful selection and correct installation of these finish materials are essential, because they directly affect the structure and beauty of the home and its ultimate market value. Also, if they are not properly installed, the interior and its inhabitants will be subjected to drafts, water leakage, and invasion by insects and rodents. Modern manufacturing methods have produced a wide selection of materials suitable for exterior finish, in various colors and textures. Most windows, doors, and frames are factory-made today. They are delivered to the jobsite ready for installation and can now be built faster, better and in more distinctive styles (*Fig. 6–1*).

Wood doors must be handled and stored carefully to prevent warpage, decay, and other damage. They should be stored flat on a level surface, in a clean, dry, well-ventilated building. When they are moved from one location to another, they should be carried, not dragged. If it is absolutely necessary to drag a door, do so only on the bottom end, which should be protected by a scuff strip or skid shoes. Deliver the door to the jobsite only after the plaster and cement is dry, and store it under cover until it is ready for hanging.

Figure 6–1 A modern home finished with rough-sawn siding.

Door Frames

Exterior door frames in homes are generally of wood construction, 1⅛ inch or thicker. Softwood is commonly used except for the sill, which is generally made of a hardwood species such as white oak. When softwood sills are used, they should be protected with metal wear strips.

The frames are usually factory-made and are furnished either assembled or knocked down (KD). Assembled units are shipped with bracing so they will stay intact, but they should nevertheless be checked for squareness before they are installed. *Fig. 6–2* shows the details of a typical door frame. The head and side jambs are rabbeted ½ inch to receive the door.

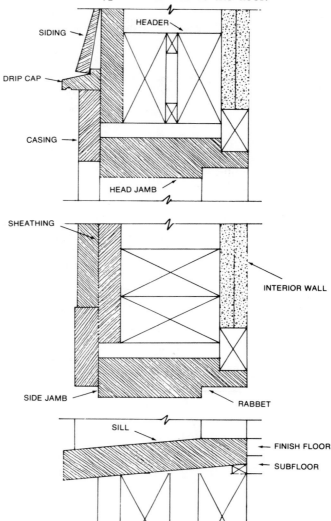

Figure 6–2 Detail of door frame.

Since the wall thickness may vary, most door manufacturers design doorframes with reversible extension strips (*Fig. 6–3*). These can be installed in various positions to attain the proper thickness.

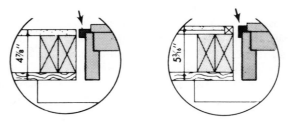

Figure 6–3 Reversible extension strip permits jambs to be converted for various wall thickness.

Installing the Door Frame. The rough opening should be ready to receive the frame. Check the size of both to make sure they will fit. Cut away a portion of the header joist so that the door sill will rest on it firmly. Trim the joist sufficiently so that the top edge of the sill will be level with the finish floor (*Fig. 6–4*).

Before placing the frame in the opening, line the edges of the rough opening with plastic sheeting or building paper. Center the frame in the opening, using wedges cut from tapered shingles for a snug fit. Locate the

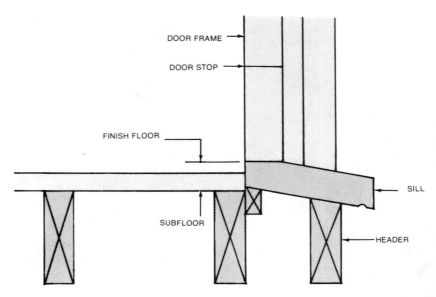

Figure 6–4 Exterior door sill detail. Note how header joist is trimmed to accommodate the sill.

wedges along the top and bottom of the frame, at the hinge locations, and also at the strike-plate location. The frame should be firmly installed in the opening with the front and rear edges set flush with the interior and exterior wall surfaces. Since the interior walls have not yet been installed, be sure to make allowances for them when you are setting the frame.

Use a straightedge and level when you plumb the frame. When you are satisfied that the frame is level and plumb, fasten it with 16d casing nails through the casing and wedges. Use care not to dent the wood with the hammer. Allow the nailheads to project slightly, then sink them with a nail set. Space the nails about 16 inches apart.

Exterior Windows

Windows, like doors, are usually factory-made and delivered fully assembled, ready for hanging. The three basic types are sliding, swinging, and fixed but many variations are available (*Fig. 6–5*). Sliding windows may slide horizontally or vertically. The vertical sliding type are called *double-hung* windows. Those that swing horizontally are *awning* types, and vertical swinging windows are called *casements*.

The materials used for windows are wood, steel, aluminum, and a combination of plastic and wood. Wood windows are usually more costly than metal; however, they have good insulating qualities, which is an important factor in these times of energy shortages. Metal windows are easier to manufacture and less costly than wood and generally are easier to install. Another advantage of metal windows is the minimal upkeep required. They are not subject to insect attack and decay, and if aluminum, they do not require painting.

Figure 6–5 Basic window types. *Left:* Sliding; *Center:* Swinging; *Right:* Fixed.

In cold climates, heat loss through metal frames and sashes is very high. To overcome this, some metal window frames are made with insulation within the frame. Other developments in window design include all-plastic frames and sash. These are lightweight, corrosion-free, and do not require painting.

Double-Hung Windows. The double-hung window consists of two sliding sashes which bypass each other in separate grooves. The partial cut-away drawing (*Fig. 6–6*) shows the components of a double-hung window and frame. This window has sealed double glass and removable grille. The double glass with a dead air space between panes provides excellent insulating qualities to the window. The details of a conventional double-hung window are shown in *Fig. 6–7.*

Various devices are used to hold the windows in place. Springs, balances, and compression weather stripping are commonly used.

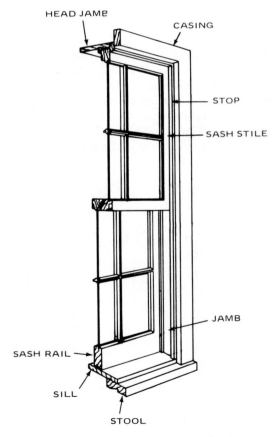

Figure 6–6 Components of a double-hung window and frame.

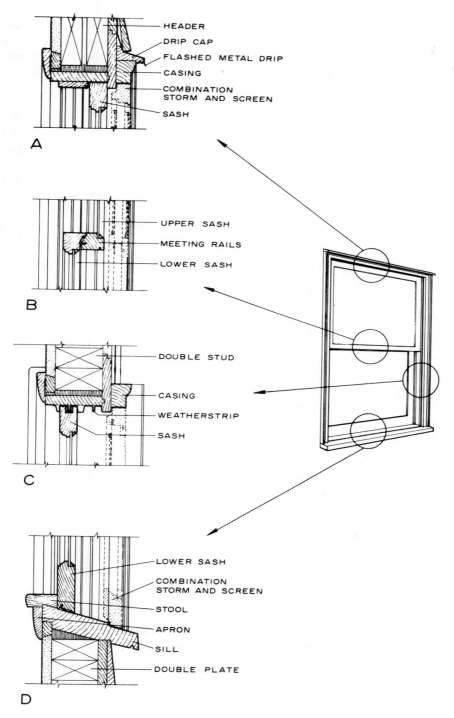

A

- HEADER
- DRIP CAP
- FLASHED METAL DRIP
- CASING
- COMBINATION STORM AND SCREEN
- SASH

B

- UPPER SASH
- MEETING RAILS
- LOWER SASH

C

- DOUBLE STUD
- CASING
- WEATHERSTRIP
- SASH

D

- LOWER SASH
- COMBINATION STORM AND SCREEN
- STOOL
- APRON
- SILL
- DOUBLE PLATE

Figure 6–7 Detail of a conventional double-hung window.

The jambs are made of nominal 1-inch lumber. The widths are made to fit drywall or plaster construction. Some manufacturers provide jamb extensions which permit adjustments for various wall thicknesses (*Fig. 6–8*). The sills are generally made of 2-inch lumber sloped for drainage.

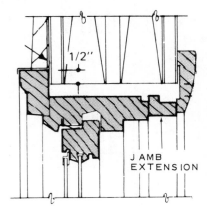

Figure 6–8 Detail of jamb extension.

Muntins are moldings which divide the sash into small sections. In the past, it was difficult to manufacture glass in large pieces, so muntins were a necessity. Today, producing large sheets of glass is not at all difficult; therefore, to keep costs down, some manufacturers provide preassembled dividers which simply snap into place. Aside from the economy factor, such windows are easy to clean and paint because the muntins are readily removable (*Fig. 6–9*).

Figure 6–9 Window unit with removable muntins.

There are several ways in which double-hung windows may be arranged. They can be used singly, doubled, or in groups of three or more. *Fig. 6–10* shows how three units are used in a bay window application. They can also be used at the sides of fixed windows and in other combinations.

Figure 6–10 Three double-hungs in a bay.

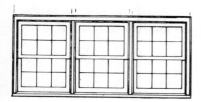

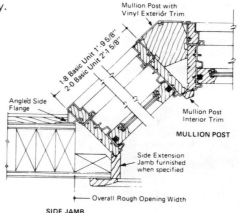

Casement Windows. The sash of casement windows is hinged at the sides. Most casement windows swing outward, but inward-swinging casements are also made. A crank mechanism is the usual method for controlling the opening and closing of the window. One advantage of the casement is that its ventilation area is 100 percent, as compared to the double-hung, which has a maximum ventilation area of only 50 percent. Another advantage is that it can be used in areas of limited access, such as over a kitchen sink. For outward-swinging casements, screens and storm sash are placed on the room side of the window. Insulated glass is often used in cold climates. Casements are available completely assembled including weatherstripping, ready to install.

Awning Windows. These are similar to casement windows except that they are hinged at the top. A similar type, called the *hopper*, is hinged at the bottom, with the sash swinging inward. A combination of both types incorporated in one window is shown in *Fig. 6–11.*

Figure 6–11 Frame with awning at top and hopper at bottom.

Fixed Windows. These are used mostly for effect as picture windows. They provide light and an attractive view of the outdoors. They may be used alone or flanked with double-hung, casement, or awning units. Large windows may be set in sash or directly into rabbeted frames. Large windows require thicker sash members. A fixed window is shown in *Fig. 6–12.*

Figure 6–12 Casement angle bay with fixed center section.

Horizontal Sliding Windows. This type of window is similar to the double-hung except that it slides horizontally instead of vertically. Such windows are available with two or more sashes. In the two-sash units, both are usually movable. In the three-sash unit the center unit is usually fixed. The section view, *Fig. 6–13*, shows details of this window.

Jalousies. Jalousies consist of transparent slats, usually glass, mounted horizontally within a frame. The slats slope downward toward the outside. They move in unison when activated by a crank or other type of control. When fully closed, the slats overlap each other slightly. They cannot be weatherproofed efficiently and therefore their use is limited to warm climates.

Basement Windows. These are single-sash windows of simple design. They are used mostly in foundation walls and are usually hinged at the top and swing inward.

Skylights. Skylights are used to admit light and air to attics, interior rooms, and hallways. One effective type, shown in *Fig. 6–14*, consists of an insulated double-domed plastic bubble appropriately framed. Skylights are easily installed on flat or sloping roofs.

CONSTRUCTION DETAILS

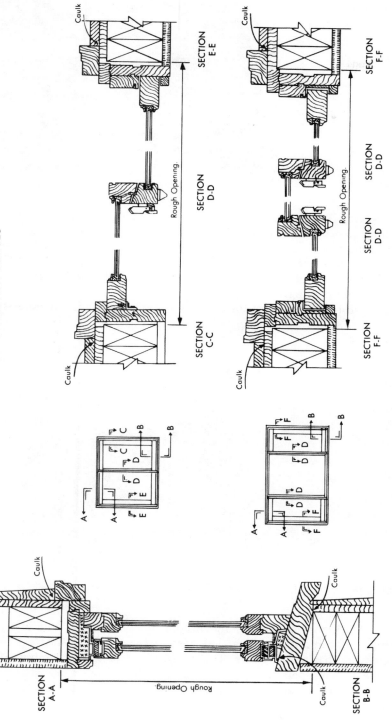

Figure 6–13 Details of horizontal sliding windows.

Figure 6–14 A double-domed attic skylight.

Installing Windows. Windows are generally delivered to the jobsite primed and braced and ready to install. If they are subjected to rough handling in transit, they may become distorted. Check each unit for squareness. If necessary, renail the braces. Check the sizes also. There should be at least ½ inch of space between jambs and rough opening studs to permit adjustment of the frame within the opening.

Before the frame is installed, the area around the opening should be flashed with a 10-inch strip of heavy building paper or plastic, as shown in *Fig. 6–15*.

Figure 6–15 Flashing around window opening.

Mark the location of the head jamb on a story pole and use it to properly position the frame. The tops of the windows should be the same height as that of the doors. This is usually 6 feet 8 inches high. Place the frame in the opening, then use wedge blocks under the sill to raise the frame to its proper height (*Fig. 6–16*). Use a level and adjust as necessary.

Figure 6–16 Wedge-shaped blocks are used to fit window frame into rough opening.

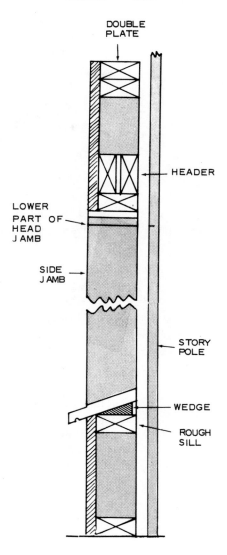

DOUBLE PLATE

HEADER

LOWER PART OF HEAD JAMB

SIDE JAMB

STORY POLE

WEDGE

ROUGH SILL

Drive nails part of the way into the lower part of the casing. Next, plumb the side jambs, then drive nails (again only partway in) near the top of the window. Now remove the frame braces and check the sash for proper operation. Make the necessary adjustments, then complete the nailing. Use a nail set to avoid marring the wood.

Cornices

The cornice, or eave, is the exterior finish and trim at the point where roof projections and side walls meet. In a hip roof, the cornice is continuous and extends completely around the structure (*Fig. 6–17*). In the gable roof, the cornice is formed along the side walls at the rafter ends. In a gable roof with a rake extension the sloping section is called a *rake cornice*.

Figure 6–17 The cornice of a hip roof extends equally around the structure.

The cornice adds to the appearance of a structure and protects it from rain, snow, and sun. It also provides a surface for installing gutters. A typical cornice with its various members is shown in *Fig. 6–18*. The soffit, which is also called a plancier, is the underside of a cornice overhang. The fascia is a finish trim member; usually a flat board applied to the ends of rafters or the outer face of a cornice.

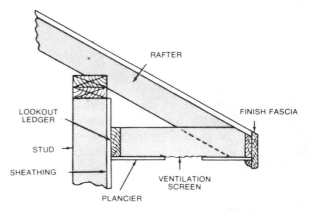

Figure 6–18 Section through a cornice.

Although many variations are possible, there are two basic types of cornices—the open and the closed (*Fig. 6–19*). The closed type is also referred to as a *box cornice*.

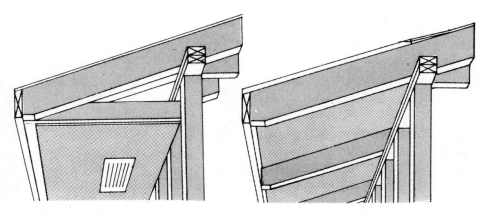

Figure 6–19 *Left:* The commonly used box cornice. *Right:* The open cornice without soffit.

Closed Cornice (Narrow).   The closed or box cornice may be either wide or narrow. Both conceal the rafters. In the narrow cornice, the rafter projections serve as the nailing surface for the soffit and the fascia (*Fig. 6–20*). The soffit is nailed to the underside of the rafters. This type of cornice is suitable with relatively short projections, usually between 6 and 12 inches.

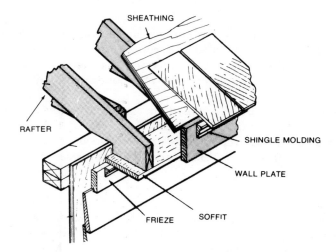

Figure 6–20 Construction detail for narrow box cornice.

Closed Cornice (Wide).   This cornice is similar to the narrow type but additional nailing surfaces for the soffit are provided. These additions are called lookouts (*Fig. 6–21*). The lookouts are toenailed to a 2 × 2 or 2 × 4 ledger and face-nailed to the rafter extensions. Lookouts are generally cut from 2 × 4s. If a 2 × 2 ledger is used, the ends of the lookouts are notched to fit (*Fig. 6–22*).

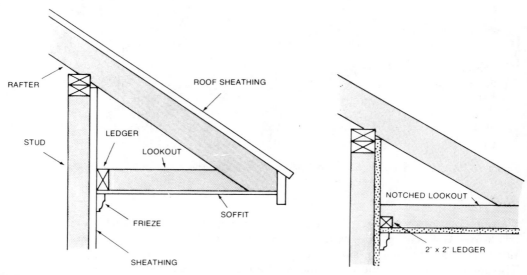

Figure 6–21 Construction detail of wide cornice. Lookouts are nailed to ledger.

Figure 6–22 Lookout is notched to fit over 2 × 2 ledger.

Stretch a line to check the tail cuts and lower corners of the rafters; these must be in alignment. If necessary, trim any irregularities. Next, install the ledger strip. Be sure it is level and that its lower edge is aligned with the lower edge of the rafter ends. The lookouts are nailed as described previously.

Plywood, hardboard, fiberboard, and metal are some of the materials used for soffits. The thickness of the soffit depends on the support spacing. For 16- or 24-inch rafter spacing, ⅜-inch plywood or ½-inch fiberboard are generally used for soffits. Heavier material is needed for 48-inch rafter spacing.

For thin soffit materials, a nailer may be used along the inside of the fascia. A nailer is a wood board serving primarily as a nailing base. Another method is to nail a header to the rafter ends. This will serve as a backer for the fascia as well as a nailer for the soffit. If a grooved fascia is used, the nailer can be eliminated (*Fig. 6–23*). The fascia should extend at least ⅜ inch below the soffit, where it serves as a drip. Ventilators are required in

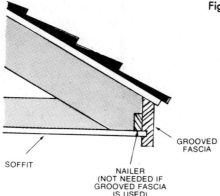

Figure 6–23 Grooved fascia supporting soffit.

GROOVED
FASCIA

SOFFIT

NAILER
(NOT NEEDED IF
GROOVED FASCIA
IS USED)

the soffit to permit a flow of air in the cornice and attic space. The ventilator may be a screened opening placed at intervals along the soffit (*Fig. 6–24*), or it can be a narrow continuous slot along the length of the soffit.

If a roof has considerable overhang and a steep slope, the soffit may fall too close to the window tops. In such cases the lookouts are eliminated and the soffit is slanted, as shown in *Fig. 6–25*.

Figure 6–24 Screened soffit ventilator.

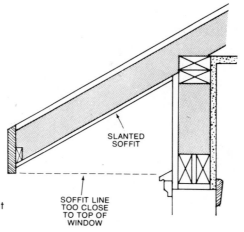

SLANTED
SOFFIT

Figure 6–25 Wide overhang may prevent use of horizontal soffit.

SOFFIT LINE
TOO CLOSE
TO TOP OF
WINDOW

Open Cornice.   This is similar to the slanting cornice except that the soffit is eliminated (*Fig. 6–26*). Since the underside of the roof boards are exposed, the overhanging section should be of good-quality material. If made of plywood, it should have an exterior glueline.

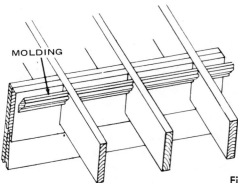

Figure 6–26 Open cornice trimmed with bed molding.

Rake Cornice.   The rake is the extension at the gable end of the roof. It can be very close or it may extend several feet. When the projection is close, the trim material may consist of only a fascia board and frieze (*Fig. 6–27*). For normal gable overhang, up to 8 inches, short lookout blocks can be toenailed to the head rafter. The soffit and fascia are then nailed to the block (*Fig. 6–28*).

For wider projections, up to 20 inches, a nailing block and fly rafter are used to support the rake section. The roof sheathing is nailed to the fly rafter and lookouts. The soffits, fascia, and other trim are then added (*Fig. 6–29*).

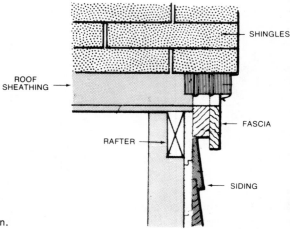

Figure 6–27 Detail of close rake projection.

Figure 6–28 Nail short lookouts to head rafter for normal gable overhang.

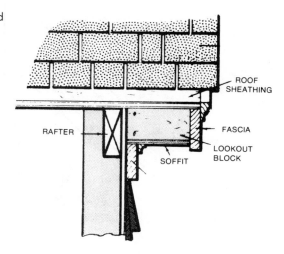

ROOF SHEATHING

RAFTER

FASCIA

LOOKOUT BLOCK

SOFFIT

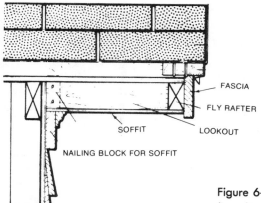

FASCIA

FLY RAFTER

SOFFIT

LOOKOUT

NAILING BLOCK FOR SOFFIT

Figure 6–29 Moderate projections require a fly rafter to support the rake section.

For very wide projections, a ladder-type frame is required, as shown in *Fig. 6–30*. It is constructed either in place on the roof or on the ground and then hoisted into place. It consists of a header and fly rafter nailed into the ends of lookouts. The head rafter is spiked to the end rafter and the lookouts rest on the plate of the end wall.

Cornice Return.   The cornice of a hip roof is continuous, but in a gable roof the cornice must be terminated at the gable end. The cornice may be returned as shown in *Fig. 6–31*. The cornice turns the corner and continues a short distance; it then turns again, where it terminates against the end wall. Great care must be exercised in cutting and fitting the various members. A more simple return is shown in *Fig. 6–32*. Here, the rake end fascia is widened to close the end of the cornice box.

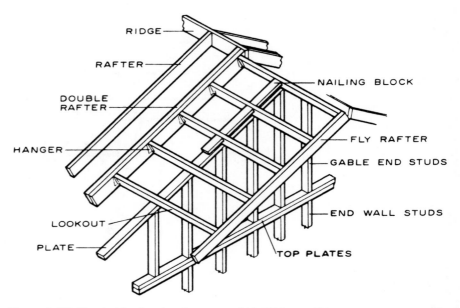

Figure 6–30 Use ladder framing for extra-wide gable extension.

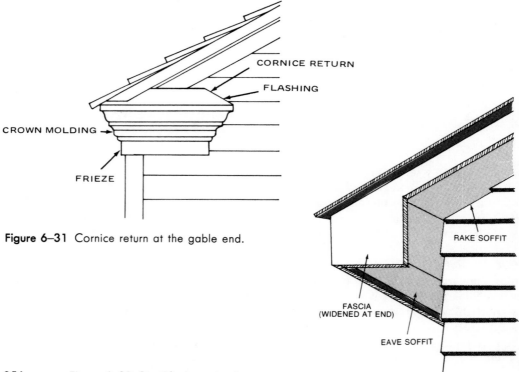

Figure 6–31 Cornice return at the gable end.

256   Figure 6–32 Simplified cornice box.

Prefabricated Soffits. Various materials are used in factory-made soffits and cornice trim. Steel, aluminum, and fiberboard are widely used. Their use greatly speeds up construction, because the parts are cut to size and prefinished ready to install. Fiberboard soffits are furnished with channel stiffeners to be assembled at the jobsite. Metal soffits are made in modular sizes and are usually ribbed for added stiffness. They are held with special channels and the ends are made to interlock, thus giving the appearance of a long, continuous soffit (*Fig. 6–33*). Another advantage of these soffit systems is that they require little maintenance.

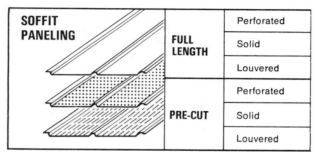

Figure 6–33 Metal soffits.

Soffit Ventilation. Improperly or nonventilated cornices can be the cause of considerable damage. One common problem caused by insufficient ventilation is the formation of ice dams at the cornice. Poor ventilation causes the attic to warm sufficiently so that the roof surface melts the snow on it. The melted snow runs down onto the colder cornice surfaces and freezes. The resulting ice dam causes water to back up, working its way into the walls and ceilings (*Fig. 6–34*). When the attic is vented so that its temperature is kept lower, the roof snow is kept from melting. Louvered openings in the gable end wall will ventilate an attic, but soffit openings will greatly increase circulation of air.

Finish Roofing

The finish roofing, when properly chosen and applied, adds beauty to the structure (*Fig. 6–35*) and also protects the interior from rain, snow, dust, and wind. When properly installed, the roof will give years of trouble-free service.

Many types of roofing are available, including asphalt, wood, asbestos, tile, slate, asphalt shingles, and wood shingles. The roofing used depends upon cost, locality, local codes, roof design, and other contributing factors. Roofs with shallow slopes generally require the watertight qualities of roll or built-up roofing. Roofs of tile or slate require stronger framing to

Figure 6–34 Lack of soffit ventilation can cause ice dams.

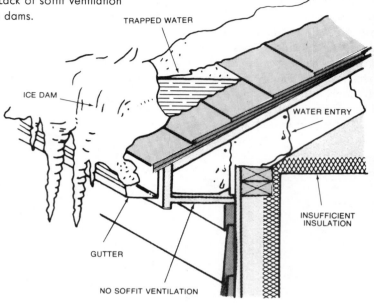

TRAPPED WATER

ICE DAM

WATER ENTRY

INSUFFICIENT INSULATION

GUTTER

NO SOFFIT VENTILATION

Figure 6–35 An asphalt shingle roof.

support their heavier weight. Generally, when the slope of a roof is under 4 inches per foot, roll or built-up roofing is recommended. Shingled roofs on slopes above 4 inches per foot are ideal.

Roofing Terms.   The application of tile, slate, metal, and similar roofs are handled by specialists in the roofing trade. However, roll and shingle roofing are often done by the carpenter, and they are therefore covered in this section. The following roofing terms should be thoroughly understood by the carpenter:

Coverage: Measures the weather protection provided to a roof and is based on the number of layers of roof-covering materials applied. For example, one layer of roll roofing provides single coverage. Shingles are usually installed in two or three layers and therefore give double or triple coverage respectively.

Exposure: The distance from the butt of one shingle to the butt of the one above it.

Square: A roofing unit of measure. One square is the amount of material needed to cover 100 square feet of roof area.

Flashing: A waterproof material used to protect joints at roof intersections and around chimneys and make them watertight.

Butt: The leading or lower edge of a shingle.

Boston Ridge: A method of applying shingles at the ridge or hip of a roof.

Fig. 6–36 illustrates some of the terms used.

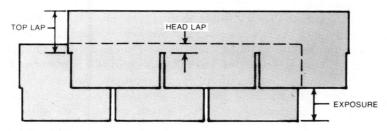

Figure 6–36 Illustrations of major roofing terms.

Deck Preparation.   Regardless of the type of roof being installed, the deck or roof surface must be sound, smooth, and free of defects. All boards or plywood sheathing must be nailed securely. Pay special attention to make sure all nailheads are flush. Board sheathing should be at least ¾ inch thick and no wider than 6 inches, and well seasoned. The use of green boards would cause the covering to buckle if shrinkage were to occur after the finish roofing is applied.

Asphalt Roofing Products. The products used in roofing can be broadly classified into three groups: saturated felts, roll roofing, and roofing shingles.

Saturated felts are made of dry felt impregnated with asphalt or coaltar saturants. They are used as an underlayment for shingles, for sheathing paper, and as the laminates in built-up roofing. Although made in various weights, the most commonly used are the #15 and #30, which are 15 and 30 pounds per square, respectively. Saturated felt is available in rolls 36 inches wide and 144 feet long.

Roll roofing consists of a saturated felt to which weather-resistant asphalt has been added. The surface may be coated with colored mineral granules, which in addition to being decorative also help fireproof the roof.

Asphalt shingles are surfaced with mineral granules. They are made in various weights and patterns in single and strip form. The table shown in *Fig. 6–37* lists some of the most commonly used shingles. The 235-pound weight per square is quite popular and easily applied.

The square butt strip shingle is widely used. It measures 12 × 36 inches and may have one, two, or three tabs. Some shingles are made with self-sealing tabs. These consist of adhesive spots on the underside of the tab. Heat from the sun activates the spot, thus eliminating the tedious work of sealing each tab by hand. To prevent dead-air pockets, a space is left between each seal. These shingles are recommended for use in high-wind localities.

Underlayment. Roofing underlayment consisting of 15-pound felt is used on roof decks with a pitch of 4 inches per foot or greater. It protects the deck from rain before the shingles are applied, and it prevents the resinous areas of the deck from direct contact with the shingles. It also protects the deck from getting wet should any shingles be lifted during a storm. The use of 30-pound felt or tar-saturated products or any other vapor barrier materials is not recommended, because they prevent the roof from "breathing." (Moisture in the attic in the form of vapor will move through roof boards and dissipate harmlessly into the atmosphere; this movement of vapor must not be restricted by a barrier, because it would have a deleterious effect on all wood structural members.)

Lay the underlayment horizontally with top and side laps of 2 inches and 4 inches, respectively, as shown in *Fig. 6–38*. Lap the felt 6 inches from both sides on all hips and ridges and use metal drip edges at the eave and rake. Note that felt is placed *under* the drip edge at the rake end but *over* it along the eaves. The drip edge prevents water from entering along the edges of the sheathing.

Low-Slope Deck. Roofs with slopes under 4 inches per foot will have a slower water runoff. Therefore, the underlayment procedure for such roofs

PRODUCT	Per Square			Size		
	Approximate Shipping Weight	Shingles	Bundles	Width	Length	Exposure
Wood Appearance Strip Shingle Single Thickness Per Strip 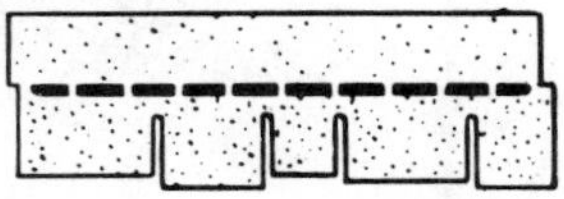	Various 250# to 350#	78 to 90	3 or 4	12" or 12-1/4"	36" or 40"	4" to 5-1/8"
Self-Sealing Strip Shingle 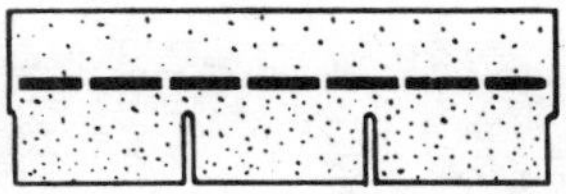	205#- 235#	78 or 80	3	12" or 12-1/4"	36"	5" or 5-1/8"
	Various 215# to 325#	78 or 80	3 or 4	12" or 12-1/4"	36"	5" or 5-1/8"
Self-Sealing Strip Shingle 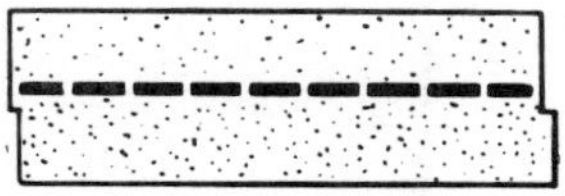 ... No Cut Out	Various 215# to 290#	78 to 81	3 or 4	12" or 12-1/4"	36" or 36-1/4"	5"

Figure 6–37 Some of the most commonly used shingles.

is different from that outlined above. The deck must first be covered with a double layer of 15-pound roofer's felt (*Fig. 6–39*). Begin with a 19-inch starter strip along the eave with the leading edge placed over a metal drip edge. Follow with a 36-inch-wide sheet placed directly over the starter course. Lay each succeeding strip with an overlap of 19 inches, thus exposing 17 inches of each preceding course.

When conditions warrant, such as in very cold climates, a flashing strip should be employed along the eave to prevent the backup of water. A course of roll roofing (not less than 50-pound) should be installed to overhang the metal drip edge and underlayment from ¼ to ⅜ inch. Allow it to extend up the roof to a point 24 inches inside the interior wall line (*Fig. 6–*

40). For normal slope roofs above 4 inches per foot, the distance should be a minimum of 12 inches inside the wall line.

Figure 6–38 Application of roof underlayment.

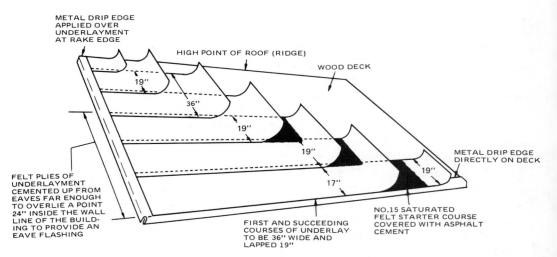

METAL DRIP EDGE
APPLIED OVER
UNDERLAYMENT
AT RAKE EDGE

HIGH POINT OF ROOF (RIDGE)

WOOD DECK

19"

36"

19"

19"

METAL DRIP EDGE
DIRECTLY ON DECK

19"

17"

FELT PLIES OF
UNDERLAYMENT
CEMENTED UP FROM
EAVES FAR ENOUGH
TO OVERLIE A POINT
24" INSIDE THE WALL
LINE OF THE BUILD-
ING TO PROVIDE AN
EAVE FLASHING

FIRST AND SUCCEEDING
COURSES OF UNDERLAY
TO BE 36" WIDE AND
LAPPED 19"

NO.15 SATURATED
FELT STARTER COURSE
COVERED WITH ASPHALT
CEMENT

Figure 6–39 Application of underlayment for low sloping roof.

Figure 6–40 Application of underlayment to prevent water backup along eaves.

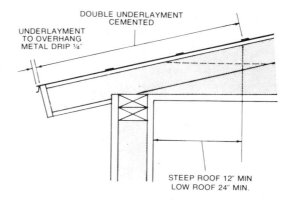

Flashing

Valley Flashing. Valley flashing is required to make the area between two intersecting roofs watertight. It is also used at chimneys, dormers, and other roof intersections. The materials used for flashing include various metals, asphalt shingles, and roll roofing. When properly installed, the flashing will divert water off the roof without allowing leakage under the shingles. Valley flashing may be either open, closed, or woven.

Open valley flashing is illustrated in *Fig. 6–41* and consists of 90-pound mineral-surfaced asphalt roll roofing applied in two layers. The first layer

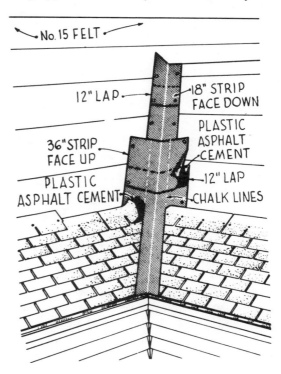

Figure 6–41 Open valley flashed with roll roofing in two layers.

is 18 inches wide and is placed face down on the roof. The second layer is 36 inches wide and is applied face up. The color of the flashing should match the shingles. When joints are necessary, they should be lapped 12 inches and secured with roofing mastic.

The width of the valley should increase toward the lower part of the slope. The minimum at the top is 5 inches and it should widen at the rate of ⅛ inch per foot (*Fig. 6–42*). Snap chalk lines to indicate the edges of the shingles.

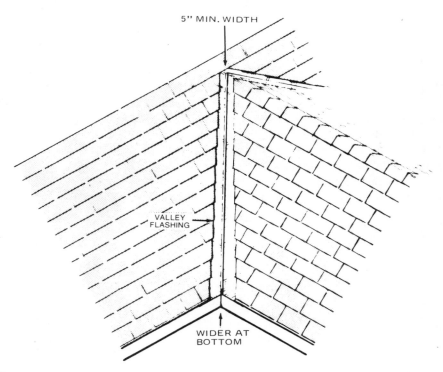

Figure 6–42 Install shingles at valley along tapering lines.

For the sake of appearance, valleys may be installed with shingles in a woven pattern (*Fig. 6–43*). Apply the shingles over the lining, working both surfaces of the roof at the same time. Lay each course alternately and interlock it to form a woven pattern.

The *closed valley flashing* is shown in *Fig. 6–44*. Here the shingles from one roof slope are made to extend at least 12 inches past the valley center line. Cut the shingles of the second slope at the valley center line and make the ends fast with plastic asphalt cement.

When adjacent roofs have varying slopes, a metal flashing with a standing seam may be used (*Fig. 6–45*). The seam acts as a barrier to pre-

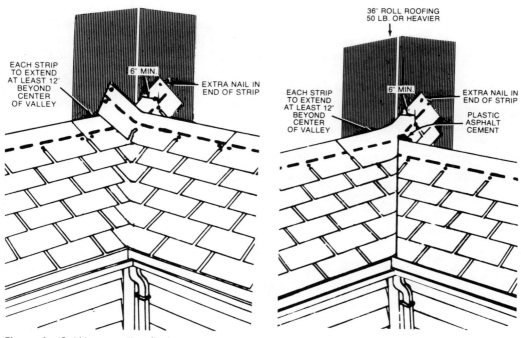

Figure 6–43 Woven valley flashing.

Figure 6–44 Closed valley flashing.

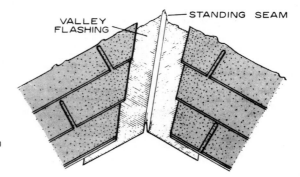

Figure 6–45 Flashing has standing seam on intersecting roofs with different slopes.

vent water from the steeper slope from overrunning the valley and forcing water under the shingles.

Flashing at a Vertical Wall. When a sloping roof abuts a vertical wall, metal flashing shingles are widely used to prevent water infiltration

through the intersecting joints (*Fig. 6–46*). The metal shingles are rectangular in shape, about 6 inches long and 2 inches wider than the exposed face
of the roof shingles. For example, when used with strip shingles laid 5
inches to the weather, the flashing shingles would measure 6 × 7 inches.
Bend them so they extend 2 inches onto the roof deck. The remainder extends up along the vertical wall. Install the strips as each course of shingles
is laid. Place one nail at the top of each metal shingle. As the shingling
progresses, each metal shingle will be covered by a regular roof shingle.

Figure 6–46 Watertight flashing at
intersecting surfaces.

Chimney Flashing.   Flashing around a chimney must be installed carefully. Since chimneys and the structures around them are subject to stresses,
some movement is bound to take place at the joints. It is therefore important to install the flashing in a manner that will permit movement without
affecting the water seal. The flashing is installed in two sections—the base
flashing, which is fastened to the roof, and the cap flashing, which is secured to the chimney. The base flashing should be installed first.

Lay the shingles up to the base of the chimney, then cut flashing sections from 90-pound mineral-surfaced roofing. Cut the pieces to size, using
the patterns in *Fig. 6–47* as a guide.

Start with the front section, then follow with the sides and top. Bed
each piece in mastic as it is applied. The back piece is added last. If the
chimney is large, a saddle to divert water is recommended. Install the saddle (also called a cricket) on the deck before you apply the underlayment.
The saddle is made of wood and has a ridge with sloping sides (*Fig. 6–48*).
Cut the flashing for the saddle and apply it as shown in *Fig. 6–49*. Add the
front section first, then the sides and finally the top. Be sure that all flashing pieces are cemented to one another and to the roof.

The cap flashing is applied next. It consists of metal sheets set into
the mortar joints and bent down over the base flashing. The front flashing

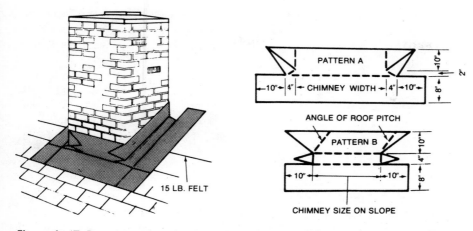

Figure 6–47 Patterns and application of chimney-base flashing.

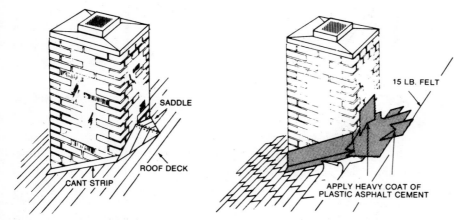

Figure 6–48 Wide chimneys require a saddle to divert water.

Figure 6–49 Details of saddle flashing.

is usually one long piece, but the side pieces are stepped to correspond to the roof slope. They are installed by the mason at the time the chimney is built. If the chimney is built without the cap flashing, the mortar joints will have to be cleared out to a depth of 1 inch. The flashing must then be inserted and the cap secured with lead plugs. The joints should be filled with caulking compound (*Fig. 6–50*).

Stack Flashing. Vent stacks and other projections must be flashed carefully. Prefabricated vent flashing boots, made of metal, rubber, or plastic,

Figure 6–50 Use metal cap flashing to waterproof the roofing around chimney.

are readily available. These are slipped over the stack and caulked with mastic. A one-piece vent collar is shown in *Fig. 6–51*. The method of flashing the vent with shingles is shown in *Fig. 6–52*. The shingles are applied up to and around the vent. A flange is then cut out and placed over the stack. It is cemented in place and the shingles are then laid over it.

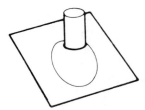

Figure 6–51 One-piece collar made to slip over vent pipe.

Applying Shingles

Roofing materials are sold by the square (100 square feet). The number of squares needed will be determined by the area of the roof in square feet. *Fig. 6–53* shows the method of determining the area of various styles of roofs.

Shingles can be laid from either end on small roofs, but on roofs that are over 30 feet, it is better to work from the center out. This will assure greater accuracy. Since there may be slight variations in the length of the strip shingles (plus or minus ¼ inch), the use of chalk lines is recommended. Strike the chalk lines parallel to the eave and space them at about every fourth or fifth course. This will assure a straight line along the butt edges of the shingles.

The application of shingles can begin after the necessary preparations are made. The underlayment must be swept clean and the surface must be checked for protruding nailheads. A starter course is applied to the roof along the eave. This can be a 9-inch or wider strip of roll roofing or a row

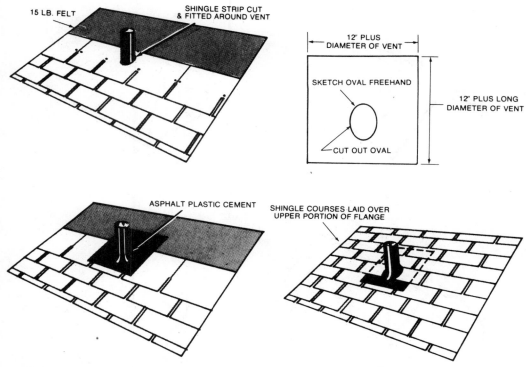

Figure 6–52 Step-by-step method of installing flashing around vent.

of strip shingles inverted so that the tabs point upward toward the ridge. Install with the leading edge overhanging the metal drip edge by about ⅜ inch. The starter strip backs up the first course of shingles and fills the space between the tabs. Nail the strip securely, placing the nails about 4 inches above the edge.

The nails used for asphalt shingles should be noncorrosive, with large flat heads. The shanks may be barbed or have annular or spiral threads (*Fig. 6–54*). Use 1¼-inch nails for new work and 1¾-inch nails when reroofing. Special staples applied with pneumatic staplers may also be used (*Fig. 6–55*). The nailing pattern is very important. For three-tab shingles in high-wind areas, use six nails for each 12 × 36-inch shingle (*Fig. 6–56*). Otherwise, use four nails per strip. Be sure to drive the nails squarely and flush with the shingle. Do not attempt to set the nails.

First and Subsequent Courses.   Start the first course with a full shingle and the following courses with full or cut strips, depending on the style of shingle used and the pattern desired. For a three-tab butt shingle with the

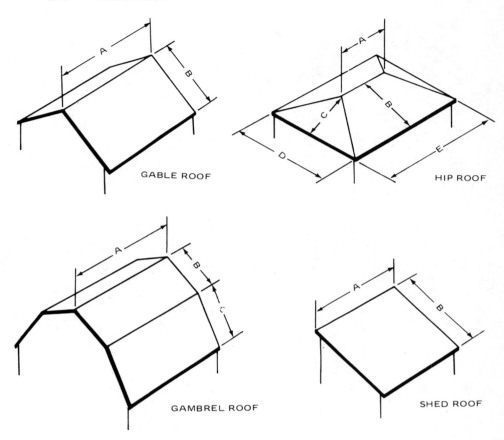

Figure 6–53 Method of determining area of roof: *Gable roof*—Multiply roof length (A) by rafter length (B). Multiply by 2. *Hip roof*—Step 1: Add roof length (A) and eaves length (E). Divide the sum by 2. Multiply the quotient by rafter length (B). Multiply by 2. Step 2: Multiply longest rafter length (C) by eaves length (D). Step 3: Add figures obtained in steps 1 and 2 for total roof area of hip roof. *Gambrel roof*—Add rafter lengths (B) and (C). Multiply sum by roof length (A). Multiply by 2. *Shed roof*—Multiply roof length (A) by rafter length (B).

Figure 6–54 Two commonly used asphalt shingle nails. *Top:* Plain barbed. *Bottom:* Annular thread.

Figure 6–55 Applying asphalt shingles with pneumatic stapler.

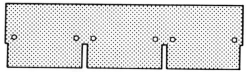

Figure 6–56 Nailing pattern for 3-tab shingle.

joints breaking on halves, remove 6 inches from the first tab on the second row of shingles. On the third row remove 12 inches. For the fourth row, remove half a strip, and so on. The result will be cutouts in each row centered in the course directly below (*Fig. 6–57*).

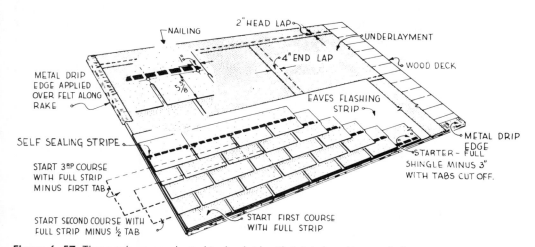

Figure 6–57 Three-tab square butt shingles laid with joints breaking on halves.

For random spacing, different amounts are removed from the tabs of succeeding courses. These rules should be followed: The width of the starting tab should be at least 3 inches; the cutout center lines of any course should be at least 3 inches to either side of the cutouts in the courses directly above or below; the starting tab widths should vary sufficiently in each row (*Fig. 6–58*).

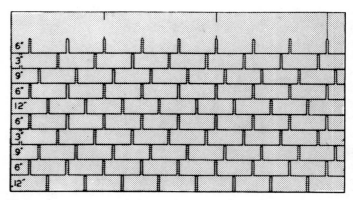

Figure 6–58 Random spacing of three-tab square butt strips. Pattern repeats each five courses.

Hips and Ridges. Special shingles are used for finishing hips and ridges. They are available ready-made, or they can be cut from 90-pound mineral-surfaced roll roofing or from the shingles used on the main roof. Cut pieces 9 × 12 inches and bend them lengthwise along the center to obtain equal exposure on each side of the hip or ridge. Start at either end of a ridge or at the bottom of a hip. Apply with a 5-inch exposure and secure with a nail at each side 5½ inches from the leading end and one inch from the edges (*Fig. 6–59*). Thus, each overlapping row of shingles conceals the nails of the row below. Be sure to warm the shingles in cold weather to prevent cracking. Do not use metal ridges with asphalt shingles, because discoloring of the shingles may result.

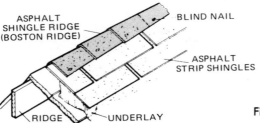

Figure 6–59 Apply ridge pieces with a 5-inch exposure.

Reroofing. It is not usually necessary to remove the old shingles when you are reroofing with asphalt shingles, as long as the old roof is strong enough to support the new dead and live loads. Also, the sheathing must be sound enough to provide good anchorage for the roofing nails. If these conditions cannot be met, the old roofing should be removed completely and any defects corrected.

Prepare the existing roofing by nailing down all loose shingles and replacing any missing ones. Remove all loose or protruding nails. If the old shingles are wood, split the badly warped ones so that the resulting segments can be nailed flat.

To insure that the new roofing installed over old wood shingles will be sound, especially in high-wind areas, cut back the shingles at the rake and eaves so that 1 × 4 or 1 × 6 boards can be installed at the edges (*Fig. 6–60*). Let the boards overhang the edge the same amount as the old wood shingles. Feather strips installed at the butts of old wood shingles are recommended.

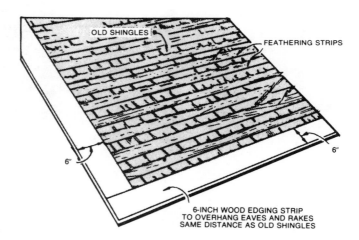

Figure 6–60 Preparation of old roof for reshingling.

Wood Shingles.   Wood shingles have been in use on dwellings for many years. They add charm and rustic beauty to a structure and because of the way they are installed they add considerably to its strength. However, because they have little fire resistance, some localities prohibit their use. The species of wood most commonly used for shingles are red cedar, redwood, and cypress; these woods are highly resistant to decay. When they are machine-sawed, the product is called shingles; when hand-split, the result is called shakes (*Fig. 6–61*).

Wood shingles are made in random widths in lengths of 16, 18, and 24 inches. They are tapered, with the butt or thicker end measuring about ½ inch in thickness. Various grades and specifications are available (*Fig. 6–62*). The better grades are cut with the annular rings perpendicular to the surface (*Fig. 6–63*). Shingles are packed in bundles. In normal usage, it takes four bundles to cover 100 square feet of roof area.

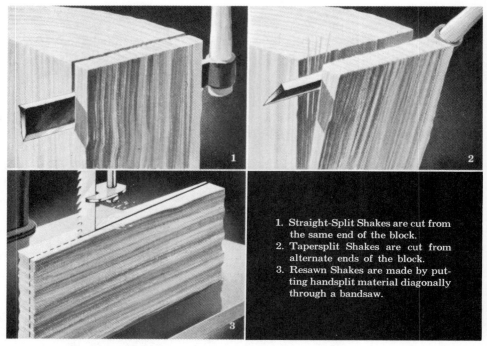

1. Straight-Split Shakes are cut from the same end of the block.
2. Tapersplit Shakes are cut from alternate ends of the block.
3. Resawn Shakes are made by putting handsplit material diagonally through a bandsaw.

Figure 6–61 Wood shingles and shakes.

GRADE	Length	Thickness (at Butt)	No. of Courses Per Bundle	Bdls/Cartons Per Square	Description
No. 1 BLUE LABEL	16″ (Fivex) 18″ (Perfections) 24″ (Royals)	.40″ .45″ .50″	20/20 18/18 13/14	4 bdls. 4 bdls. 4 bdls.	The premium grade of shingles for roofs and sidewalls. These top-grade shingles are 100% heartwood . . . 100% clear and 100% edge-grain.
No. 2 RED LABEL	16″ (Fivex) 18″ (Perfections) 24″ (Royals)	.40″ .45″ .50″	20/20 18/18 13/14	4 bdls. 4 bdls. 4 bdls.	A proper grade for some applications. Not less than 10″ clear on 16″ shingles, 11″ clear on 18″ shingles and 16″ clear on 24″ shingles. Flat grain and limited sapwood are permitted in this grade.
No. 3 BLACK LABEL	16″ (Fivex) 18″ (Perfections) 24″ (Royals)	.40″ .45″ .50″	20/20 18/18 13/14	4 bdls. 4 bdls. 4 bdls.	A utility grade for economy applications and secondary buildings. Not less than 6″ clear on 16″ and 18″ shingles, 10″ clear on 24″ shingles.
No. 4 UNDER-COURSING	16″ (Fivex) 18″ (Perfections)	.40″ .45″	14/14 or 20/20 14/14 or 18/18	2 bdls. 2 bdls. 2 bdls. 2 bdls.	A utility grade for undercoursing on double-coursed sidewall applications or for interior accent walls.
No. 1 or No. 2 REBUTTED-REJOINTED	16″ (Fivex) 18″ (Perfections) 24″ (Royals)	.40″ .45″ .50″	33/33 28/28 13/14	1 carton 1 carton 4 bdls.	Same specifications as above for No. 1 and No. 2 grades but machine trimmed for exactly parallel edges with butts sawn at precise right angles. For sidewall application where tightly fitting joints are desired. Also available with smooth sanded face.

Figure 6–62 Wood shingle specifications.

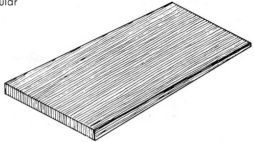

Figure 6–63 Better shingles have annular rings perpendicular to surface.

Wood shingles are installed over solid sheathing or over spaced boards (*Fig. 6–64*). In either case, it is important that the attic be well ventilated to prevent a buildup of moisture or heat.

The type of nails used for shingling is important. They must be a corrosion-resistant type, such as hot-dipped galvanized steel or spiral-threaded aluminum nails, and they must have good holding qualities. Although shanks can be smooth or threaded, threaded nails are recommended for plywood sheathing. Nail sizes are shown in *Fig. 6–65*. For hips and ridges the nail size should be 2d larger than shown.

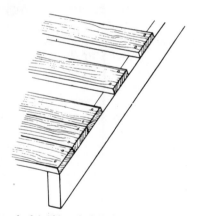

Figure 6–64 Wood shingles may be installed over solid sheathing or spaced boards.

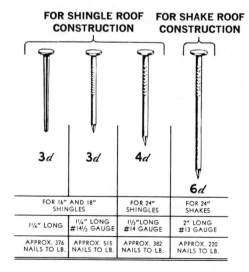

FOR SHINGLE ROOF CONSTRUCTION		FOR SHAKE ROOF CONSTRUCTION
3d	3d	4d	6d
FOR 16" AND 18" SHINGLES		FOR 24" SHINGLES	FOR 24" SHAKES
1¼" LONG	1¼" LONG #14½ GAUGE	1½" LONG #14 GAUGE	2" LONG #13 GAUGE
APPROX. 376 NAILS TO LB.	APPROX. 515 NAILS TO LB.	APPROX. 382 NAILS TO LB.	APPROX. 220 NAILS TO LB.

Figure 6–65 Nail sizes recommended for application of wood shingles.

Tools for Shingling.   A special hatchet is used for shingle installation. It should be well balanced, lightweight, and sharp. Some have a gauge that can be adjusted to check the shingle exposure (*Fig. 6–66*). A trimming saw and straightedge are also needed.

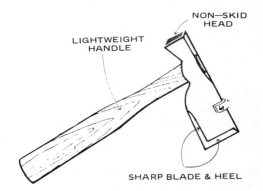

NON—SKID
HEAD

LIGHTWEIGHT
HANDLE

SHARP BLADE & HEEL

Figure 6–66 Shingler's hatchet.

Slope and Exposure.   The exposure of wood shingles depends upon the roof slope. For roofs with slopes of 5 in 12 or steeper, standard exposures of 5, 5½ or 7½ inches are used for 16-, 18-, or 24-inch-size shingles, respectively. By using less exposure, the shingles may be used on slopes as low as 3 in 12. For example, a 4 in 12 roof requires exposures of 4½, 5, and 6¾ inches. For a 3 in 12 slope, the figures drop to 3¾, 4¼, and 5¾ inches, respectively. This will result in four layers of shingles over the entire roof area.

Roof Sheathing.   As mentioned previously, sheathing for wood shingles may be solid or spaced. Solid sheathing may be plywood or nominal 1-inch boards. The advantage of spaced sheathing is the reduced cost. Also, spaced sheathing permits the shingles to dry out faster. The center-to-center spacing of open sheathing should be equal to the shingle exposure (*Fig. 6–67*).

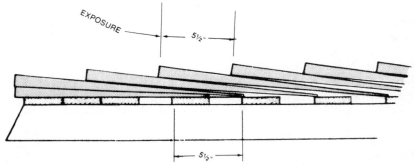

EXPOSURE

5½"

5½"

Figure 6–67 The spacing of open sheathing should be the same as the exposure.

Other methods of spacing may also be used, depending on the preference of the carpenter.

In areas of high wind, snow, and rain, solid sheathing is recommended. Underlayment is not needed with wood shingles, but it may be desirable to prevent air infiltration. Use a nonsaturated felt.

Flashing. Eaves flashing is highly recommended in areas where temperatures drop as low as zero degrees Fahrenheit. Follow the same procedure as used for asphalt shingles.

Metal flashing is generally used with wood shingles. The type of metal is important. Do not use copper flashing with red cedar, because the wood may react with the copper, causing deterioration of the metal. Galvanized steel or aluminum is preferable. When using galvanized steel flashing, be careful not to crack the zinc coating.

On roofs with a pitch of ½ (12-inch rise in 12-inch run) or steeper, the valley flashing should extend at least 7 inches to either side of the valley center line. For lower slopes, the measurement should be 10 inches to each side (*Fig. 6–68*). If the two slopes leading to a valley are unequal, the valley flashing should contain a vertical waterstop along its center line.

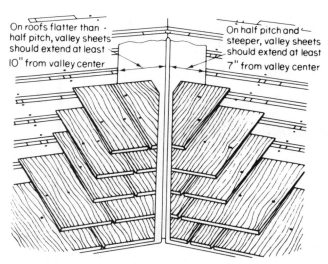

Figure 6–68 Flashing specifications for shingle roofs.

Applying the Shingle. Double or triple the first course of shingles at the eaves. Let them project beyond the sheathing 1 inch at the rake and 1½ inches at the eave. Leave ¼ inch of space between all shingles to allow room for expansion. When laying down the second row of shingles in the first course, offset the joints by 1½ inches (*Fig. 6–69*). Use two nails per shingle regardless of its width. Place the nails ¾ inch from the edge and 1 inch above the exposure line. Drive the nails flush without crushing the wood fibers (*Fig. 6–70*). To divert rainwater away from the gable end, use a cant strip made of beveled siding (*Fig. 6–71*). This will direct the water away from the edge and toward the eaves, where it will run off into the rain gutters.

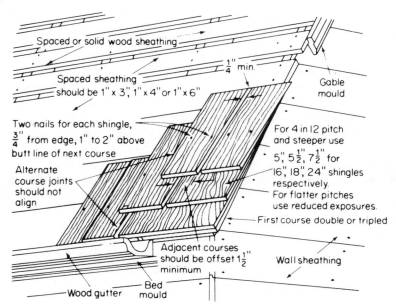

Figure 6–69 Typical shingle application.

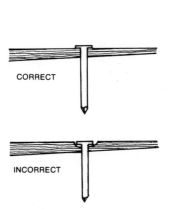

Figure 6–70 Drive nails until the heads meet the shingle surface.

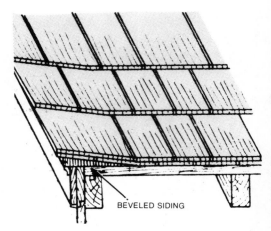

Figure 6–71 Application of shingles at gable end.

Hips and Ridges.   Good tight joints are important. The width of the hip or ridge should match the exposure. Start the course with a double shingle and use the "Boston" method of alternate overlapping. (To keep the ridge line straight, some workers tack a guide strip on each side of the ridge.) Cut and bevel each piece, letting them overlap as shown in *Fig. 6–72*. Position the nails so they are concealed by the succeeding shingles. Use nails long enough to penetrate the sheathing.

Figure 6–72 Alternate hip and ridge shingles to produce watertight joint.

Wood Shakes.   Wood shakes are hand-split shingles with a textured surface. They have a distinct character unlike that of other roofing materials (*Fig. 6–73*). Shakes are available in three types: hand-split and resawn, straight-split, and taper-split (*Fig. 6–74*). They are produced in three lengths —18, 24, and 32 inches—and in widths varying from 5 to 18 inches. The resawn hand-splits are made by diagonally sawing a straight-split shingle to produce two pieces. They are applied in a manner similar to wood shingles. Because of their thickness, however, greater exposures are possible. Typical exposures are 7½, 10, and 13 inches.

Shakes should not be used on slopes of less than 4 in 12. Installation procedure is similar to that used for regular wood shingles. Apply a

Figure 6–73 Application of wood shakes.

Grade	Length and Thickness	20" Pack		18" Pack		Shipping Weight	Description
		# Courses Per Bdl.	# Bdls. Per Sq.	# Courses Per Bdl.	# Bdls. Per Sq.		
No. 1 HANDSPLIT & RESAWN	15" Starter-Finish	8/8 10/10	5 4	9/9	5	225 lbs.	These shakes have split faces and sawn backs. Cedar logs are first cut into desired lengths. Blanks or boards of proper thickness are split and then run diagonally through a bandsaw to produce two tapered shakes from each blank.
	18" x ½" to ¾"	10/10	4	9/9	5	220 lbs.	
	18" x ¾" to 1¼"	8/8	5	9/9	5	250 lbs.	
	24" x ⅜"	10/10	4	9/9	5	225 lbs.	
	24" x ½" to ¾"	10/10	4	9/9	5	280 lbs.	
	24" x ¾" to 1¼"	8/8	5	9/9	5	350 lbs.	
	32" x ¾" to 1¼"	6/7	6			450 lbs.	
No. 1 TAPERSPLIT	24" x ½" to ⅝"	10/10	4	9/9	5	260 lbs.	Produced largely by hand, using a sharp-bladed steel froe and a wooden mallet. The natural shingle-like taper is achieved by reversing the block, end-for-end, with each split.
No. 1 STRAIGHT-SPLIT	18" x ⅜" True-Edge*	14 Straight	4			120 lbs.	Produced in the same manner as tapersplit shakes except that by splitting from the same end of the block, the shakes acquire the same thickness throughout.
	18" x ⅜"	19 Straight	5			200 lbs.	
	24" x ⅜"	16 Straight	5			260 lbs.	

Figure 6–74 Grades and sizes of wood shakes.

36-inch-wide strip of 30-pound felt along the eaves. Double or triple the first course. (A doubled course is sufficient, but a triple course may be used for design effect.) As each course is completed, apply an 18-inch-wide strip of 30-pound roofing felt over the top part of the shakes. Place the felt so that the bottom edge is positioned a distance equal to twice the weather exposure. For example: 24-inch shakes with a 10-inch exposure would require the felt to be placed 20 inches above the shake butts (*Fig. 6–75*).

Figure 6–75 Apply felt strips across top of shakes in each course.

Spacing between shakes should be about $5/16$ inch to allow for expansion. Offset the joints at least 1½ inches in adjacent courses. The use of felt is not required in snow-free areas.

Hip, ridge, and valley shakes are applied like the regular wood shingles. Select uniform shakes and apply them over an 8-inch-wide strip of 30-pound felt. Use 6d or longer hot-dipped zinc-coated nails.

The open valley is recommended for shakes. To avoid leaks, the metal valley sheets should be at least 20 inches wide and preferably edge- and center-crimped.

Gutters for use on roofs with hand-split shakes should be at least 5 inches wide. The wider gutter is needed because shakes tend to shed water in a wider zone.

Other Roofing

Carpenters are not required to install specialized roofing but they should be familiar with the various types, some of which are briefly described here.

Mineral Fiber.   Mineral-fiber shingles are made by combining asbestos fibers and Portland cement. They are practically indestructible. They resist burning, rotting, and decay, and are unaffected by water or ice. Made in various sizes, shapes, and colors, they can be used on new or old roofs alike. The material is very hard and brittle; therefore, special tools are required to cut and punch the shingles. Nails cannot be driven through the shingles, so they are furnished with prepunched nail holes. *Fig. 6–76* shows one method of making cutouts in the shingles. The perimeter of the cutout is outlined with a series of punched holes. A hammer blow is then struck at the center to knock out the waste. Straight cuts are made by scoring the shingles with a carbide-tipped blade, then snapping them apart much like glass is cut. Roof framing must be sufficient to support the load of the shingles, which is substantial.

Figure 6–76 Punching hole in asbestos shingle.

Slate Roofing.   The beauty of slate roofs is unexcelled. Slate is expensive, but it is fireproof and is unaffected by the elements. Made by splitting natural rock, slate is available in various colors and sizes. However, slate tiles are brittle and must be handled with care. Fastening is accomplished with copper or slater's nails.

Tile Roofing.   Tile roofs are commonly used in mission and Spanish architecture. The tiles are made of fired clay and are available in various patterns and colors. Since they are made from molds, uniformity is assured. The tiles are applied over felt underlayment and fastened with nails. Matching hip and ridge tiles are available. Sheet copper is generally used for the valleys.

Metal Roofing.   Roof coverings are available in various metals. These include galvanized iron, copper, lead, zinc, and aluminum. They are used mostly on commercial structures.

Drainage

Gutters and downspouts are used to collect and dispose of water from roofs. The gutter collects water from the edge of the roof and directs it to the downspout, which may direct the water either onto the ground and away from the structure or into a drainage system (*Fig. 6–77*). Gutters and downspouts may be fabricated of wood, aluminum, copper, steel, or plastic. Regardless of the material used, the gutters and downspouts should be large enough to carry the water away without overflowing.

Figure 6–77 A good drainage system will direct rainwater away from the structure.

Wood Gutters. These are generally made of cypress, red cedar, or red-wood. Other species of wood may be used, but they must be treated. Unaffected by corrosive fumes, heat, or moisture, wood gutters will give many years of service. Unlike metal gutters, which are supported by hangers, wood gutters are fastened to the fascia board with screws. To provide an air space between the gutters and fascia, small blocks are spaced 24 inches on center (*Fig. 6–78*). Whenever possible, the gutter should be in one continous section. If splices are necessary, they should be doweled.

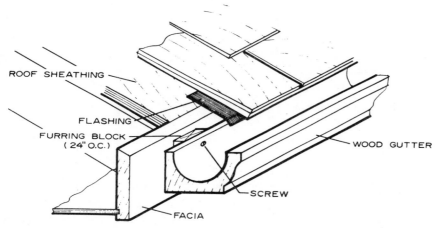

ROOF SHEATHING

FLASHING

FURRING BLOCK
(24" O.C.)

WOOD GUTTER

SCREW

FACIA

Figure 6–78 Wood gutters should be installed with an air space at the rear.

Gutters are installed with a slight pitch toward the downspout. Standard practice is to slope the gutter 1 inch for every 20 feet. On runs over 35 feet, use two downspouts, one at each end. Install the gutters with the front edge about 2 inches below the roof edge. This will prevent water from flowing between the shingles in case of an overflow.

Metal Gutters. These are generally made of aluminum, copper, or steel. They come in a variety of sizes and shapes. The most common are the half-round and the box-type (*Fig. 6–79*). Aluminum and copper gutters are not as strong as steel and must be handled carefully- to prevent denting. Aluminum gutters are available either finished or unfinished. Copper gutters have soldered joints and therefore are more difficult to install than the aluminum and steel types, which are assembled with slip-joint connectors and caulking.

The component parts of the gutter system, shown in *Fig. 6–80*, include inside and outside mitered corners, gutter sections, connectors, hangers, end caps, elbows, and downspouts.

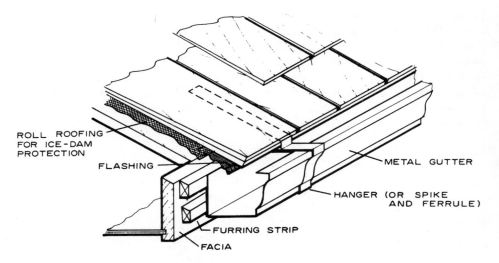

ROLL ROOFING
FOR ICE-DAM
PROTECTION

FLASHING

METAL GUTTER

HANGER (OR SPIKE
AND FERRULE)

FURRING STRIP

FACIA

Figure 6–79 Conventional metal gutter.

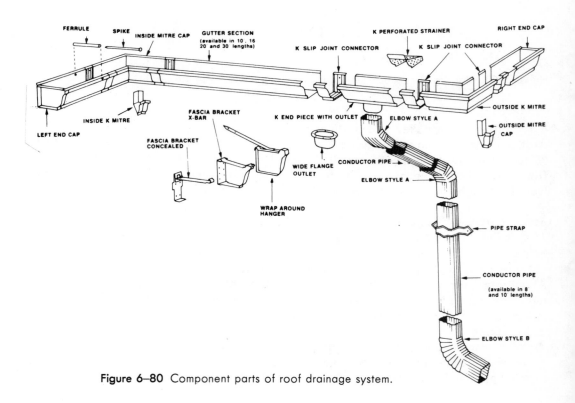

FERRULE   SPIKE   INSIDE MITRE CAP   GUTTER SECTION
(available in 10 , 16
20 and 30 lengths)

K PERFORATED STRAINER   RIGHT END CAP

K SLIP JOINT CONNECTOR   K SLIP JOINT CONNECTOR

INSIDE K MITRE

LEFT END CAP

FASCIA BRACKET
X-BAR

FASCIA BRACKET
CONCEALED

K END PIECE WITH OUTLET   ELBOW STYLE A

OUTSIDE K MITRE

OUTSIDE MITRE
CAP

WIDE FLANGE
OUTLET

CONDUCTOR PIPE

ELBOW STYLE A

WRAP AROUND
HANGER

PIPE STRAP

CONDUCTOR PIPE

(available in 8
and 10 lengths)

ELBOW STYLE B

Figure 6–80 Component parts of roof drainage system.

The size of the gutter depends on the roof area. Use a 4-inch gutter for roofs with an area up to 750 square feet. Use a 5-inch gutter for areas over 750 square feet, and a 6-inch gutter for anything over 1,500 square feet.

Downspouts are installed with pipe straps, which are placed at the top and bottom of the pipe and at 6-foot intervals in between. The size of the downspouts is governed by the roof area. A 3-inch downspout will handle roof areas up to 1,000 square feet. Slip connections permit the sections to move when the temperature changes.

When downspouts are made to flow onto the ground, splash blocks should be used to direct the flow away from the foundation (*Fig. 6–81*).

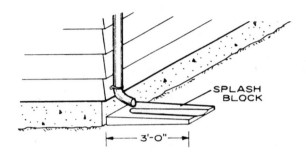

SPLASH BLOCK

3'-0"

Figure 6–81 Splash block diverts water flowing from downpipe.

Exterior Wall Finish

The exterior wall covering of a house must be selected and installed with great care. Although its chief function is to protect the structure and its interior from the elements, other factors must be considered in choosing the most appropriate wall-covering material, such as appearance, durability, insulating value, and labor and material cost. When properly chosen and applied, exterior wall products will give years of trouble-free service.

Sheathing Paper. Wall-sheathing paper, also called building paper, is used to protect against air infiltration and moisture. It is applied over sheathing boards or directly over studs when sheathing is not used. It is not needed when walls are sheathed with plywood, treated gypsum board, or fiberboard (*Fig. 6–82*).

The paper should be water-resistant but not waterproof. It must not act as a vapor barrier, because this could cause condensation problems. A low-vapor-resistant asphalt-saturated felt is recommended. It should be applied as soon as possible after the walls have been sheathed. Use 4-inch side and top laps and 6-inch laps at corners. When they are not protected by wide overhangs, drip caps over doors and windows should be flashed

(*Fig. 6–83*). Although other materials are available, aluminum, copper, or galvanized steel are the usual flashing materials used.

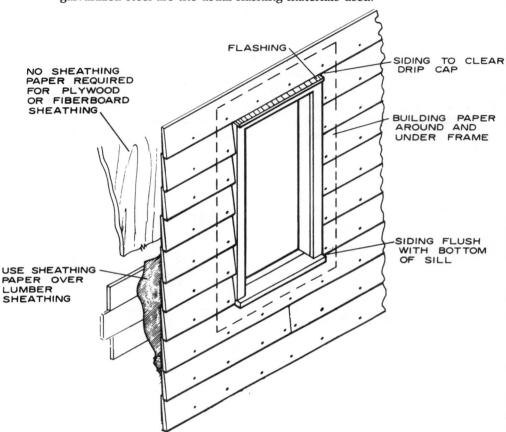

NO SHEATHING PAPER REQUIRED FOR PLYWOOD OR FIBERBOARD SHEATHING

FLASHING

SIDING TO CLEAR DRIP CAP

BUILDING PAPER AROUND AND UNDER FRAME

USE SHEATHING PAPER OVER LUMBER SHEATHING

SIDING FLUSH WITH BOTTOM OF SILL

Figure 6–82 Application of building paper over sheathing boards.

Figure 6–83 Application of drip-cap flashing.

Wood Siding. Wood sidings are available in many styles for application horizontally or vertically, or a combination of both (*Fig. 6–84*).

The most popular horizontal siding is the bevel, or tapered, siding (*Fig. 6–85*). Bevel sidings have a narrow end of 3/16 inch, tapering to ½ or ¾ inch at the thick edge. The widths vary from 4 to 12 inches. Bevel sidings are made by sawing plain-surfaced boards diagonally. The surfaced face is smooth, while the other side is rough and shows the saw cuts. Either surface may be placed outward, depending on the finish desired. The smooth side is usually exposed when the siding is to be painted. The rough-sawn surfaces are normally stained for a rustic appearance. Bevel siding with a rabbeted edge is widely used, because it lies flat and tight against the wall sheathing.

Figure 6–84 Horizontal wood siding.

Figure 6–85 Bevel siding.

Drop siding is also applied horizontally. It is heavier than bevel siding and is often used on low-cost houses and garages where sheathing has been eliminated. The matched joints form a strong, tight wall, highly resistant to wind and water.

The two patterns shown in *Fig. 6–86* can be used horizontally or vertically. Vertical sidings (*Fig. 6–87*) are usually limited to one-story houses. Sometimes they are used in combination with horizontal siding for a pleasing effect. Wide boards are applied vertically, with 1 × 2 battens used to cover the joints. They are economical and easy to install. Reverse battens (*Fig. 6–88*) comprise another method of applying boards and battens. In this application, the air space behind the panel adds a slight degree of insulation to the structure.

Figure 6—86 Two wood siding patterns suitable for horizontal or vertical application.

Figure 6—87 Vertical wood siding.

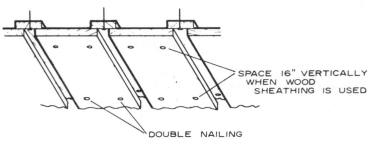

SPACE 16" VERTICALLY
WHEN WOOD
SHEATHING IS USED

DOUBLE NAILING

Figure 6—88 Detail of reverse batten.

Exterior siding should be of high quality and free of knots, pitch pockets, and other defects. It should have the correct moisture content. Wide boards are especially troublesome if excessive shrinkage occurs after installation. If used in southern states, the moisture content of the siding should be 9 percent; 12 percent is desirable for the rest of the country.

Most sidings are available prefinished; some are furnished with a prime coat of alkyd resin. Siding is applied after doors and windows are in place. Bevel siding must be lapped at least 1 inch so it will be weathertight (*Fig. 6–89*). However, there are occasions when slight variations are permissible, especially when it is necessary to align the siding with the bottom of windowsills and drip caps at the tops of doors and windows. Butt joints should be staggered in adjacent courses and made over studs (*Fig. 6–90*).

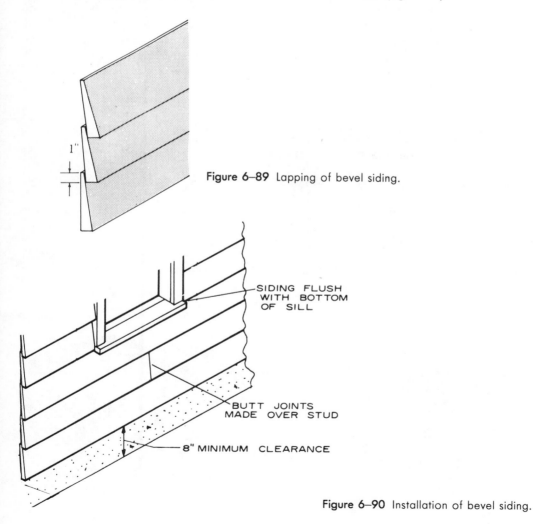

1"

Figure 6–89 Lapping of bevel siding.

SIDING FLUSH WITH BOTTOM OF SILL

BUTT JOINTS MADE OVER STUD

8" MINIMUM CLEARANCE

Figure 6–90 Installation of bevel siding.

Applying Wood Siding.   A story pole is used to aid in the installation of siding. Its use will insure that the siding will be applied level and aligned around the perimeter of the building. Use a flat, straight piece of 1×2 stock for the story pole. The siding should start 1 inch below the top of the foundation wall, and it usually extends up to the soffit. Mark these locations on the pole (*Fig. 6–91*). Now divide this distance into equal parts, representing the exposure of the siding. Make slight adjustments to align the bottom of the windowsills and the drip caps so they can be installed without notching (*Fig. 6–92*). Transfer the marks from the pole to the house corners and to the window and door casings (*Fig. 6–93*). Snap a chalk line on the lower part of the wall to locate the top edge of the first piece. A starter strip, with the same thickness as the thin edge of the siding, is usually used with bevel siding. This should be placed slightly above the lower edge of the first piece (*Fig. 6–94*). If the design calls for the use of a water table, it is constructed and flashed as shown in *Fig. 6–95*. (The water table serves as a miniature shed to direct water away from the structure.)

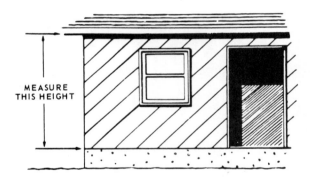

MEASURE THIS HEIGHT

Figure 6–91 Story pole should include foundation and soffit lines.

Figure 6–92 Layout siding courses to align with tops and bottoms of windows.

Figure 6–93 Transfer story-pole markings to house corners.

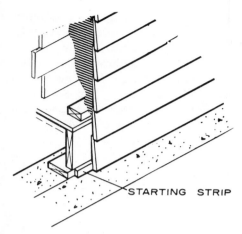

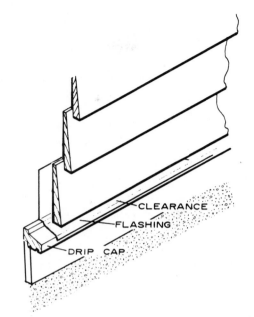

Figure 6–94 Use starter strip under first course of bevel siding.

Figure 6–95 Details of flashed water table.

Various corner treatments are possible. The most difficult method calls for mitered joints. If not properly cut and installed, these joints have a tendency to open, which detracts from their appearance. If mitered corners are used, the miter must be cut with the siding held at the same angle it will have when installed. A simple jig set up in the miter box will aid in cutting the miters. Corner boards and metal corners are more widely used than mitering. They are easily installed and thus save on labor costs. Metal corners are provided with tabs to align the bottom of the siding. *Fig. 6–96* shows a typical metal corner.

Figure 6–96 Metal corners add finished appearance to siding.

Corner boards are installed before the siding is applied. The siding is then cut to fit snugly against them. *Fig. 6–97* shows how inside and outside corners are installed. To insure properly fitting joints, use a siding gauge. Make the gauge by notching a piece of 1 × 4 stock as shown in *Fig. 6–98*. The double notching of the gauge assures that the siding is held level while it is being marked. Use a pencil or knife to mark the cutting line. For a neat appearance, cut the siding with a fine-tooth saw. Before installing each piece, brush the ends of the siding with a preservative (*Fig. 6–99*). Joints at doors and windows should be caulked.

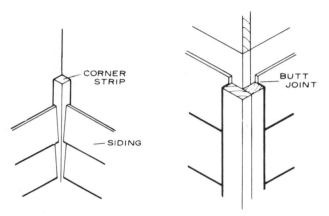

Figure 6–97 Corner treatment. *Left:* Inside corner. *Right:* Outside corner.

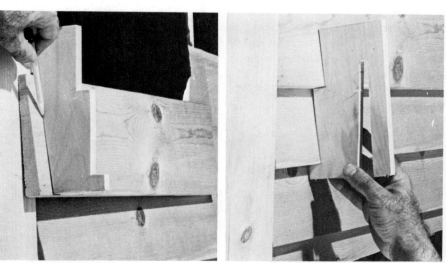

Figure 6–98 *Left:* Use siding gauge to mark length of siding. *Right:* Use notches at rear of gauge to obtain required exposure.

Figure 6–99 Treat siding cut on the job with a water repellent.

Tips on Nailing. Proper nailing is important. Many walls have been marred because the wrong nails were used. Rusting nails not only discolor the siding, but they will eventually corrode to the point where they lose their holding power. Use only noncorrosive fasteners such as aluminum, hot-dipped galvanized, or stainless steel nails. For ½-inch siding over plywood or board sheathing, use 6d nails. Use 8d nails over gypsum or fiberboard sheathing. For ¾-inch siding, use 7d nails over plywood and board sheathing and 9d nails over gypsum and fiberboard. Nailing details are shown in *Fig. 6–100*. Note that nails are placed so they just clear the tops of the preceding courses. This is especially important when you are using wide boards, because it allows expansion clearance, thus preventing cupping of the siding.

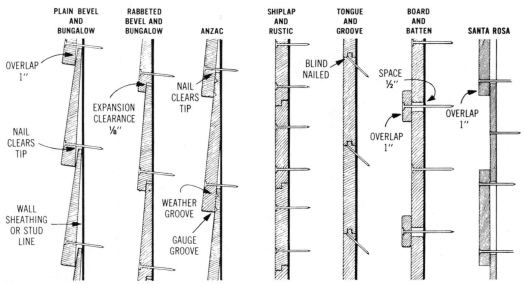

Figure 6–100 Nailing details for different kinds of siding.

Wood Wall Shingles.   Wood shingles are widely used for side-wall cover-ing. They are durable and require little in the way of maintenance, especially when used in their natural state. They are available painted or stained with a grooved surface for side-wall application. They are made in uniform or random width in several grades. Installation is similar to that for roofs but greater exposure is possible. Wood shakes are also used for wall covering, but their cost and difficulty of installation has prevented their widespread use.

Wood shingles are applied either single- or double-coursed. The max-imum weather exposure for single-course work is 7½ inches for 16-inch shingles, 8½ inches for 18-inch shingles, and 11½ inches for 24-inch shin-gles. *Fig. 6–101* shows a single-course application. For double-course appli-cation, the permissible exposures are 12, 14, and 16 inches for the 16-, 18-, and 24-inch shingles, respectively. A lower-grade undercoursing shingle is used on double-coursed work. When it is double-coursed, the insulation value of that wall is increased. *Fig. 6–102* shows the detail of a double-coursed wall.

Applying Shingles.   The entire wall must be covered with building paper before wood shingles are installed. If the walls have been sheathed with wood or plywood, the shingles may be applied directly over the building paper. However, for non-wood sheathing and open studding, nailers are re-quired to support the wood shingles. The nailers are 1 × 3 or 1 × 4 wood strips installed over the building paper with proper spacing to receive the courses of shingles as they are nailed to the wall. On open studding, the paper is applied to the framing members, as shown in *Fig. 6–103.*

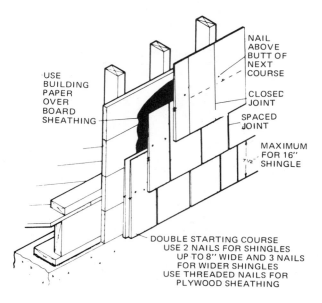

USE
BUILDING
PAPER
OVER
BOARD
SHEATHING

NAIL
ABOVE
BUTT OF
NEXT
COURSE

CLOSED
JOINT

SPACED
JOINT

MAXIMUM
FOR 16"
SHINGLE
7½"

DOUBLE STARTING COURSE
USE 2 NAILS FOR SHINGLES
UP TO 8" WIDE AND 3 NAILS
FOR WIDER SHINGLES
USE THREADED NAILS FOR
PLYWOOD SHEATHING

Figure 6–101 Single-course shingle application.

Figure 6–102 Double-course shingle application.

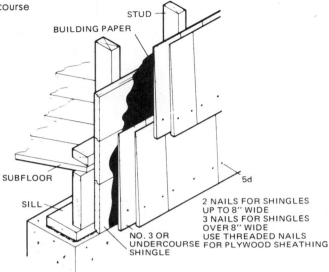

STUD

BUILDING PAPER

SUBFLOOR

5d

SILL

2 NAILS FOR SHINGLES
UP TO 8" WIDE
3 NAILS FOR SHINGLES
OVER 8" WIDE
USE THREADED NAILS
FOR PLYWOOD SHEATHING

NO. 3 OR
UNDERCOURSE
SHINGLE

Use the story pole to indicate exposures as well as soffit and base lines. To align the shingles across the wall, snap a chalk line on the building paper. A better method, which will speed up installation, consists of tacking a 1 × 4 straightedge across the wall. The butts of the shingles are then aligned with it. As each course is completed, the straightedge is repositioned (*Fig. 6–104*).

For single-course application, start the first course by doubling the shingles at the foundation line. Apply the second row of shingles over the starter course with just enough overlap to give the desired exposure. Each following course should have the same overlap.

The nails in single-course work are concealed. Use 3d rust-resistant nails, placing them ¾ inch from each edge and 1 inch above the butt line

Figure 6–103 Apply building paper and nailers to open studding before applying shingles.

Figure 6–104 Straightedge aids in aligning each shingle course.

of the following course. If the shingle is wider than 8 inches, use a third nail at the center and aligned with the others.

For double-course applications, make the starter row with three shingles, the first two being of undercoursing grade. To allow for expansion, lay the first two courses with spaces between the joints, then apply the top course with tight joints. Set the butts of the outer course ½ inch below the butt edge of the undercourse. A shiplap straightedge with a ½-inch rabbet will serve to align the under and top courses (*Fig. 6–105*).

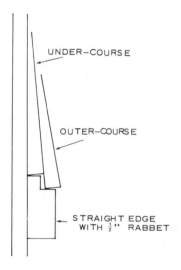

Figure 6–105 Rabbeted straightedge provides proper spacing for shingles.

The nails in double-coursed work are exposed. Place them 2 inches above the butt and ¾ inch from the edge. Use a third nail on shingles wider than 8 inches. The outside corners are alternately lapped or mitered. Inside corners are alternately butted.

In both single- and double-course application of shingles, the joints in adjacent courses should be offset by at least 1½ inches.

Overwalling. Shingles and shakes can be applied over old walls with nailable surfaces. On masonry and other non-nailable surfaces, nailer strips must be fastened to the wall (*Fig. 6–106*). Before fastening the strips, apply building paper to the old wall surface. Because of the added thickness of the new wall, it will probably be necessary to add new molding strips around doors and windows.

Plywood Siding. Plywood is ideally suited for exterior wall covering. Because of the large sheet sizes available, installation time is reduced considerably. Plywood can be applied directly to the framing, thus eliminating the need for sheathing (*Fig. 6–107*). Standard-size panels are available in

Figure 6–106 Application of wood
shingles over masonry walls.

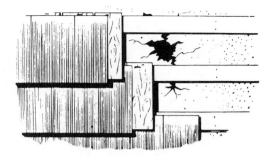

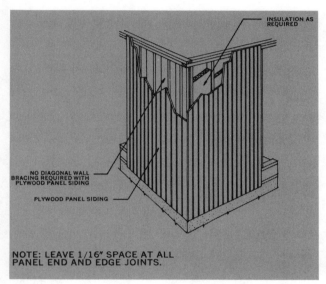

NO DIAGONAL WALL
BRACING REQUIRED WITH
PLYWOOD PANEL SIDING

PLYWOOD PANEL SIDING

INSULATION AS
REQUIRED

NOTE: LEAVE 1/16" SPACE AT ALL
PANEL END AND EDGE JOINTS.

Figure 6–107 Installation of plywood
siding without sheathing.

⅜-inch, ½-inch, and ⅝-inch thickness in 4-foot width, and in lengths of
8, 9, and 10 feet.

On open studding, ⅜-inch panels are used for 16-inch stud spacing;
⅝-inch panels are used for 24-inch stud spacing. All joints should be over
studs. Horizontal joints require solid blocking. Various joint treatments are
shown in *Fig. 6–108*. Before installing panels, paint all edges. Install the
panels with a ¹/₁₆-inch space at all edges and ends. Use caulking at all joints
except where joints are matched. Place nails at 6-inch intervals around the
perimeter and 12-inch intervals along intermediate supports.

Plywood is also made in 12-, 16-, and 24-inch widths for use as lap
and beveled siding in many textures and finishes. It is installed with a ⅜
× 1½-inch starter strip behind the lowest panel to bring it away from the
wall (*Fig. 6–109*). Use wedges as backing under all vertical joints. Lap
succeeding courses 1 inch and fasten with 8d galvanized nails. Place nails
½ inch from the bottom. This will insure that they go through both siding
courses at the lap.

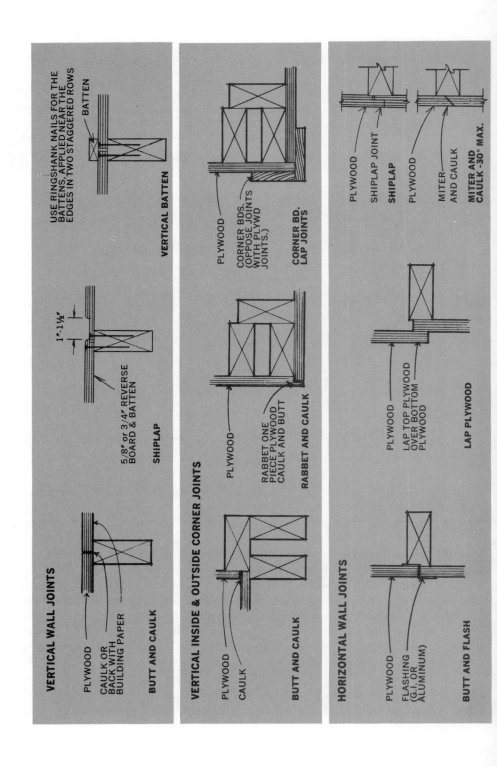

VERTICAL WALL JOINTS

PLYWOOD

CAULK OR BACK WITH BUILDING PAPER

BUTT AND CAULK

5/8" or 3/4" REVERSE BOARD & BATTEN

1"-1½"

SHIPLAP

USE RINGSHANK NAILS FOR THE BATTENS, APPLIED NEAR THE EDGES IN TWO STAGGERED ROWS

BATTEN

VERTICAL BATTEN

VERTICAL INSIDE & OUTSIDE CORNER JOINTS

PLYWOOD

CAULK

BUTT AND CAULK

PLYWOOD

RABBET ONE PIECE PLYWOOD CAULK AND BUTT

RABBET AND CAULK

PLYWOOD

CORNER BDS. (OPPOSE JOINTS WITH PLYWD JOINTS.)

CORNER BD. LAP JOINTS

HORIZONTAL WALL JOINTS

PLYWOOD

FLASHING (G.I. OR ALUMINUM)

BUTT AND FLASH

PLYWOOD

LAP TOP PLYWOOD OVER BOTTOM PLYWOOD

LAP PLYWOOD

PLYWOOD

SHIPLAP JOINT

SHIPLAP

PLYWOOD

MITER AND CAULK

MITER AND CAULK - 30° MAX.

298

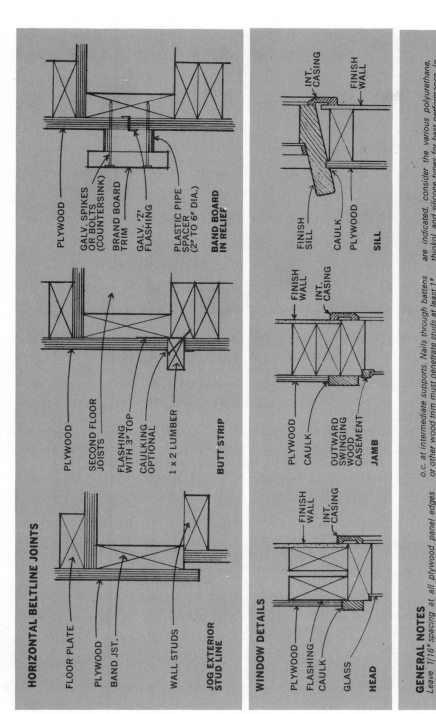

HORIZONTAL BELTLINE JOINTS

FLOOR PLATE
PLYWOOD
BAND JST.
WALL STUDS
JOG EXTERIOR STUD LINE

PLYWOOD
SECOND FLOOR JOISTS
FLASHING WITH 3" TOP
CAULKING OPTIONAL
1 x 2 LUMBER
BUTT STRIP

PLYWOOD
GALV. SPIKES OR BOLTS (COUNTERSINK)
BRAND BOARD TRIM
GALV. 'Z' FLASHING
PLASTIC PIPE SPACER (2" TO 6" DIA.)
BAND BOARD IN RELIEF

WINDOW DETAILS

PLYWOOD
FLASHING
CAULK
GLASS
HEAD

FINISH WALL
INT. CASING

PLYWOOD
CAULK
OUTWARD SWINGING WOOD CASEMENT
JAMB

FINISH WALL
INT. CASING

INT. CASING
FINISH SILL
CAULK
PLYWOOD
SILL

FINISH WALL

GENERAL NOTES

Leave 1/16" spacing at all plywood panel edges and ends.
Nailing: General nailing requirements for plywood panel siding call for 6" o.c. edge nailing and 12" o.c. at intermediate supports. Nails through battens or other wood trim must penetrate studs at least 1". To prevent staining of siding, use hot-dip galvanized. aluminum, or other non-staining nails.
Caulks and sealants: Where caulks or joint sealants are indicated, consider the various polyurethane, thiokol, and silicone types for best performance. In some cases, a foam rod or other type filler material maybe used behind the sealants as recommended by the manufacturer.

Figure 6–108 Joint details for plywood siding.

299

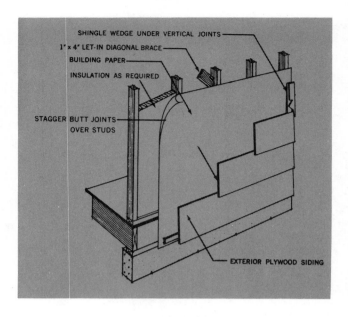

SHINGLE WEDGE UNDER VERTICAL JOINTS

1" x 4" LET-IN DIAGONAL BRACE

BUILDING PAPER

INSULATION AS REQUIRED

STAGGER BUTT JOINTS OVER STUDS

EXTERIOR PLYWOOD SIDING

Figure 6–109 Use a starter strip when applying plywood siding.

Hardboard Siding.   Because of its uniformity, durability, and moisture-resistant surface, hardboard is a popular siding material. It is made in many sizes and textures in both panel form and as lap siding. Special embossing techniques yield surfaces with a remarkable resemblance to wood, masonry and other materials. *Fig. 6–110* shows a hardboard lap siding embossed to look like wood.

Standard panels are available $7/16$ inch thick and 4 feet wide in lengths of 8, 9, and 10 feet (*Fig. 6–111*). Lap siding is made in thicknesses from $3/8$ inch to $9/16$ inch and up to 12 inches wide and 16 feet long. The top edge is usually made with a 1-inch rabbet.

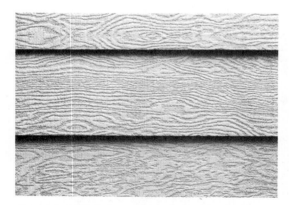

Figure 6–110 Embossed hardboard lap siding.

Figure 6–111 Application of standard-size hardboard panel siding.

Hardboard siding is applied to open or sheathed walls. Use building paper on open wall framing. For lap siding, stud spacing should be 16 inches on center. For panel siding, either 16- or 24-inch spacing is permissible.

The siding is applied following the same procedure used for other siding materials. *Fig. 6–112* shows nailing details and joint treatment. Weatherproof nails must be used. For prefinished panels, matching battens are used to conceal the joints. *Fig. 6–113* shows a system using batten backer strips with snap-on covers. There are numerous patented products manufactured for use in applying siding, such as special moldings, mounting strips, and various metal fasteners. *Fig. 6–114* shows how mounting strip and joint molding are used.

Mineral-Fiber Shingles. Mineral-fiber shingles are manufactured by combining asbestos fibers with Portland cement, and are made with either

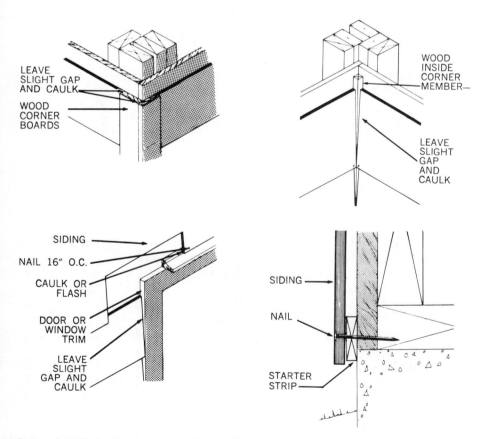

Figure 6–112 Application details for hardboard siding.

Figure 6–113 Hardboard batten snaps into place.

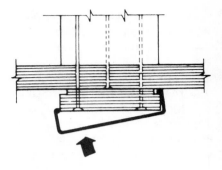

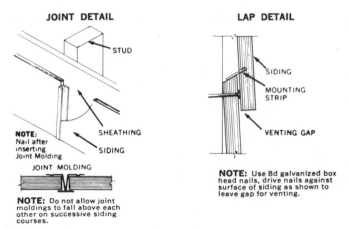

JOINT DETAIL

STUD

NOTE:
Nail after
inserting
Joint Molding

SHEATHING

SIDING

JOINT MOLDING

NOTE: Do not allow joint
moldings to fall above each
other on successive siding
courses.

LAP DETAIL

SIDING

MOUNTING
STRIP

VENTING GAP

NOTE: Use 8d galvanized box
head nails, drive nails against
surface of siding as shown to
leave gap for venting.

Figure 6–114 Joint and lap details for hardboard siding.

straight or wavy butts. Various sizes are available, the most popular being 12 × 24 inches. Thicknesses range from $^5/_{32}$ to $^3/_{16}$ inch. Because of its excellent weather resistance and durability, this material is highly suited for exterior wall finish.

Apply the material to flat dry walls. If the wall sheathing will not support nails, use wood nailer strips as shown in *Fig. 6–115*. Use only corrosion-resistant nails. The proper size and type of nails are usually specified by the siding manufacturer.

The shingles may be installed over backer boards to provide additional insulation and deep shadow lines (*Fig. 6–116*). The backer boards should be at least $^5/_{16}$ inch thick and ¼ inch narrower than the shingle, thus forming a drip edge at the shingle butt. Shingles are installed with the lower edge of the first row placed against a wood starter strip. The butts of the first row of shingles should overhang the starter strip by ½ inch. For

Figure 6–115 Nailing strips are required when sheathing will not support nails.

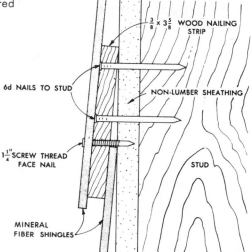

$\frac{3}{8}$ x $3\frac{5}{8}$ WOOD NAILING STRIP

6d NAILS TO STUD

NON-LUMBER SHEATHING

$1\frac{1}{4}''$ SCREW THREAD FACE NAIL

STUD

MINERAL FIBER SHINGLES

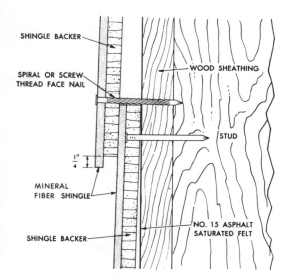

SHINGLE BACKER

SPIRAL OR SCREW THREAD FACE NAIL

WOOD SHEATHING

STUD

$\frac{1}{4}''$

MINERAL FIBER SHINGLE

SHINGLE BACKER

NO. 15 ASPHALT SATURATED FELT

Figure 6–116 Backer-board shingle application.

12 × 24-inch shingles, the exposure should be 10½ inches. Use asphalt backer strips behind each vertical joint (*Fig. 6–117*).

Corners may be backed or caulked as shown in *Fig. 6–118*. If caulking is used, be careful not to smear it on the face of the shingles. Flashing over doors and windows should be done as illustrated in *Fig. 6–119*. Corners may be covered with wood or matching mineral-fiber strips. Prefinished metal corners are also available (*Fig. 6–120*).

Figure 6–117 Vertical joints must have backer strips.

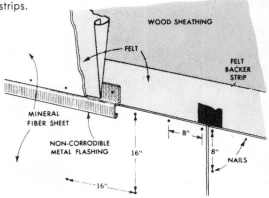

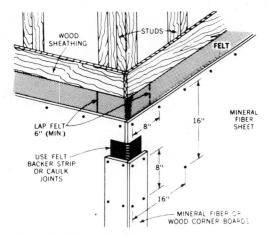

Figure 6–118 Backed corner.

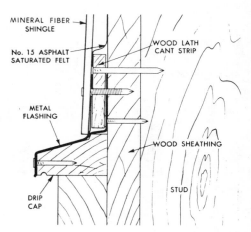

Figure 6–119 Flashing details for doors and windows.

Metal Siding. Factory-finished aluminum and steel sidings are made in numerous styles and sizes. Surfaces may be smooth or textured and coated with plastic or baked-on finishes which require a minimum of maintenance. Installation details are illustrated in *Fig. 6–121.*

Plastic Siding. Extruded vinyl sidings with embossed or smooth surfaces have been recently developed. They are not affected by weather or pests. They are made in widths of 8 inches and in lengths of 12 feet 6 inches. They are available with backer boards which add to the rigidity and insulat-

Figure 6–120 Metal corners for asbestos shingles.

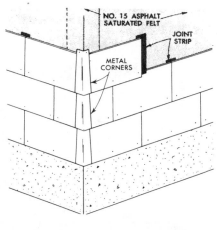

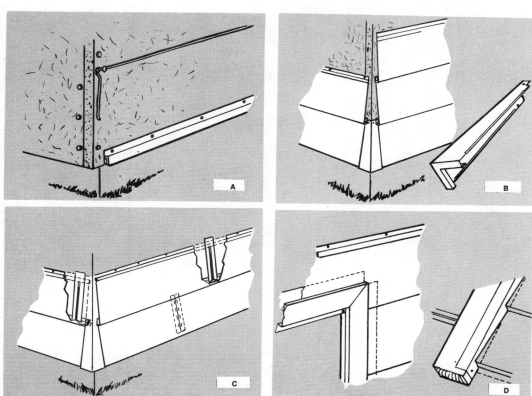

Figure 6–121 Installation details for aluminum siding: A. Attach first panel to starter strip; use chalk line to assure levelness. B. Apply outside corner caps for 8-inch horizontal siding one at at time. C. Use backer strips under ends of siding to provide stiffening. D. Various accessories are available for building breaks and corners.

ing value. Nailing slots enable the panels to expand and contract without buckling. Interlocking lips permit the sections to be snapped together. *Fig. 6–122* shows some of the accessories that are available for vinyl sidings.

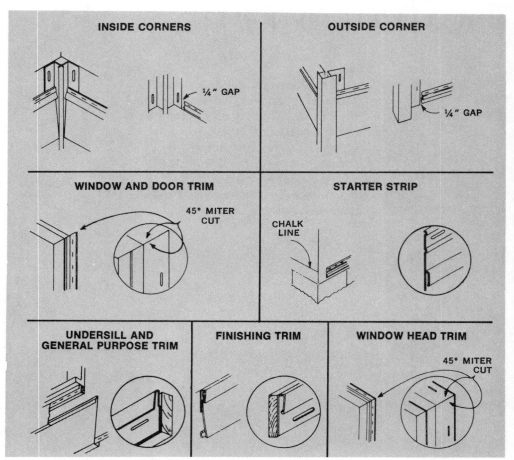

INSIDE CORNERS

¼" GAP

OUTSIDE CORNER

¼" GAP

WINDOW AND DOOR TRIM

45° MITER CUT

STARTER STRIP

CHALK LINE

UNDERSILL AND GENERAL PURPOSE TRIM

FINISHING TRIM

WINDOW HEAD TRIM

45° MITER CUT

Figure 6–122 Accessories for vinyl sidings.

Brick Veneer. Walls made with a single layer of brick or stone applied to a wood-frame structure are called veneer walls. They are non–load-bearing and serve only as decoration. Noncorrosive metal ties, spaced 15 inches vertically and 32 inches horizontally, are used to hold the wall to the frame. Such walls must have a 1-inch space between the veneer and sheathing. Weep holes spaced about 4 feet apart are placed along the bottom course of brick. These allow moisture to escape (*Fig. 6–123*). Sheathing should be covered with building paper and base flashing added (*Fig. 6–124*).

Figure 6–123 Removing weep-hole plug.

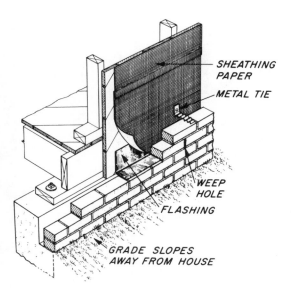

SHEATHING PAPER

METAL TIE

WEEP HOLE

FLASHING

GRADE SLOPES AWAY FROM HOUSE

Figure 6–124 Details of brick veneer wall.

Brick or masonry veneers are often used in combination with other siding materials. Flashing is required where changes in exterior finish occur, such as at the juncture of a wood siding with a brick veneer wall (*Fig. 6–125*). Flashing is also needed under a window opening where masonry or brick veneer is used (*Fig. 6–126*).

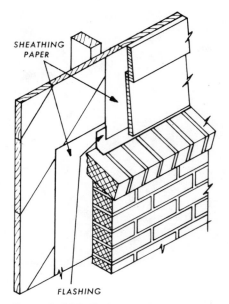

SHEATHING
PAPER

FLASHING

Figure 6–125 Flashing is required where wood siding and brick veneer meet.

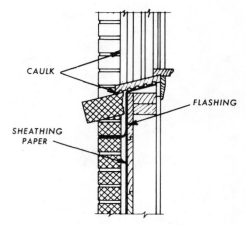

CAULK

FLASHING

SHEATHING
PAPER

Figure 6–126 Flashing under window in exterior wall.

Insulation and Moisture Control 7

Condensation takes place when moisture changes from a vapor to a liquid. It is the cause of many problems in homes located in the colder regions of the United States. Condensation can cause paint to peel, siding to stain, and ice to form at cornices. It is also responsible for the decay of wood-frame members. In severe cases it will cause sheathing and siding to bow or buckle. Insulation can also be affected by condensation. When wet, insulating material offers less resistance to heat loss. Unfortunately, the better a home is insulated, the greater is the potential for condensation. Other factors that contribute to condensation are better weatherstripping, storm sash, and better building practices. All of these restrict the escape of moisture generated in the house by cooking, bathing, mopping, and laundering. Houseplants and people also add considerably to the vapor in a home. *Fig. 7-1* shows an example of condensation. The window on the right is of insulated glass; the one on the left is single-glazed.

Figure 7-1 Condensation in window at right has been eliminated by the use of insulating glass in wood sash.

Vapor Barriers

The condensation problem can be prevented by the proper use of vapor barriers, insulation, and good ventilation practices. Water vapor will always migrate toward colder surfaces owing to differences in vapor pressure. Wood paneling, plywood, plaster, and sheetrock will not restrict the passage of moisture under most conditions. During the heating season, water vapor within a house, if unrestricted, can move through walls and ceilings and will condense upon contact with any colder surface. Condensation problems due to formation of water, ice, and frost, on the framing members may result (*Fig. 7–2*).

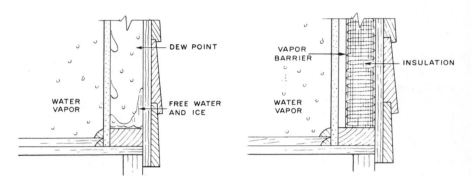

Figure 7–2 Vapor barriers avoid moisture problems in walls. *Left:* Water vapor from inside house moved out through walls. When vapor met outside cold air, moisture condensed and froze. As outside temperature rose in spring and summer, ice melted, and moisture was free to move through siding and destroy paint coating. *Right:* Vapor barrier (on warm side of wall) has prevented moisture from getting into walls.

To prevent the passage of moisture through the walls, a satisfactory vapor barrier must be used. Some building materials used in construction may effectively restrict the passage of moisture. Oil-base paints, foil-backed gypsum board, plastic-faced panels, and the vapor-barrier face of some insulating products fall within this category (*Fig. 7–3*). Materials used as vapor barriers include asphalt-coated papers, kraft-backed aluminum foil, and plastic films. The barriers are used in walls, ceilings, and floors. They should always be placed as close as possible to the interior of walls, ceilings, and floors (*Fig. 7–4*). For example, on walls it should be placed on the inside edge of the studs (*Fig. 7–5*). At the ceiling it is placed on the underside of the joists of the uppermost floor. In floors, vapor barriers are placed between subfloor and finish floor, except in unheated crawl spaces, where it should be under the subfloor.

Houses constructed over a concrete slab are subject to soil moisture which can penetrate the slab. In such cases, a vapor barrier should be in-

Figure 7–3 The foil face of the insulation serves as a vapor barrier.

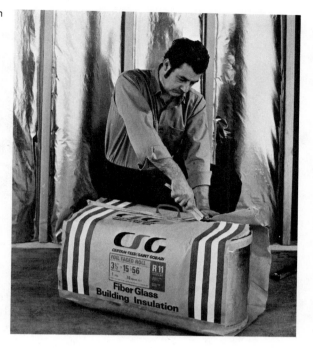

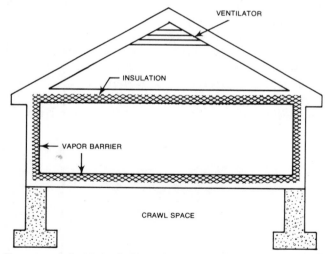

Figure 7–4 Place vapor barriers close to the interior walls.

stalled on the soil before the concrete is poured. The barrier will slow the curing of the concrete. Do not punch holes in the barrier to hasten curing; this will defeat the purpose of the barrier.

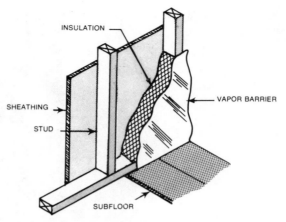

INSULATION

VAPOR BARRIER

SHEATHING

STUD

SUBFLOOR

Figure 7–5 Place vapor barrier on inside edge of studs.

Finished rooms in basements with exposed walls are treated much the same as framed walls. If drainage in the area is poor, special treatment may be required. A drain tile should be used at the perimeter of the footing (*Fig. 7–6*). In addition, a waterproof coating should be applied to the inner face of the exterior wall. A vapor barrier is then used either as part of the insulation or in addition to it (*Fig. 7–7*).

Other excellent methods for controlling moisture within a construction include the use of metal flashings in roof valleys, around chimneys and dormers, and at roof eaves; cap flashing over doors and windows; and the use of water-repellent preservatives on sidings and other framing members. More efficient use of these controls will produce more durable, more comfortable homes. For more information on flashing, refer to Chapter 6.

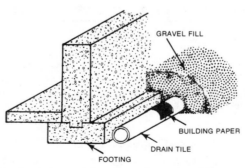

GRAVEL FILL

BUILDING PAPER

DRAIN TILE

FOOTING

Figure 7–6 Use drain tile at footing in areas where drainage is poor.

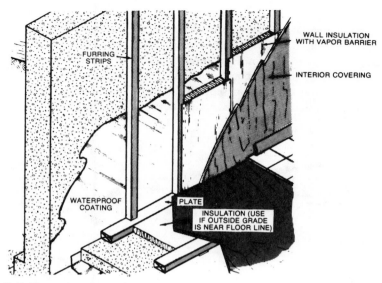

Figure 7–7 Vapor barrier on foundation wall.

Thermal Insulation

Heat will flow into a building in the hot weather and out in cold weather. The result of this heat transfer in or out will have an effect on the comfort of the occupants. To provide adequate comfort and to reduce heating and cooling costs, good insulation properly installed is a necessity, especially to-day, when conservation of natural resources is so important. To some de-gree, all materials used in building construction have insulating value. Even the space between studs will offer some resistance to the passage of heat. If these stud spaces are filled with material highly resistant to the transmis-sion of heat, the insulation value of the wall will be greatly increased. Bet-ter thermal insulation will save the consumer money and conserve energy as well.

So serious is the energy problem that the FHA has issued new insula-tion requirements for one- and two-family homes. These insulation re-quirements vary, depending on the size of the home and the local weather conditions. The map in *Fig. 7–8* is a rough guide to minimum requirements for heated but not air-conditioned homes in different areas of the United States. The following FHA regulations must be observed:

   1. Smaller homes (including split-level and two-story homes) usual-ly require better insulation because of a greater exposed exterior wall area relative to the floor area. In cold climates, double glazing and storm doors are required. In moderate areas, R–11 wall insulation and storm doors or partial double-glazing should be specified. In warmer areas, side-wall insulation is sufficient.

2. Increased insulation is also appropriate where glass and door areas exceed 15 to 20 percent of the exterior walls.

3. When the home is air-conditioned, full insulation in ceilings and walls is usually required, regardless of location.

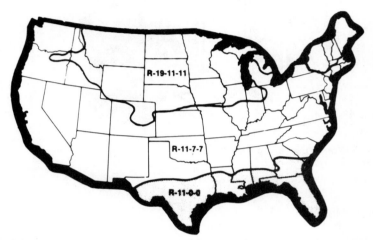

Figure 7–8 A guide to determine insulation requirements throughout the United States.

Insulating Materials.   Insulation is made in a variety of forms and types. There are four general classifications: flexible, loose fill, reflective, and rigid. Among the materials used for insulation are the mineral fibers, glass fibers, organic fibers, foamed plastics, and reflective materials. Regardless of the type used, it should be efficient, permanent, odor-free, vermin-free, and economical.

The thermal properties of commonly used building materials are known and the rate of heat flow is easily calculated. The measure of heat flow between the warm side and the cool side of the construction unit is known as the *U-value*. The U-value represents the heat loss in Btu's (British thermal units) per hour per square foot per degree Fahrenheit difference between inside and outside temperature. In other words, it represents the heat loss through a building section. The lower the U-value, the higher the insulating value. By comparing U-values, it is possible to choose the best combination of materials for a particular installation.

Another term used in describing insulation is *R-value*. This is the measured resistance of a material to the flow of heat. The higher the R-value, the more efficient the insulation. The R-value is always expressed as a numeral, such as R–7, R–11, and so forth (*Fig. 7–9*).

Figure 7–9 All insulation is identified with its R-value.

A typical wood-frame wall without insulation has a U-value of .22. When an insulating material with an R–7 value (such as 2¼-inch insulation) is used, the U-value is lowered to .10. If we use an insulation with an R–11 value (3½-inch fiberglass insulation), the U-value drops to .07. Thus, to lower the U-value, increase the R-value. Regardless of the type or thickness of the material used, all products with the same R number will have the same insulating value. The recommended R-values are as follows: ceilings R–19, walls R–11, floors R–11. An insulating material with an R–11 value (3½-inch fiberglass) provides the same insulation as 8 feet of face brick or 1 foot of solid wood or plywood.

The "cold wall effect" is a phenomenon caused by cold surfaces such as floors, walls, and ceilings. It causes a person to feel cold even when the temperature of the room is fairly high and otherwise comfortable. The effect is caused by the body's radiating, or losing its heat to a cooler surface. This effect is also noticeable to persons sitting at a metal desk. Although the upper part of the body is comfortable, the legs will feel cold. Most office workers will use an electric heater to warm the legs, but the same effect can be achieved by lining the knee opening of the desk with corrugated paper or even several layers of newspaper. Proper insulation is the answer, whether in the walls of a room or the kneehole of a desk. The various types of insulation are described below:

Flexible Insulation: Available in blankets and batts. Blanket insulation is produced in rolls or packages in widths to fit 16-inch or 24-inch stud spacing. Usual thicknesses are 1½, 2, and 3 inches. The blanket is made up of fluffy mats of insulating fibrous materials such as rock, glass, wool, cotton, or wood. The organic fibers are treated to resist fire, vermin, and decay. The blanket may have paper covers on one or both sides. The inner cover facing the room is usually treated to serve as a vapor barrier. If the outer face is covered, the material must be porous. Sometimes the inner covering is aluminum foil. Continuous tabs along the sides of the blanket provide a means for fastening the material to joists or studs.

Batt insulation is also made of fibrous materials, usually in 3½-inch and 6-inch thicknesses and in widths to fit 16-inch and 24-inch spacings. Lengths are usually 48 inches. Available with or without vapor barrier, they are generally used to insulate unfinished attic floors where the joists are exposed.

Fill-type Insulation: Bulk-packed and delivered in bags, or bales to be poured in place or packed by hand into small spaces (*Fig. 7–10*). It can be used to fill the space between vertical walls or to fill horizontal spaces to any desired thickness. It can be blown into an area with special equipment. It is made of various materials similar to those used for batt and blanket insulation.

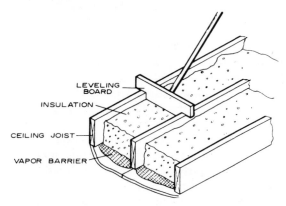

LEVELING BOARD
INSULATION
CEILING JOIST
VAPOR BARRIER

Figure 7–10 Installation of loose-fill ceiling insulation.

Reflective Insulation: Usually made of metal foil or foil-backed materials. Its efficiency is not dependent on thickness but rather on the number of reflective surfaces. To be effective, the reflective surfaces must be exposed to an air space. One product is made in an accordion-type manner. It consists of several layers of foil arranged so that they can be stretched to form numerous reflective layers. The bright surfaces of reflective insulation reflect the heat back to its source. Thus, in summer, the hot outside air is reflected back outside. In the winter, the room heat is reflected back into the room.

Rigid Insulation: Available in various types. One type is made from some form of wood or cane fibers. They are formed into large lightweight boards which combine thermal and acoustical properties. Sizes vary from 12-inch squares to panels 4 feet wide by 12 feet long. Thicknesses range from ½ to 1 inch.

Other materials used in the manufacture of rigid insulation include polyurethane, polystyrene, cork, and mineral wool.

Other types of insulation include lightweight aggregates and sprayed urethane foam.

Roof insulation is often referred to as "block" or "slab" insulation. It is nonstructural and is manufactured in thicknesses from ½ inch to 3 inches and in panels 2 × 4 feet in size.

Insulation must be chosen with care, because some insulating materials present serious problems. Cellulose insulation is made from reprocessed paper and is highly flammable unless treated with fire-retardant chemicals. Even when it has been treated, the chemical can separate from the cellulose, leaving the material dangerously flammable again. In such a state, a spark from a wall switch or outlet can cause a fire. There are strict standards governing the flammability of cellulose insulation and the installer should use only those materials which meet the standards.

Urea-formaldehyde insulation (foam) is another material that must be chosen and used with extreme care. If improperly installed, it can be injurious to health. Exposure to light, open air, high temperatures, and humidity can cause the product to deteriorate and produce an obnoxious formaldehyde odor that can permeate the house. In addition to the sickening odor, it is believed that serious health problems can also result from prolonged exposure to the product. Although the material will burn, it is not likely to do so within the wall cavity. However, when it does burn, it releases noxious fumes. In some states its use is outlawed. Check the local building codes before using any of these materials.

Where to Insulate. All areas that separate heated from unheated areas should be insulated; this includes walls, ceilings, roofs, and floors (*Fig. 7–11*).

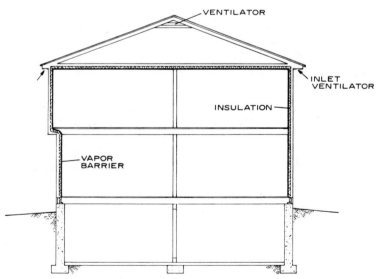

Figure 7–11 Location of vapor barriers and insulation in full two-story house with basement.

The insulation should be placed as close as possible to the room used as living quarters. Basements below grade which are to be used as living quarters should also be insulated. In houses with unheated but vented crawl spaces, the insulation should be placed between the floor joists. If the insulation used is the flexible type, be sure to support it between the slats with wire mesh or with tight-fitting sticks bowed into place (*Fig. 7–12*). Place it with the vapor barrier facing the subfloor. If the space is unvented, the insulation can be placed around the perimeter of the foundation wall. Use a vapor barrier to cover the ground and the wall area around it up to the sill. Place one edge of the insulation on the wall so it will be held by the sill (or header, if a sill is not used). The vapor-barrier side of the insulation must face the room. Let the remainder drape over the wall as shown in *Fig. 7–13*.

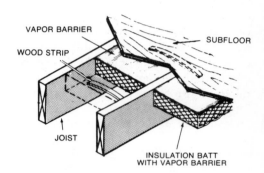

Figure 7–12 Batts are held between joists with sticks or wire mesh.

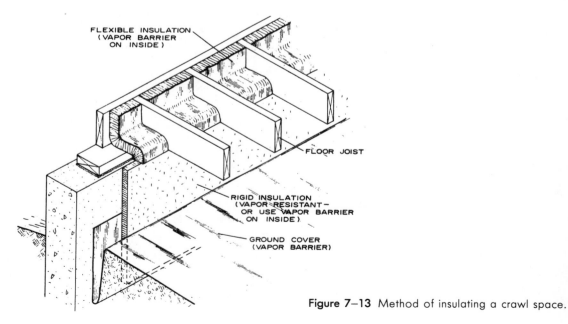

Figure 7–13 Method of insulating a crawl space.

Slab Insulation

Besides the usual insulation required for floors, walls, and ceilings, a house built over a concrete slab must be protected from soil moisture and heat loss by the use of a vapor barrier and insulation around the perimeter of the slab. These must be applied before the slab is poured. The perimeter insulation is in the form of rigid slabs and may be applied vertically, horizontally, or both ways (*Fig. 7–14*).

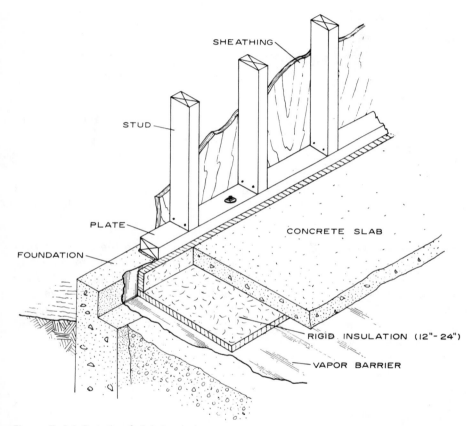

Figure 7–14 Details of slab insulation.

In the horizontal application, as in *Fig. 7–15*, a vapor barrier is first laid down over the soil. Thermal insulation, in the form of a rigid slab 24 inches or wider, is installed over the vapor barrier around the perimeter, and covered with polyethylene film. Then the concrete slab is poured over the insulation.

For a vertical installation, *Fig. 7–16*, a rigid insulation, 1 inch thick or more, should be installed on the inside of the foundation wall, extending vertically from the top of the slab to a minimum of 12 inches or more below the outside finish grade. The depth of the perimeter insulation depends on the climate of the region in which the house is built.

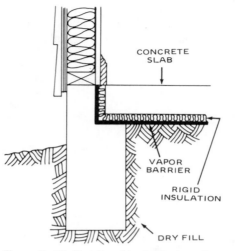

Figure 7–15 Details of horizontal application of vapor barrier and insulation.

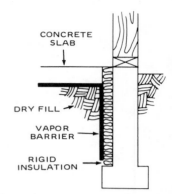

Figure 7–16 Vertical application of vapor barrier and rigid insulation.

Ceiling Insulation.   If the insulation is installed before the ceiling is covered, apply flexible blankets from below, stapling the flanges to the joists. Be sure the blankets extend over the top plate at the exterior wall. Do not let them overhang past the plate, because this could block eaves ventilation. Be sure to fill any gaps that may occur between the plate and insulation. If pressure-fit batts are used, simply wedge them between the joists. Otherwise staple them to the joist sides (*Fig. 7–17*).

Figure 7–17 Installation of insulation batts between joists.

Wall Insulation.   Install blankets in stud spaces, letting the back of the blanket touch the sheathing or siding. Work from the top down, stapling the flanges every 8 inches. The flanges should fit snugly against the wall without gaps or openings. Cut the blankets so that they fit square against the bottom plate. Some workers prefer to cut the end a trifle longer, then staple it through the vapor barrier (*Fig. 7–18*). Both methods are satisfactory. If pressure-fit blankets without a vapor barrier are used, wedge them into place, then cover the entire wall with a 2-mil-thick polyethylene vapor barrier (*Fig. 7–19*).

The areas around doors and windows require special attention. Fill in all the spaces between the rough framing and headers, sills, and jambs. Do not pack tightly, because the insulating value drops when the material is compressed. After the spaces are filled, staple a strip of vapor-barrier material over the openings (*Fig. 7–20*). If the wall is over a cantilevered floor, be sure to insulate the section of floor that projects beyond the wall.

Until recently, masonry walls of concrete block were seldom insulated, but because of the current high cost of energy, the trend is now to insulate. This is accomplished by pouring loose insulation into the core spaces of the block (*Fig. 7–21*). Similarly, the space between brick walls may be filled with vermiculite or cellular glass.

To sum up, keep the following points in mind when you are installing insulation:

1. A vapor barrier consisting of a sheet of foil, plastic, or treated paper must be installed facing the interior of the room. This will minimize any condensation of moisture that could soak the insulation and thus lower its insulating value. Moisture could also cause rotting of joists, studs, and sheathing.

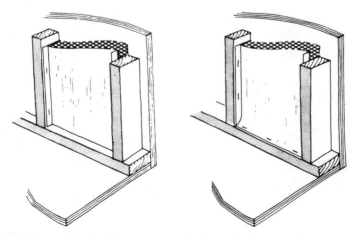

Figure 7–18 Two methods of fastening insulation at the bottom plate.

Figure 7–19 Install vapor barrier over pressure-fit blankets.

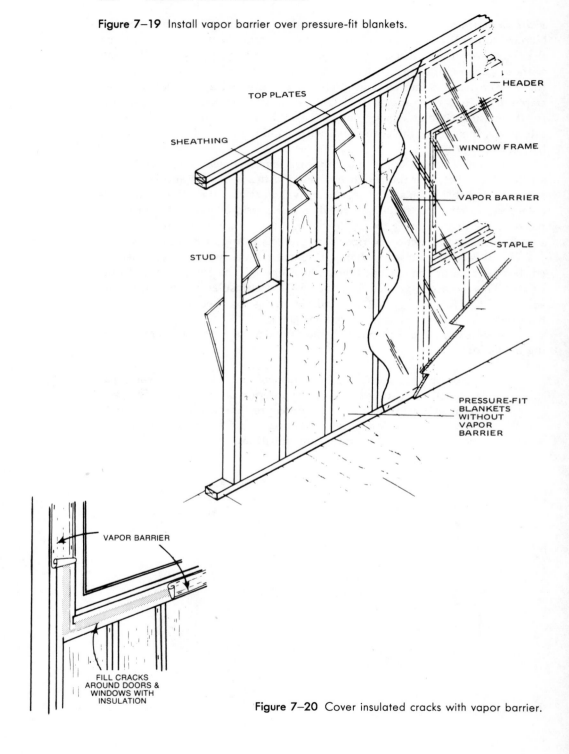

TOP PLATES

SHEATHING

STUD

HEADER

WINDOW FRAME

VAPOR BARRIER

STAPLE

PRESSURE-FIT
BLANKETS
WITHOUT
VAPOR
BARRIER

VAPOR BARRIER

FILL CRACKS
AROUND DOORS &
WINDOWS WITH
INSULATION

Figure 7–20 Cover insulated cracks with vapor barrier.

2. Keep vents in eaves and crawl spaces free of insulation because they must provide movement of air to control moisture.

3. Keep insulation away from recessed lights and heaters in ceilings. If covered with insulation, these units could overheat and cause a fire.

4. Dress properly when installing insulation to keep dust and glass fibers away from your eyes, skin, and lungs. Goggles and a dust mask are essential when you are working in an enclosed area.

Figure 7–21 Pouring insulation into core spaces of cement block.

Ventilation

Condensation may occur in attic spaces even when vapor barriers are used. This problem is overcome by using ventilators to prevent the buildup of moisture vapor. When properly installed, the ventilators will cause a flow of air to circulate in and out of the attic space (*Fig. 7–22*).

ATTIC
VENT

Figure 7–22 A properly vented attic will prevent condensation.

The unoccupied attic space should be kept at a low temperature to prevent snow next to the roof from melting and forming ice dams. This is accomplished by both insulating the attic ceiling and installing ventilators. The use of ventilators also provides an excellent means of reducing attic temperatures in hot weather.

Ventilators in the end walls of gable roofs require wind pressure to move the air within. Their efficiency can be increased by adding soffit vents along the eaves. A soffit vent is shown in *Fig. 7–23*. The difference in

Figure 7–23 Soffit vents increase efficiency of attic gable vent.

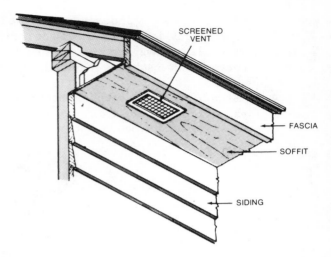

temperature between the attic and the outside will then cause the movement of air through the attic.

The use of attic fans will provide a more efficient method of moving air through an attic. The size of the fan required is determined by the volume of air to be moved. Fans are rated accordingly.

Crawl spaces under houses also require ventilators to remove the moisture-laden air which rises from the soil, especially when a vapor barrier is not used. A crawl-space ventilator is shown in *Fig. 7–24*.

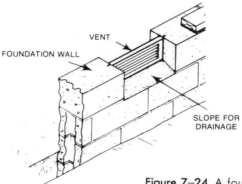

Figure 7–24 A foundation vent in a crawl space masonry wall.

The vents in gabled roofs should be placed as close to the ridge as possible. When soffit vents are added, the gable vents can be reduced in size. For hip roofs, each soffit should be vented with an outlet vent placed near the peak. Flat roofs require a greater ratio of venting because air movement is less positive.

Outlet Vents.   Many types of outlet vents are made to blend with any style of architecture (*Fig. 7–25*). Regardless of the type used, they should be screened to keep out insects. Do not use too fine a mesh, however, because dust and lint can clog it easily. Louvers are required to keep out rain and snow. The louvers must slope downward toward the outside of the building. When you install vents, be sure they are properly fitted into the rough opening.

Figure 7–25 Various types of outlet ventilators.

Inlet Vents.   Inlet soffit vents may consist of a series of continuous slots or prefabricated inserts. Both types are shown in *Fig. 7–26*. The insert type is available in numerous sizes. They are fitted snugly into place and secured with screws. The openings can be sawed out before the soffits are installed.

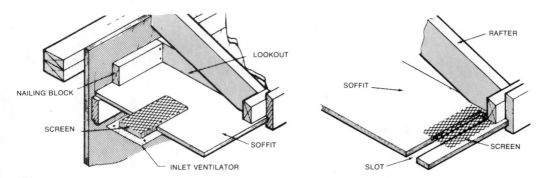

Figure 7–26 Two types of inlet soffit vents. The slotted vent should be placed close to the fascia.

Continuous slot vents should be placed near the outer edge of the soffit toward the fascia. The slots should be interrupted with a bridge at regular intervals to prevent weakening of the soffit. (*Fig. 7–27*).

Weatherstripping

Because of the energy shortage, greater attention has been given to weatherstripping in recent years. Studies show that the amount of heat loss

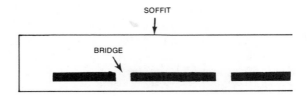

Figure 7–27 Slotted soffits require bridges at intervals to prevent weakening.

resulting from inadequate or absent weatherstripping is considerable. When properly installed, weatherstripping will cut down on heat loss in winter and reduce heat gain in the summertime. Additionally, it helps keep out dust, soot, and other airborne contaminants. *Fig. 7–28* shows three weatherstripping devices that are widely used in modern home construction.

Figure 7–28 *Left:* Aluminum and vinyl door weatherstripping with slotted holes for adjustment. *Center:* Aluminum and felt weatherstripping can be formed to fit around door or window. *Right:* Bronze weatherstripping for doors.

Sound Control

Modern living is very noisy, thanks to such sources as TV, radio, dishwashers, dryers, vacuum cleaners, jet planes, road traffic, and so forth. Not only is noise annoying, it can be harmful as well. It causes fatigue, irritability, inefficiency, and sometimes damage to sensitive hearing nerves.

Many construction methods and materials contribute to the noise problem. The hard surfaces of walls and ceilings reflect or bounce back about 90 percent of the sounds that strike them. This is called *reverberation.* The sound continues to bounce back and forth, building up to an irritating level. Even if the sound source is stopped, the sound does not stop at the same instant. Instead, it continues to bounce back and forth until its energy is expended. Some of the sound is absorbed by walls, ceilings, and floors and some is transmitted through these surfaces. If the sound strikes a surface that absorbs it 100 percent, then only the original sound will be heard without reverberation. This can be just as annoying as too much sound.

Speaking in a completely soundproofed room would give one an eerie feeling, however. The aim of sound control is to simply lower the buildup of sound by using sound-absorbing materials, and by reducing the transmission of sound. Acoustical materials are designed to absorb more than 50 percent of the sound striking them. These materials are highly porous. The movement of air particles in and out of the pores causes friction, which dissipates the sound energy as heat (*Fig. 7–29*).

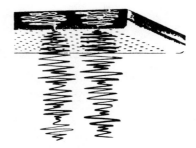

Figure 7–29 Sound is dissipated as heat upon entering sound-absorbing material.

Sound is measured in decibels (db), which indicate the degree of loudness. Sound reductions up to 10 decibels in a room are considered sufficient. Reductions greater than this are impractical and uneconomical. In an average-size room with an acoustical tile ceiling, the sound reduction will be between 5 and 7 decibels. This reduction is obvious and comfortable.

Although sound is reflected back and forth in the room in which it originates, it is also transmitted through the construction as the walls, ceilings, and floors vibrate. The vibrations are caused by sound waves striking these surfaces. They in turn cause the air molecules in the adjoining rooms to vibrate, resulting in the original sound being reproduced at lower intensity (*Fig. 7–30*).

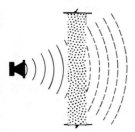

Figure 7–30 Sound waves being transmitted through a wall.

Building materials vary in their ability to absorb and transmit sound. The type of use made of materials also affects these qualities. Materials and their application are thus rated according to their sound resistance. The STC, or sound transmission class, was established to rate the noise qualities of various materials. The higher the number, the better the sound barrier. For example, ½-inch gypsum wallboard nailed to 2 × 4 studs has an STC rating of 32. This number can be increased to 46 by the use of ½-inch sound-deadening board together with the ½-inch gypsum wallboard. *Fig. 7–31* illustrates how various materials and their application to a stud wall affect the STC rating.

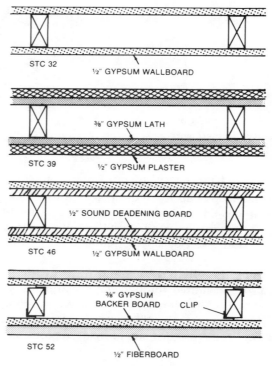

Figure 7–31 Sound transmission class (STC) ratings for partition walls with various combinations of materials.

Separating Opposite Surfaces. Two wall surfaces mounted to the same frame member will vibrate equally to sounds produced in a room. By staggering the stud frame, sound transmission to the second wall can be reduced considerably. A double wall with two rows of studs on separate plates is even more effective (*Fig. 7–32*). Both can be improved by the addition of thermal insulation to one of the wall members.

Adjoining rooms that are relatively quiet do not require soundproof walls or partitions. Measures should be taken, however, to soundproof walls or partitions that contain plumbing.

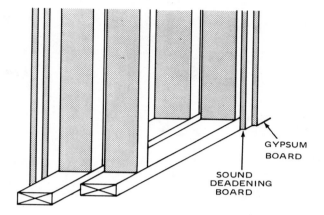

GYPSUM
BOARD

SOUND
DEADENING
BOARD

Sound transmission between rooms may be increased by placing medicine chests and outlet boxes back to back. This practice should be avoided if possible.

Floors and Ceilings.   Floors are subject to airborne and impact sounds caused by footsteps, dropped objects, and furniture movement. One effective way to control impact sounds is by cushioning the floor. This is easily done with carpeting and heavy padding. It must be noted, however, that carpeting is not very effective for airborne sound insulation.

Floors may be floated or separated from the structural subfloor to lessen the impact noises transmitted to the room below. No nails or other fasteners are used to attach the finish floor to the structure. The finish floor is fastened to a plywood underlayment which rests on sound-deadening board. This in turn rests on the structure subfloor (*Fig. 7–33*).

One method of lessening both airborne and impact sounds is to float the ceiling in the room below. This is done by using resilient clips or channels to which the ceiling is mounted (*Fig. 7–34*). The use of mineral-fiber insulating blanket between the joists will also help.

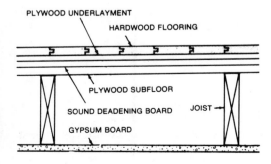

PLYWOOD UNDERLAYMENT

HARDWOOD FLOORING

PLYWOOD SUBFLOOR

SOUND DEADENING BOARD   JOIST →

GYPSUM BOARD

Figure 7–33 Resilient underlayment reduces impact sound transmission.

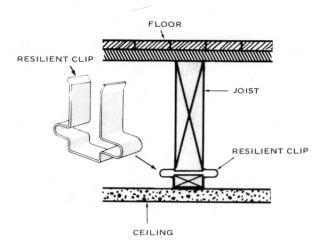

Figure 7–34 Resilient clips spaced 24 inches apart separate the ceiling finish from the framing.

Doors and Windows. Doors and windows, as well as the cracks and holes around them, create paths for the transmission of sound. The door and window are like large holes through which sound leaks. To be effective barriers to sound transmission, doors and windows should be airtight and heavy enough to minimize transmission. Double glazing is also helpful.

Hollow doors transmit more sound than solid doors, but much sound leaks through the edges. Weatherstripping helps, especially the soft foam types. They not only seal the door but also absorb impact sounds created by slamming doors.

Sound Absorption. Acoustical ceiling tiles are widely used to reduce noise levels within rooms. They are commonly made of wood fiber or similar materials. Lightweight and easily installed, they range in size from 12 × 12 inches to 24 × 48 inches and are available in thicknesses of ½ inch and ¾ inch. The tiles may be applied with adhesives to smooth, flat surfaces, stapled to furring strips, or suspended by means of one of many patented systems. *Fig. 7–35* shows an acoustical tile ceiling being installed on furring.

REVIEW QUESTIONS

1. How does moisture condensation affect building structures?
2. What is a vapor barrier? Where should it be placed in the house frame?

Figure 7–35 Installation of an acoustical tile ceiling.

8 Interior Finish

Interior finish includes ceilings, walls, floors, windows, doors, and trim. Many materials are used in this work, both wood and non-wood (*Fig. 8–1*). Walls and ceilings are generally covered with plaster or drywall. Although plastering is a specialized trade practiced by skilled artisans, the carpenter must do the preliminary work to prepare the walls and ceilings for the plasterer.

Plastering

Plaster Base. Plaster is made from a white mineral called gypsum. It is mined in the earth much like coal. The rock is crushed and ground and then heated (calcined) to drive off its chemically combined water. The oxidized powder that results is mixed with sand and a binder to make plaster. This is mixed with water into a paste and spread thinly on a flat surface, where it dries hard and smooth.

The surface to which the plaster is applied must hold the plaster securely. In the past, wood lath was used extensively for plaster application. The wet plaster was made to penetrate the spaces between the lath, forming keys which held the dried plaster securely. Wood lath is seldom used in modern construction; instead, metal and gypsum laths are utilized. *Fig. 8–2* shows several forms of metal lath. These are made by slitting sheet steel and expanding it.

Gypsum lath consists of a plaster core sandwiched between paper. One paper face is porous and becomes filled with plaster crystals, resulting in a perfect bond between the two materials. Another form of gypsum lath contains ¾-inch holes spaced 4 inches apart throughout its surface. The perforations supplement the surface bond. The perforated type is generally used in commercial applications. *Fig. 8–3* shows three types of gypsum lath.

Gypsum lath is made in several sizes. The standard size is 16 × 48 inches, except on the Pacific coast, where the standard width is 16⅕ inches. Extra-length lath, 24 inches wide by 12 feet long, is also available. Thicknesses are ⅜, ½, and ⅝ inch. For residential construction the ⅜-inch thickness is used on 16-inch stud spacings, and the ½-inch thickness is used on 24-inch stud spacings. The lath is applied horizontally across framing members (*Fig. 8–4*). Joints must be staggered and must not be made on jamb lines. Fasten the lath with nails or staples. The recommended nails for ⅜-inch lath are 1⅛-inch blued with a ¹⁹⁄₆₄ head. Use four nails per bearing. For ½-inch lath use 1¼-inch nails, five per bearing. If staples are used, the recommended size is 16 gauge with ⅞-inch divergent legs for ⅜-inch lath and 1-inch divergent legs for ½-inch lath.

Figure 8–1 Interior finish requires skilled workmanship in many materials.

Figure 8–2 Four types of expanded metal lath.

Figure 8–3 Three types of gypsum lath bases.

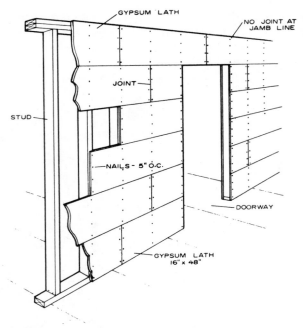

Figure 8–4 Gypsum lath should be laid out with staggered joints.

Certain areas of a wall are subject to stresses that can cause cracks in the finished plaster wall. The areas around doors and windows are especially vulnerable. To minimize the development of such cracks, reinforcing strips are utilized. Narrow pieces of metal lath serve this purpose well. Install the mesh diagonally as shown in *Fig. 8–5*. Other problem areas are inside and outside corners and the area under flush beams. Metal lath is used under flush beams (*Fig. 8–6*). Wire fabric "cornerites" are generally used for inside corners (*Fig. 8–7*). Metal beads are used for reinforcing outside corners (*Fig. 8–8*). Metal lath is seldom used in residential construction except in the shower or tub area of bathrooms.

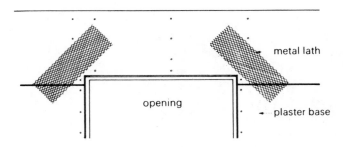

Figure 8–5 Reinforcement of plaster at openings.

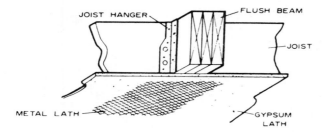

Figure 8–6 Area under a flush beam must be reinforced before plastering.

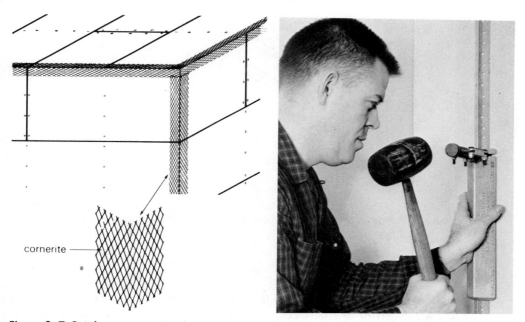

Figure 8–7 Reinforcement at inside corners.   Figure 8–8 Installation of outside corner beads.

Plaster Grounds.   These consist of 1 × 1-inch or 1 × 2-inch wood strips placed around doors and windows and at the base of the walls as a guide for plasterers. They insure a true, flat plaster surface and they provide a nailing surface for finish trim. The grounds may, however, be installed temporarily, then removed when the plastering is completed. *Fig. 8–9* shows a typical application of plaster grounds. Often the jambs of doors and the jamb extensions of windows serve as plaster grounds. When you install the grounds at the floor line, be sure to allow a clearance for the finish floor. Place the grounds so they will be slightly above the finish floor.

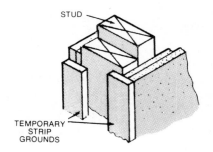

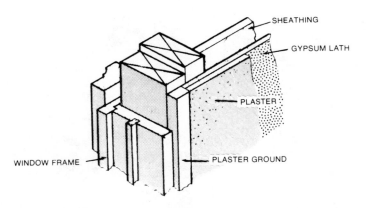

Figure 8–9 Plaster grounds. *Top:* Temporary application. *Bottom:* Permanent ground around window.

Plaster Applications. Plaster is applied in two or three coats. Two-coat work consists of one base coat and a top or finish coat. The base coat contains plaster combined with sand, wood fibers, perlite or vermiculite. Scratch and brown is the first coat used in all plastering. In two-coat work, both are applied at almost the same time, in what is called "double-up work." The brown coat is applied doubled back within a few minutes. This coat fills the space between the plaster base and the grounds. The second and final coat gives the wall its finish surface.

In three-coat work the first coat is a scratch coat. It is applied directly to the plaster base and scratched after it sets slightly. The scratched surface provides a good bond for the following brown coat. After the scratch coat dries, the brown coat is applied, then brought up to the surface of the grounds and leveled. The final coat is then troweled on as in the two-coat method (*Fig. 8–10*).

Figure 8–10 Applying plaster to scratch coat.

Gypsum Drywall

Gypsum wallboard, also known as plasterboard and sheetrock, consists of a core of gypsum plaster sandwiched between two sheets of heavy paper. The sheets are 4 feet wide and range in length from 6 to 16 feet. Thicknesses are ¼, ⅜, ½, and ⅝ inch. The ¼-inch boards are normally used for re-covering old walls and ceilings. The ⅜-inch material is used in two-ply construction and the ½-inch and ⅝-inch boards are used in single-ply work. The ⅝-inch boards provide better fire resistance and sound control. Because of its low cost and ease of installation, drywall is used extensively in residential construction. The plain papered face provides an excellent surface for paint or wallpaper.

The edges of gypsum wallboard are made in square, tapered, or beveled form (*Fig. 8–11*). The tapered edges along the length are made with a slight depression which allows for a filled and taped joint.

Single-Ply Construction. Cover the ceiling first, placing the sheets at right angles to the joists. Next, cover the walls by applying the sheets vertically or horizontally. The horizontal application is preferred by most installers, because it requires less taping. To assure a good joint at the ceiling line, apply the top panel first in the horizontal method (*Fig. 8–12*). Irregularities, if any, will be concealed by the baseboard at the floor line.

Figure 8–11 Common wallboard edges.
Top: Square;
Center: Tapered;
Bottom: Beveled.

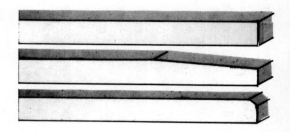

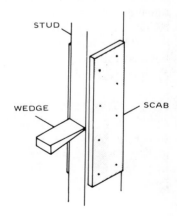

Figure 8–12 In horizontal application, install wallboard first.

The wall framing must be square and true. If some of the studs are bowed, correct this by making a cut on the hollow side of each such stud. Drive a wedge into the kerf until the stud straightens, then nail 1 × 4 scabs on each side and cut the wedge flush (*Fig. 8–13*).

Cut the panels by scoring the face side with a sharp blade. Then snap the core by bending it back along the cut line. Complete the cut by running the blade through the back of the board. Fasten the panels with nails

STUD

WEDGE

SCAB

Figure 8–13 Method of straightening bowed stud.

or screws. For ceilings, place the nails 7 inches on center; for walls, 8 inches on center. The nails should be driven at least ⅜ inch from panel ends and edges. Press down on the panel near the nailing area to assure good contact between the panel and framing member. Use 1¼-inch annular ring nails with a ¼-inch head. Drive the nails with a crowned-head hammer to produce a uniform dimple (*Fig. 8–14*). The use of screws will virtually eliminate "popping," which is caused by drying out of frame members. Popping can also be reduced by double-nailing. Place the nails 2 inches apart on 12-inch centers. Use single nailing around the perimeter. After the second nail is applied, restrike the first nail to seat it tightly (*Fig. 8–15*).

Figure 8–14 Properly driven nail leaves a dimple in the surface of the wallboard.

Figure 8–15 Double-nailing reduces popping.

Cracked joints and nail-popping are often due to movement of the wallboard as the framing members dry out. This problem can be minimized if you use the floating angle method, illustrated in *Fig. 8–16*. In this system, one sheet at each interior angle is not nailed; thus the wallboard can "move" under stress.

Panels can also be installed by means of adhesives, the advantage being a sturdier wall with fewer nails. Apply beads of adhesive to each stud in a continuous strip (*Fig. 8–17*). When two panels join on the same framing member, apply the adhesive in a zigzag fashion. If possible, brace the panels until the adhesive sets. Use supplemental nails or screws at the edges. If prefinished wallboard is used, install the panels vertically and apply nails along the top and bottom edges only.

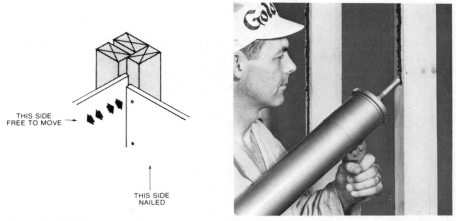

THIS SIDE
FREE TO MOVE →

THIS SIDE
NAILED

Figure 8–16 Floating angle application permits wallboard to move under stress.

Figure 8–17 Fastening wallboard adhesive.

Joint Finishing.   Joints are finished by the application of joint compound and perforated tape. Spread the compound into the depression formed by the tapered edges of the wallboard. Apply tape the full length of the joint, embedding it in the compound. The compound will be forced out through the perforations of the tape. Apply additional compound, feathering the outer edges. Allow to dry, then apply a second coat of compound. Feather the second coat about 2 inches beyond the first. Repeat this procedure for the final coat and feather it 2 to 4 inches beyond the previous coat. For best results the finishing tool (broad knife) should be held at a 45-degree angle (*Fig. 8–18*). The details of a typical taped joint are illustrated in *Fig. 8–19*. Fastener heads are concealed with three applications of compound, each coat feathered slightly. Taping at the wall-ceiling line can be eliminated if moldings are used (*Fig. 8–20*).

Two-Ply Construction.   Two-ply or double-layer applications offer more fire protection and greater sound-deadening. Nail-popping is also eliminated because the second layer is fastened with adhesive. The first layer is fastened with nails, screws, or a combination of nails and adhesives. The first layer need not be gypsum board. It can be backer board or sound-deadening board. Apply adhesive to the entire surface of the panel, using a notched spreader or mechanical applicator. Then apply the top layer at right angles to the first (*Fig. 8–21*). Face joints should offset those of the base by 10 inches. In some applications the adhesive may be applied in strip form.

Figure 8–18 Applying compound to taped joint.

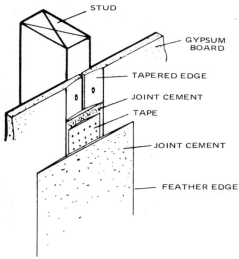

STUD

GYPSUM BOARD

TAPERED EDGE

JOINT CEMENT

TAPE

JOINT CEMENT

FEATHER EDGE

Figure 8–19 Detail of taped joint.

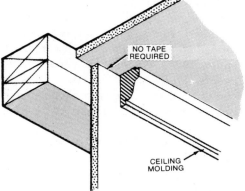

NO TAPE REQUIRED

CEILING MOLDING

Figure 8–20 Molding used at celing line eliminates the need for taping.

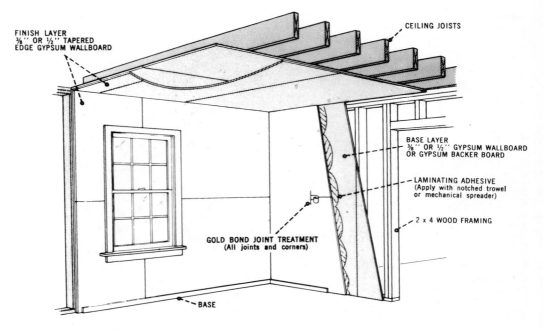

Figure 8–21 Double-layer construction.

Wall Paneling

Board Paneling.   Solid lumber paneling comes in various random widths from 4 to 12 inches, and in lengths from 6 to 16 feet. Thicknesses range from ⅜ to ¾ inch. Edges may be square, shiplapped, or tongue-and-grooved. Some patterns have molded edges. *Fig. 8–22* shows several tongue-and-groove designs. Boards may be installed vertically, horizontally, or even diagonally for some interesting effects (*Fig. 8–23*). When they are used vertically, furring strips are required at the top and bottom, and in the intermediate spaces on 24-inch centers. Furring is also needed at the corners and around window and door frames. For horizontal applications, the boards can be nailed directly to the studs. Cut the inside corners to length and butt them against the paneling on the adjacent wall. The joints in boards on adjacent walls must be aligned; therefore, use boards of matching widths.

Tongue-and-groove panels under 6 inches wide must be blind-nailed. This will eliminate the need for sinking nail-heads and filling. Use one 6d nail per bearing. For boards that are 8 inches or wider, use two nails per bearing. Face-nail boards that are 8 inches or wider. Sink the heads and fill.

Figure 8–22 Various tongue-and-groove boards with molded edges.

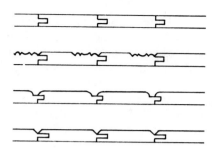

Figure 8–23 Board paneling being installed in a herringbone pattern.

Plywood Paneling. Plywood is produced in many styles, patterns, and textures. Made with prefinished surfaces, it is tough and durable. Because the large sheets cover surfaces fast, installation time is minimal. Plywood panels are made in 4-foot widths and in lengths of 7, 8, 9, and 10 feet. Standard thickness is ¼ inch. Although plywood is available ungrooved, most prefinished panels have V-grooves or channels spaced to give the panel a random-plank effect. Two of the intermediate grooves are always spaced 16 inches on center so they will bear on a framing member when the studs are on 16-inch centers. In most cases this will eliminate the need for nails on the face of the panel. Mark the backs of the panels at 16-inch intervals so they can serve as a guide for applying adhesive when it is used.

Several days before installation is to start, the panels should be conditioned in the room in which they are to be used. Stack the panels horizontally with full lengths of furring between each sheet. This will enable air to circulate freely around each panel and allow the panels to stabilize to the temperature and humidity of the surrounding area.

Because the grain patterns and color characteristics of each panel differ, the panels should be arranged into a pleasing sequence. Number the backs of the panels with chalk to indicate the order in which they are to be installed.

Panels can be applied to existing walls or to open studding. The studs should be straight and sound. If necessary, plane down high spots and shim the low ones. Fastening can be done with nails or adhesives. The panel edges must join on a bearing surface. If the studs will not permit direct application because of misalignment, furring will be necessary. Furring is also needed when paneling is applied to a masonry wall. Apply furring strips at right angles to the panel application. Furring can be ⅜ × 1⅞ plywood or 1 × 2 lumber. Space the furring 16 inches on center (*Fig. 8–24*).

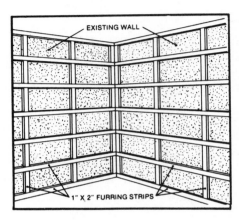

Figure 8–24 Furring applied to masonry wall.

Measure and cut the panels accurately. Start the installation at a corner, following this procedure: Plumb the first panel using a level or a plumb line. Place two nails at the top and drive them partially into the top plate. Mark the panel edge parallel to the corner. If the corner is rough or irregular, use a compass to scribe the cutting line (*Fig. 8–25*). After cutting, refit the panel into the corner, maintaining the plumb line, and nail. Allow ¼ inch clearance at top and bottom for expansion. Butt the second panel to the first and continue around the room.

Figure 8–25 Scribing irregular corner.

For stud application, use 3d finish nails, spaced 6 inches along the panel edges and 12 inches elsewhere (*Fig. 8–26*). Over furring, use 3d finish nails at 8-inch spacings along edges and 16-inch spacings into horizontal furring elsewhere. Sink all nail-heads carefully and fill with matching putty stick.

Contact adhesives may be used instead of nails. Each panel must be cut and fitted perfectly before the adhesive is applied. Manufacturers' directions should be followed for the correct use of their products.

Figure 8–26 Whenever possible, nails should be applied in grooves.

Hardboard Paneling. Hardboard is widely used for interior wall finishing. It is made from small wood chips that are "exploded," cleaned, and refined. Further processing converts them into a dense, rigid board, free of knots or other imperfections. The grain-free boards are then finished with a variety of surfaces and colors, including wood grains. Grooved boards have the appearance of random planks. Other surfaces include marble, brick, stone, leather, and many others. *Fig. 8–27* shows just a few of the many surfaces available.

Hardboard panels are made in various thicknesses and sizes. Wall paneling is usually ¼ inch thick. The panels can be applied with nails or adhesives. All edges must be supported. Where humidity conditions are normal, panels should stand unwrapped around the room for at least 48 hours before application. This will permit them to stabilize to the moisture condition of the room. When hardboard is to be used below grade in areas of high humidity, it is necessary first to eliminate all dampness by waterproofing the exterior walls, and insulating, and by installing dehumidifying equipment.

Panels are installed using the same procedure as for wood paneling.

Figure 8–27 Various hardboard products.

Prefinished matching moldings of wood, metal, or plastic are readily available.

Plastic Wall Paneling. Plastic laminates are known for their use on sink and counter tops. They are also widely used for paneling, especially in bathrooms, where good appearance, durability, and resistance to stains and moisture are required. When they are used for tub and shower enclosures, a backed laminate should be used. One such product consists of a laminate backed with polystyrene foam (*Fig. 8–28*). The foam core will absorb the irregularities of the wall surface beneath. The material is applied directly over ½-inch waterproof gypsum board. It can also be applied over waterproof plywood. Although the foam backing will accept small surface irregularities, large voids should be filled. Masonry walls can also be covered, but only above grade, and they must be smooth and free of defects.

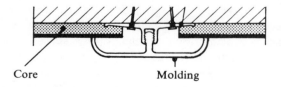

Core   Molding

Figure 8–28 Detail of foam core laminate with two-piece snap molding.

Because wall sections must fit perfectly, paper templates are usually made before actual cutting of the laminate. Cutouts for plumbing stubs and other projections or openings are cut in the template, then transferred to the laminate. The necessary cutouts and edge cuts are made with an expansive bit and saber saw fitted with a fine-toothed blade. The cut panels should be dry-fitted to be sure they conform (*Fig. 8–29*). The panels are applied with a special adhesive. Two-piece "matching" snap moldings are used for trim.

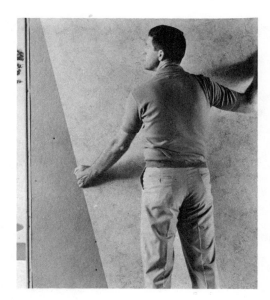

Figure 8–29 Dry-fitting cut panel before applying adhesive.

Ceiling Tiles

Ceilings may be finished with a variety of materials. Ceiling tiles are very popular, because they can be used for new and old construction. They are made in numerous sizes, styles, and materials; some are acoustical, others are purely decorative. For residential construction, the 12 × 12-inch tile is commonly used (*Fig. 8–30*). Other sizes available are 12 × 24 inches, 16 × 32 inches, 24 × 24 inches, and 24 × 48 inches. Standard thicknesses are ½ inch and ¾ inch. Edges may be square-cut or they may have interlocking seams (*Fig. 8–31*).

Tiles may be applied directly to a flat ceiling, or to furring strips, or they may be suspended. Adhesive can be used when they are applied to a flat surface; however, 1 × 3 furring is required when an old ceiling in poor condition is being tiled or when tiles are being applied to open joists. The furring should be applied at right angles to the joists (*Fig. 8–32*). Special metal channels may be used instead of furring.

Tile Layout. The tiles bordering the ceiling should be of equal width at opposite sides of the room. If the length of the room leaves less than a full tile at one end, add the length of one full tile to the odd figure and divide by 2. The result will be the width of the border tiles to be used at each

Figure 8–30 Regular 12 × 12 ceiling-tile installation.

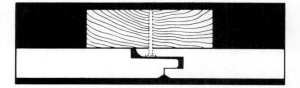

Figure 8–31 Section through interlocking ceiling tiles.

Figure 8–32 Applying 12-inch tiles to furring.

end. For example: Assume that the length of a ceiling is 14 feet 4 inches, and the tiles being applied are 12 × 12 inches. Add 12 inches to 4 inches (the excess) and divide the total (16 inches) by 2. The result is 8 inches. This is the size of the border tiles at the two opposite lengthwise ends of the ceiling. Each end will have 8-inch tiles instead of 4 inches at one end and 12 inches at the other. Do the same for the width of the room.

Installing Furring.   Nail the first furring strip against the wall, perpendicular to the joists. Place the second strip parallel to the first and with its center the width of a border tile away from the wall. If the border tiles are to be 8 inches, then the center of this second furring strip should be exactly 8 inches from the wall. Install the rest of the furring strips at 12-inch centers (for 12-inch tiles; as in *Fig. 8–33*). Use two 8d nails at each bearing, but do not drive them flush until you check the level of the strips (*Fig. 8–34*). If this is correct, drive the nails tightly. If necessary, use shims cut from shingles to bring the furring strips to an even plane (*Fig. 8–35*).

Pipes and other obstructions should be boxed (*Fig. 8–36*). If the obstruction is less than 1½ inches, double furring can be used to eliminate the projection altogether.

Figure 8–33 Installing furring onto joists.

Figure 8–34 Furring strips should be checked for levelness before they are nailed tightly to joists.

Figure 8–35 Use shims to bring furring strips out to an even plane.

Figure 8–36 Frame built of 1 × 2 stock conceals heating ducts and pipes.

Installing Tiles.   Before installing the tiles, it is necessary to establish a square reference point. On the long wall, snap a chalk line along the center of the second furring strip, line A–a (*Fig. 8–37*). This line will be one border-tile width away from the wall. Snap a second line, B–b, at right angles to the first, again one border-tile width from the wall. Since the adjacent walls may not be perpendicular to each other, it is best to use the 3-foot, 4-foot, 5-foot method to establish the square corner. (See the section on layout in Chapter 2 for a further description of this method, which is based on the theorem that the sides of a right triangle always will be in the ratio of 3:4:5.)

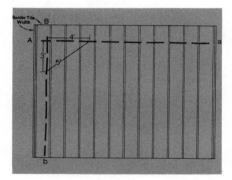

Figure 8–37 Establishing square reference point to begin tile installation.

Start installing the tile in the corner of the room where the square reference point was established. Cut away the tongue edges, as shown in *Fig. 8–38*. Staple the tile with its face matching the chalk line (*Fig. 8–39*). Cut and fit the border tiles along the reference lines (*Fig. 8–40*). Measure and cut each tile individually, because the wall may be uneven. Cut tiles from the face side, using a sharp knife and straightedge (*Fig. 8–41*). Irregular cuts can be made with a compass saw.

Working outward from the corner, fit the tiles tightly and fasten with three staples per tile. Finish the joint between the wall and the ceiling with

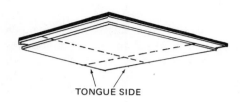

TONGUE SIDE

Figure 8–38 Cut the first border tile to fit within the border lines established. Cut away tongue sides.

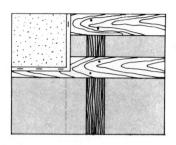

Figure 8–39 Staple first tile at corner, using the chalk lines as a guide.

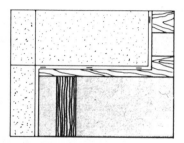

Figure 8–40 Cut and fit the remaining border tiles along the reference lines.

Figure 8–41 Cut tiles with a sharp knife.

a piece of molding. If the molding is not prefinished, paint or stain it before installation so you will not smear the wall or ceiling later.

Tiles may be applied to a flat, sound surface by means of adhesive mastic. Use the same procedure for establishing the reference point. Snap the chalk lines directly onto the ceiling. Use four daubs of adhesive, half the size of a walnut, on each tile (*Fig. 8–42*). Press the tile firmly into place.

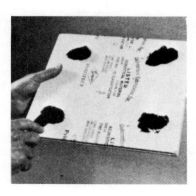

Figure 8–42 Apply mastic to back of tile.

Suspended Ceilings

Suspended ceilings are often used when obstructions such as ducts, pipes, and wiring are present. They may also be used to advantage on remodeling work, especially when you are dropping a high ceiling. The system consists of a metal framework mounted along the walls and suspended from the ceiling with wires or special hanger clips. Lay-in ceiling panels are then made to rest on the flanges of the various members.

Attach a wall angle to the wall around the room at the ceiling height desired (*Fig. 8–43*). Then position main tees with hanger wires connected to screw eyes installed in the ceiling or joists (*Fig. 8–44*). Fasten cross-tees to the main tees by inserting the ends into the slots provided (*Fig. 8–45*). When the framework is completed, install the lay-in panels (*Fig. 8–46*).

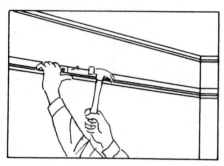

Figure 8–43 Attach wall angle around room at ceiling-tile height.

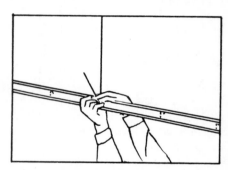

Figure 8–44 Position main tees with hanger wires.

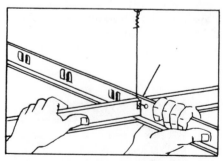

Figure 8–45 Install cross-tees between the main tees.

Figure 8–46 Install panels by tilting them into place.

Translucent plastic panels can be installed to provide light from concealed fixtures (*Fig. 8–47*).

Other systems make use of concealed channels where the tile joints are interlocked with flanges on the runners (*Fig. 8–48*).

Figure 8–47 A suspended ceiling allows the installation of concealed lighting.

Figure 8–48 Tile system using concealed interlocking slots.

Finish Floors

In residential construction, *finish flooring* refers to the final covering placed on floors—the wearing surface. The finish flooring should be installed only after all plumbing, wiring, plastering, painting, and ceilings are completed. The only work remaining should be the final trim.

The most commonly used floors are made of wood. These include strip, plank, parquet, and block. Unfortunately, because of the spiraling costs of lumber and wood products, many home builders are installing wall-to-wall carpeting directly on the subfloor and eliminating the cost of a finish floor. If this practice continues, finish flooring will become a lost art. There are some areas of a home where finish flooring other than wood is desirable, such as the kitchen, bathroom, and laundry or mudroom, where moisture is a major problem. Carpeting is desirable in the living areas, but it should be used over a substantial wood floor, and not in place of one. Some wood-finish flooring is too beautiful to be hidden by any carpeting at all.

A wood floor properly installed will give many years of service. Although they are firm and durable, wood floors have a degree of resiliency that makes them comfortable to walk on.

Strip Flooring. This type of flooring consists of long, narrow pieces of hardwood or softwood in varying thicknesses. Most are made with tongue-and-grooved joints along the sides and ends (matched flooring). Oak is almost universally used for flooring, but maple, birch, and beech are other hardwoods that are popular as well. Among the softwoods, yellow pine and fir are often used. *Fig. 8–49* shows a typical strip floor. The back or bottom side of matched strip flooring is grooved (*Fig. 8–50*). This permits the boards to bridge small irregularities in the subfloor and thus lie flat.

Figure 8–49 Side-matched strip flooring.

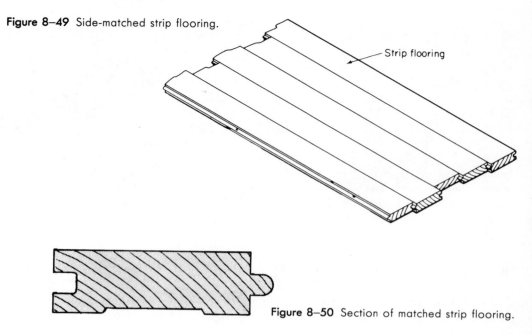
Strip flooring

Figure 8–50 Section of matched strip flooring.

Strip flooring ranges in width from 1 inch to 3½ inches and in thickness from $5/16$ inch to $25/32$ inch, except for square-edge maple, beech, birch, and pecan, which come in $25/32$- and $33/32$-inch thicknesses. Standard thicknesses are ⅜ inch, ½ inch, and $25/32$ inch for strip flooring. For residential construction the $25/32 \times 2¼$-inch strips are the most popular. Nominal and actual sizes of oak strip flooring are shown in *Fig. 8–51*. Counted size refers to the size used in determining the board feet in a shipment. The thinner flooring is generally used in remodeling work.

OAK			
Nominal	Actual	Counted	Wts. M Ft.
TONGUED AND GROOVED-END MATCHED			
25/32 x 3¼ in.	25/32 x 3¼ in.	1 x 4 in.	2300 lbs.
25/32 x 2¼ in.	25/32 x 2¼ in.	1 x 3 in.	2100 lbs.
25/32 x 2 in.	25/32 x 2 in.	1 x 2¾ in.	2000 lbs.
25/32 x 1½ in.	25/32 x 1½ in.	1 x 2¼ in.	1900 lbs.
3/8 x 2 in.	11/32 x 2 in.	1 x 2½ in.	1000 lbs.
3/8 x 1½ in.	11/32 x 1½ in.	1 x 2 in.	1000 lbs.
1/2 x 2 in.	15/32 x 2 in.	1 x 2½ in.	1350 lbs.
1/2 x 1½ in.	15/32 x 1½ in.	1 x 2 in.	1300 lbs.
SQUARE EDGE			
5/16 x 2 in.	5/16 x 2 in.	face count	1200 lbs.
5/16 x 1½ in.	5/16 x 1½ in.	face count	1200 lbs.

Figure 8–51 Standard sizes, counts, and weights of oak strip flooring.

Care should be exercised in the delivery and storage of hardwood flooring. It should not be transported or unloaded in rain or snow. In very damp weather it should be protected with a canvas covering. It should be stored in a well-ventilated area and must not be installed in a damp or cold building. If flooring is laid in winter, the building should be heated.

The flooring should be laid over a sound subfloor properly nailed as described in the section on floor framing. The subfloor adds bracing strength to the building and provides a solid base for the finish floor.

Installing Strip Flooring.   Check the subfloor for defects. Hammer down protruding nails and look for loose or warped boards. Remove blobs of plaster, mortar, or other foreign material. If necessary, scrape the floor to produce a smooth, flat surface. Finally, sweep the floor thoroughly, and if the subfloor is of board construction, lay down a good-quality asphalt-coated building paper. This will protect the room from dust and moisture that may infiltrate the seams. It is not needed when subfloors are plywood, but it is often used as a sound-deadener. In areas directly over a heating unit, double the layer of building paper to protect against excessive heat, which may dry out and shrink the floorboards. Lap all seams at least 4 inches (*Fig. 8–52*).

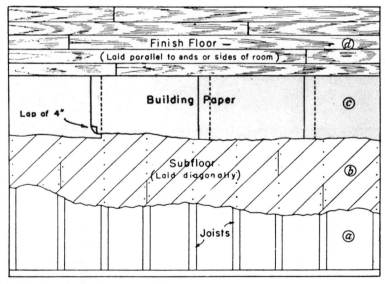

Figure 8–52 Cutaway plan view of floor.

Normally, strip flooring is laid parallel to the longer side of the room. There are times, however, when it is laid down crosswise, especially in very wide rooms. Since interior thresholds are no longer used in residential construction, the flooring should run continuously from room to room. Preferably, the strip flooring should run perpendicular to the floor joists. Use the shortest pieces of flooring in closets and other less prominent areas. Place the longer pieces at entrances and doorways. If the direction of the strips is to be changed between rooms, be sure to make the change under doors to conceal the joint.

Nailing.   Proper nailing of flooring is most important. Insufficient or improper nailing will almost always result in loose or squeaky floors.

Blind-nail tongue-and-groove flooring through the tongue at an angle of approximately 50 degrees (*Fig. 8–53*). Countersink nail-heads with a nail set (*Fig. 8–54*). Use 7d or 8d steel-cut or screw-type flooring nails, spacing them 10 to 12 inches apart.

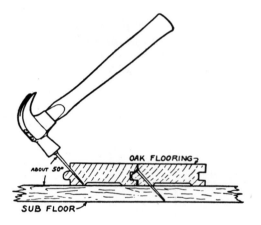

Figure 8–53 Blind-nailing is done through tongue of flooring.

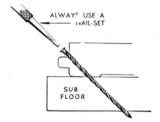

Figure 8–54 Use a nail set to sink nail-heads.

Start the installation by snapping a chalk line about a foot away and parallel to the wall (*Fig. 8–55*). If the near and far walls are not parallel to each other, adjust the chalk line to take an average (split the difference). Lay the first strip with the grooved edge facing toward the wall, leaving a space of about ½ inch between the strip and the wall to allow for expansion. This narrow space will be covered later with a shoe molding. Face-nail the edge nearest the wall, placing the nails so they will be concealed later on by the molding. Do not drive these nails all the way.

After the first strips along the wall are in place, check to be sure that they are parallel to the chalk line established earlier. If so, drive the face nails all the way, then blind-nail the tongue edge (*Fig. 8–56*). Note that the shoe molding is nailed into the subfloor. This is done so that if the finish floor were to contract, it would not pull the shoe molding away from the wall.

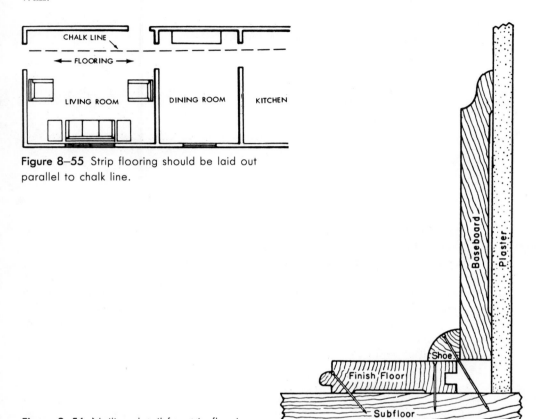

Figure 8–55 Strip flooring should be laid out parallel to chalk line.

Figure 8–56 Nailing detail for strip flooring.

Continue to blind-nail all succeeding strips. The end joints in each row must be staggered at least 6 inches (*Fig. 8–57*). Strips should be arranged so grain and color will blend harmoniously. Cut and fit a number of boards, laying them out a few rows ahead of the nailing (*Fig. 8–58*). This procedure will speed up the job and keep waste to a minimum. The strips

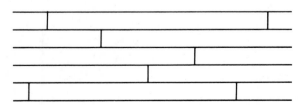

Figure 8–57 End joints in each row must be staggered at least 6 inches.

Figure 8–58 Arrange strips to blend color and grain pattern.

must be set tightly against the preceding rows. After three or four pieces have been nailed, place a scrap piece of flooring from which the tongue has been removed against the edge of the outside piece and drive the units up snugly (*Fig. 8–59*). Repeat this procedure after every three or four pieces have been laid. Nailing machines are available to draw the boards tightly as they are being nailed. Some machines even sink the nail-heads (*Fig. 8–60*).

The last row on the opposite wall should be face-nailed, as was the first row. Use a pinch bar to draw the last row tight. Place a board against the wall to protect it from being marred by the pinch bar.

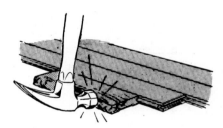

Figure 8–59 Use scrap to tap flooring strips into place and to tighten joints before nailing.

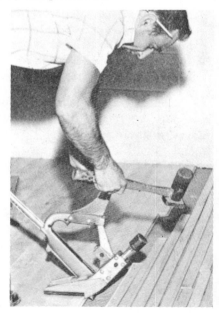

Figure 8–60 Nailing machine being used in strip flooring.

Special Problems. In some installations, the flooring may be interrupted by a partition or other projection. In such cases the flooring must be applied in two directions (*Fig. 8–61*). Lay out the main area to a line even with the projection. Snap a chalk line to extend this line into the adjoining area. Then reverse the tongue direction and continue the flooring in the opposite direction. Use a hardwood spline to fill the groove of the two opposing strips.

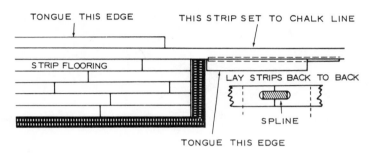

Figure 8–61 Method of laying strip flooring around a projection.

Strip Flooring over Concrete. It is possible to lay strip flooring over concrete, provided a suitable vapor barrier is used to provide resistance to ground moisture and vapor. If a vapor barrier was used under the concrete slab, another is not needed. There are several approved methods of waterproofing a slab prior to installing the flooring. In one method, the slab is coated with asphaltic mastic followed by a layer of polyethylene sheeting and another coating of mastic on top of the plastic film. Treated 2 × 4-inch screeds or sleepers, in random lengths of 18 inches to 48 inches, are laid in the top layer of mastic in rows 16 inches on center, with end joints staggered and ends lapped 4 inches. The sleepers must be impregnated with a wood preservative other than creosote, and a termite repellent. Install them flat side down and perpendicular to the length of the room with a ¾-inch space between the sleepers and the wall line. Lay another layer of polyethylene film across the tops of the sleepers, with edges overlapped 4 to 7 inches, before you install the strip flooring. Lay the strip flooring perpendicular to the sleepers and nail it into both sleepers where they are lapped.

This method is rather involved and many workers prefer the method illustrated in *Fig. 8–62*. In this second method, 1 × 2-inch strips spaced 16 inches on center are nailed to the slab, which has been coated with 2-inch-wide rivers of mastic suitable for bonding wood to concrete. Next, a .004 gauge polyethylene film is placed over the entire floor and another set of 1 × 2 sleepers, untreated this time, is nailed to the first. The finish flooring is then applied at right angles to the sleepers. *Fig. 8–63* illustrates the steps involved.

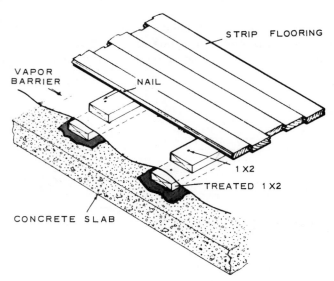

Figure 8–62 Approved method of applying strip flooring.

Figure 8–63 Installation of strip flooring over concrete slab. *Top left:* 1 X 2 sleepers are laid over ribbon of mastic and nailed into concrete slab. *Top right:* Sleepers, 16 inches on center, are covered with sheet of polyethylene film. *Bottom left:* Second course of 1 X 2 sleepers is nailed through the polyethylene film into the first. *Bottom right:* Strip flooring is nailed at right angles to the sleepers.

A third method does not use sleepers as a nailing base. Instead, exterior plywood panels, ¾ inch × 4 feet × 8 feet are installed directly on the slab, which has been covered by a polyethylene film vapor barrier. The panels are laid parallel to the length of the room, with ½- to ¾-inch spaces between panels. Joints are staggered. Powder-activated fasteners are driven through the plywood directly into the slab, spaced 12 inches apart and 2½ to 3 inches from all edges. The finish strip or plank flooring is also installed parallel to the length of the room.

Plank Flooring.   Plank flooring has been used for hundreds of years in Europe and since the early days of settlement in America. Because of its charm and rugged appearance, there is great demand for this type of flooring, especially in Colonial-style buildings (*Fig. 8–64*). Plank flooring is generally made of oak in random widths from 3½ to 8 inches wide and is available with or without a bevel at the top. Usually the pieces are tongued and grooved with square or matched ends. However, they are also produced with square sides and ends. The plugs used to conceal the heads of installation screws are made to resemble the pegs used by the colonists to fasten the flooring to the subfloor. The plugs are glued into the holes on top of the countersunk screws (*Fig. 8–65*). The planks are blind-nailed and face-nailed or screwed every 30 inches. If you choose to face-nail them, countersink the heads and fill with putty. Install the planks with a slight space between the boards to allow for expansion.

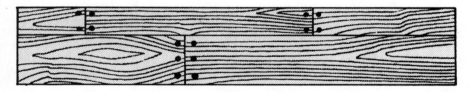

Figure 8–64 Random-width plank flooring with pegs.

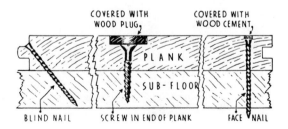

Figure 8–65 Fastening detail for plank flooring.

Block Flooring.   There are several types of block flooring. One form, the unit block, consists of short lengths of strip flooring joined at the edges to produce a square block. It has tongues at right angles to each other. This type of flooring is subject to expansion in use and therefore must be installed with a space at the wall line. Often a cork or rubber expansion strip is used around the perimeter of the room with the baseboard and shoe moldings concealing it.

Another type is the laminated block, made up of three layers of wood with the center ply placed at right angles to the face plies. These may have tongues on opposite sides or they may be made with grooves on all four edges (*Fig. 8–66*). The grooved blocks must be installed with special metal splines.

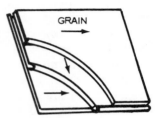

Figure 8–66 Laminated plywood block.

Generally, block flooring can be installed with nails or mastic over wood or concrete. However, some types are made for mastic application only. For installations on concrete either on or below grade, a damp-proofing membrane must be used in addition to the felt. The nailing procedure is similar to that used for plank and strip floors. Nails are driven through the tongues. When mastic is used, it is troweled on in a thin coat, then covered with 30-pound asphalt-saturated felt. This is followed by another application of mastic about ³/₃₂ inch thick. The blocks are then laid in the mastic (*Fig. 8–67*).

Figure 8–67 Installing block in bed of mastic.

Prefinished Flooring. When unfinished hardwood floors are laid down, they must usually be sanded and the finish applied after installation. However, most manufacturers can furnish the flooring completely prefinished.

When used in this form, it must be handled very carefully. Although prefinished flooring is more costly than the unfinished variety, it has several advantages. Factory-finished flooring is produced under controlled conditions with highly sophisticated equipment. The result is a finish far superior to that produced on the job. Also, the floor is ready to use as soon as installation is completed. *Fig. 8–68* shows a prefinished block floor. Since trim work will have to be done after the finish floor is installed, most carpenters protect the flooring with heavy kraft paper carefully taped in place.

Figure 8–68 A prefinished block floor.

Underlayment. Resilient flooring materials such as linoleum, vinyl, rubber, and asphalt tend to mold themselves to any irregularities over which they are installed. Open joints, cracks, and other defects in subfloors would show through the surface if a proper underlayment were not used. The underlayment provides a durable base for these materials and also prevents failure of the flooring due to movement of the subfloor during changes in temperature and humidity. The materials commonly used for underlayment over wood subfloors are hardboard, plywood, and particleboard. Mastic underlayments are generally used over concrete.

Regardless of the type of underlayment used, it is important that the material be uniform in thickness and smooth and that it be properly installed.

Plywood Underlayment. Plywood is excellent for use as underlayment. It is smooth and strong, and its high dimensional stability eliminates excessive swelling and consequent buckling. The plywood selected should be an underlayment grade, either interior or exterior type. Interior grades are usually used because they are less expensive, but the exterior grade is recommended in areas where excessive water spillage is possible. Underlayment grades of plywood have a solid surface with an inner ply construction that resists dents from concentrated loads. Panels to receive carpeting can be touch-sanded. Those to receive the conventional thin resilient flooring must be the fully sanded type.

Plywood underlayment is made in various thicknesses from ¼ inch to ¾ inch. Combined subfloor underlayment is made in thicknesses from ⅜ inch to 1¼ inches. The thickness used depends on the thickness of the resilient flooring to be installed as well as the surrounding floor thickness. If a ³/₃₂-inch vinyl tile floor is to butt against a ²⁵/₃₂-inch strip floor, the underlayment used should be ⅝ inch thick (*Fig. 8–69*). For maximum stiffness, place the face grain of the underlayment across the supports (*Fig. 8–70*).

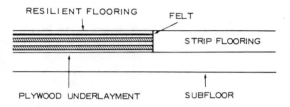

Figure 8–69 Alignment of resilient and strip flooring.

Figure 8–70 Face grain of underlayment should run across supports for greater stiffness.

Apply the underlayment just prior to laying the finish floor; otherwise, protect it against water or physical damage. Stagger the panel end joints and be sure that none of the joints align with those in the subfloor. Space the panels ¹/₃₂ inch apart to allow for expansion. Use 3d ring shank or cement-coated nails when applying ¼- or ⅜-inch underlayment. Space the nails 6 inches apart at panel edges and 8 inches apart at intermediate bearings. Installation time can be reduced by using mechanical fasteners.

Hardboard Underlayment.   Untempered hardboard is also used for underlayment. It is made in 3 × 4 and 4 × 4-foot panels with a ¼-inch nominal thickness (average caliper .215). It is applied like plywood underlayment.

Before applying, stand the panels around the room for 24 hours to allow them to stabilize to room temperature and humidity. Start installation at a corner, leaving a 1⅛-inch space at the wall line. Spacing between panels should be ¹⁄₆₄ inch. Use a matchbook cover as a spacing guide (*Fig. 8–71*). Install the panels with the smooth side down; the sanded side forms a better bonding surface for the adhesive. Some manufacturers print a nailing pattern on the top side of the panel (*Fig. 8–72*).

Figure 8–71 Panels are installed with a 1/64-inch space between joints.

Figure 8–72 Nailing pattern on underlayment simplifies installation.

Particleboard Underlayment. Because of its tough, smooth surface, particleboard underlayment provides an excellent base for resilient flooring. It is made in sheets that are 4 × 8 feet in size, in thicknesses of ⅜, ½, ⅝, and ¾ inch. It is fastened with ring-groove or cement-coated nails or with staples. The fasteners must penetrate into the subfloor at least 1 inch.

Use 6d nails on particleboard underlayment with thickness from ⅜ inch through ⅝ inch and 8d nails for particleboard underlayment that is ¾ inch thick. The nails should be not more than ¾ inch away from the edges and should be spaced no more than 6 inches on center along the edges of the panels, and 10 inches on center elsewhere.

When stapling particleboard underlayment, use the appropriate size staples for the thickness being installed. Use 1⅛-inch-long staples for ⅜-inch particleboard, 1½-inch-long staples for ⅝-inch particleboard, and 1⅝-inch-long staples for ¾-inch underlayment. Space the staples 3 inches apart along the edges of the underlayment and 6 inches apart elsewhere.

Interior Trim

Interior trim involves the installation of doorjambs, doors, casings around doors and windows, baseboards, and moldings. They are installed after the finish floors are in place. Many different woods are used in trim work, including oak, pine, redwood, birch, mahogany, cherry, maple, walnut, and others. In the past, trim members were very ornate, but the modern trend is for smooth, simple shapes that are easier to finish and maintain. Regardless of the type used, all trim must be carefully installed to insure neat, tight joints, smooth surfaces, and good appearance. Trim should be installed only after the structure has dried out sufficiently.

Door Frames.   The door frame consists of jambs, casings, and stops. The jambs cover the partition ends and support the door. They consist of two vertical members or side jambs and the top member, which is called the head or head jamb (*Fig. 8–73*). The side jambs should be dadoed near the top to receive the head jamb. Grooves or kerfs are sometimes cut on the back side to minimize warping. A slight bevel at jamb edges assures a tight joint line when the casing is applied (*Fig. 8–74*).

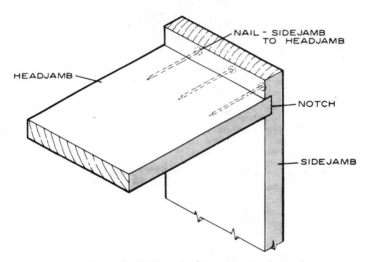

Figure 8–73 Detail of head and side jamb.

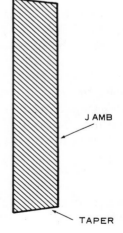

Figure 8–74 Slight taper at jamb edges improves joint between casing and jamb.

Interior doorjambs are of 1-inch nominal thickness and of the same material as the interior trim. The jambs are made 5⅛ inches wide for plaster walls and 4½ inches wide for walls with ½-inch drywall finish. Two-piece jambs which adjust to various wall thicknesses are also available; however, they are more costly than conventional jambs.

The jambs are obtainable assembled ready for installation or they may be delivered in knocked-down form. Some manufacturers offer units in which the doors are prefitted and hung in the door frame. The jambs for 6-foot 8-inch residential doors are made with a clear underhead size of 6 feet 8⅝ inches so there will be a clearance at the top and bottom of the door (*Fig. 8–75*). Some jambs are made with slightly more or less clearance.

Figure 8–75 Prehung door unit.

Frame Installation.   If the frame is received in the knocked-down form, assemble it using three 8d coated nails driven through the dado (*Fig. 8–76*). Place the frame in the rough opening, letting the side jambs rest on the finish flooring. If the finish flooring has not yet been installed, rest the jambs on spacer blocks equal to the thickness of the finish floor. Check the head jamb with a level. If it is off, trim the bottom of the jamb that is too high. Reinstall and check the head again for levelness. When it is level, cut a 1 × 6 spreader, equal in length to the distance between the side jambs measured just below the head jamb. Place the spreader at the floor line between the jambs (*Fig. 8–77*). Center the frame in the opening and secure it with doubled wedge blocks placed at each side near the top and bottom.

Figure 8–76 The side jamb must be dadoed to receive the head jamb. Fasten it with three nails.

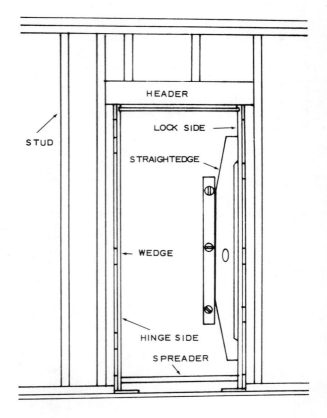

Figure 8–77 Setting the doorjamb.

On each side of the side jambs, draw a light nailing line along their length. Position this line from the door side, and locate it from the edge a distance equal to the door thickness plus one-half of the door-stop width (*Fig. 8–78*). Plumb the side jambs, using a long level or a short one in combination with a straightedge. Adjust as necessary, then drive 8d casing nails into the side jambs near the top and bottom on the nailing line. Do not drive the nails all the way.

Add three more wedge blocks to each side. On the hinge side, place two blocks at the hinge locations, normally 7 inches down from the top and 11 inches up from the bottom. Center the third block between the two. On the other jamb, place one wedge block at the lock height that is between 36 to 38 inches from the floor. Center the other two blocks in the remaining spaces.

Figure 8–78 Gauging nailing line on doorjamb.

Adjust the wedges as necessary, then complete the nailing with two 8d casing nails at each block location. Stagger the nails ⅜ inch on each side of the nailing line. The door stop will cover the nail heads.

Door Trim.   Casings are added to the door frame after the finish floor is in place. The casing covers the space between the wall and jambs and adds rigidity to the frame. Several typical casings are shown in *Fig. 8–79*.

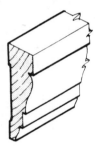
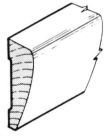

Figure 8–79 Two commonly used casings.

Since no further adjustments of the frame are possible after the casing is applied, check the frame once again before proceeding. Install the casings with the inside edge about ¼ inch back from the corner of the jambs (*Fig. 8–80*). Draw a light gauging line at the top and side jambs as a guide.

If necessary, trim the lower end of the side casings to square them. Hold the casing along the side jamb with its bottom resting squarely on the finish floor. Mark the top at the miter line, then cut, using a miter box or power miter saw (*Fig. 8–81*). Repeat for the second piece, then tack both into place. Cut and fit the head casing. When you are satisfied that the pieces fit with good miter joints, nail them permanently. Use 4d nails into the jambs and 8d nails along the outer edge into the studs. Space the nails 16 inches apart.

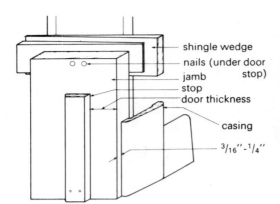

shingle wedge
nails (under door
jamb   stop)
stop
door thickness

casing

$^3/_{16}$''-$^1/_4$''

Figure 8–80 Set casing back 1/4 inch from corner of jamb.

Figure 8–81 Casing being cut on portable cutoff saw.

Doors.   There are two basic types of door—flush and panel. Flush doors consist of a plywood or hardboard face fastened to either solid or hollow cores. Panel doors are made of vertical and horizontal members called stiles and rails. Wood, glass, or other materials are enclosed within the framework. Both types are made for interior and exterior use. Exterior doors are usually 1¾ inches thick. Interior passage doors are generally made 1⅜ inches thick. A variety of interior and exterior door styles is shown in *Fig. 8–82.*

Figure 8–82 Several door styles.

Exterior flush doors are usually made with a solid core because they are more durable and can withstand the weather better. The core material may be wood or composition. The most common type of solid-core construction consists of wood blocks bonded with end joints staggered. Composition cores are usually solid. *Fig. 8–83* illustrates both types.

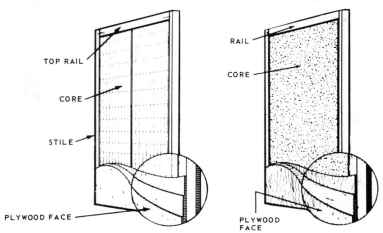

TOP RAIL

CORE

STILE

PLYWOOD FACE

RAIL

CORE

PLYWOOD FACE

Figure 8–83 Two popular flush doors. *Left:* Solid core. *Right:* Composition core.

Hollow flush doors are sometimes used as exterior doors, but they must be bonded with waterproof glue. Face panels made of three-ply ⅛-inch plywood or other materials are bonded to cores of wood or similar products. Several core patterns are shown in *Fig. 8–84.* Note that all are provided with lock blocks for lock mortising.

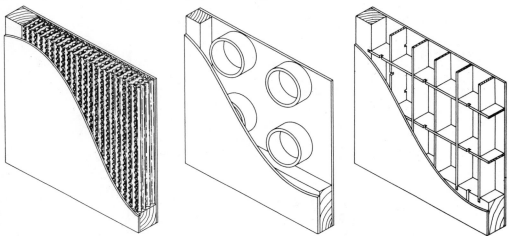

Figure 8–84 Typical hollow-door cores.

The stiles and rails of panel doors are interlocked and glued. Panel doors are usually made of solid lumber but veneered doors are also available. *Fig. 8–85* shows the principal parts of a panel door. Panels may be raised or flat. *Fig. 8–86* shows how doors are milled to receive the various panels.

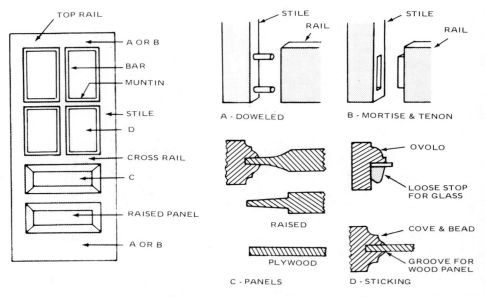

Figure 8–85 Typical construction of stile and rail doors.

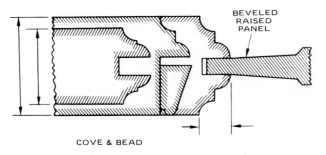

COVE & BEAD

Figure 8–86 Stiles and rails milled to accept various panels.

Door sizes vary depending upon use. Entryway doors for residential use are generally 6 feet 8 inches high and 2 feet 8 inches or 3 feet wide. The standard size for interior passage doors is 6 feet 8 inches high and 2 feet 6 inches wide. Closet doors are usually 2 feet wide.

Door Hanging. Doors may be trimmed slightly to fit an opening, but they must not be cut to smaller nominal sizes. The recommended clearance at the sides and top of a door is $1/16$ inch, as shown in *Fig. 8–87.* Cutouts for glass and louvers in flush doors must not exceed 40 percent of the face area and cutouts must be not closer than 5 inches from the door edge.

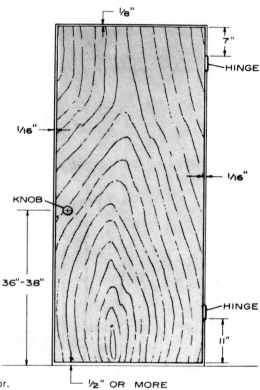

Figure 8–87 Standard clearances for typical door.

To provide swinging clearance, doors must be beveled on the lock edge (*Fig. 8–88*). This clearance is required only on doors thicker than $1\frac{3}{8}$ inches. Make a 3-degree bevel, using a hand or power plane. Hold the door firmly while planing. Special door holding jacks are available, but most carpenters build their own. A simple one you can make is shown in *Fig. 8–89.*

After trimming, place the door in the opening, resting it on a wedge. Check that the clearances are correct, then break all sharp edges with sandpaper.

Fitting Hinges. Exterior doors require three hinges; interior doors can be hung with two. The hinges used for doors are called butts or butt hinges.

Figure 8–88 Beveling lock edge of door with a power plane.

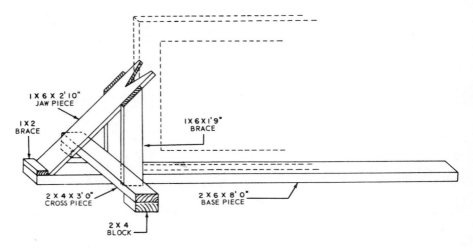

Figure 8–89 An easily made door-holding jack.

They are made with fast or loose pins. The size of the hinge used depends on the width and weight of the door and the thickness of the trim. Door hinge lengths are shown in the table in *Fig. 8–90*. A loose-pin butt is shown in *Fig. 8–91*. When installed, the barrel must clear the casing and the door should be able to swing back fully to the wall (*Fig. 8–92*).

Place the trimmed door in the opening, wedging it into place so that it has the proper clearance at the top. Lay two 4d nails about 18 inches apart at the top of the door, then force a wedge at the bottom. This will then position the door vertically with the proper clearance at the top.

Mark the position of the hinge on the edge of the jamb and on the face of the door. Holding the door in the jack, use a knife to scribe the hinge gain outline on the edge of the door. The edge of the hinge should be set back about ¼ inch from the edge of the door. Gauge the depth of the hinge along the edge of the jamb and face of the door. NOTE: The depth of the gain depends on the type of hinge used. For swaged hinges,

Figure 8–90 Hinge selection table.

FRAME THICKNESS OF DOOR	WIDTH OF DOOR	SIZE OF HINGE
1-1/8" to 1-3/8"	Up to 32"	3-1/2
1-1/8" to 1-3/8"	32" to 37"	4
1-3/8" to 1-7/8"	Up to 32"	4-1/2
1-3/8" to 1-7/8"	32" to 37"	5
1-3/8" to 1-7/8"	37" to 43"	5 extra heavy
Over 1-7/8"	Up to 43"	5 extra heavy
1-7/8"	Over 43"	6 extra heavy

Figure 8–91 A loose-pin butt hinge.

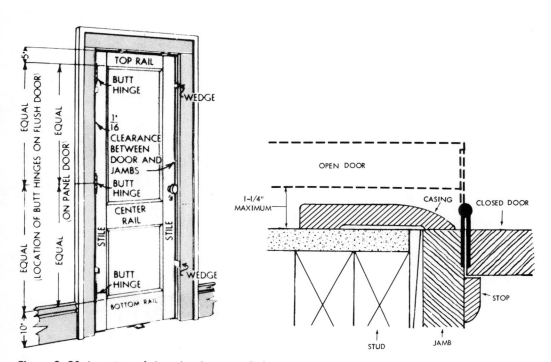

Figure 8–92 Location of door hardware and clearances.

the depth of the gain is equal to the thickness of the leaf. For straight hinges, the gain depth is a trifle less than one-half the barrel diameter (*Fig. 8–93*). A butt gauge (*Fig. 8–94*) will assure accuracy.

Butts can be mortised by hand with a chisel or with a router and hinge template. Use a sharp chisel to outline the gain, then make a series of chipping cuts spaced about ⅛ inch apart. Pare these with the flat side of the chisel to complete the mortise (*Fig. 8–95*). Repeat the procedure on the jambs.

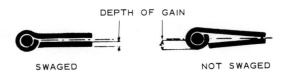

Figure 8–93 Depth of gain depends on type of hinge used.

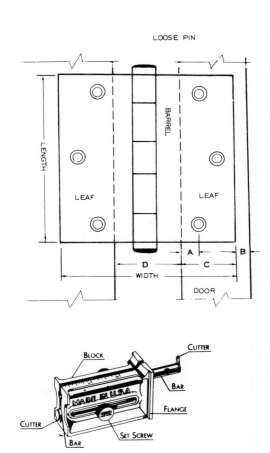

Figure 8–94 The butt gauge.

Figure 8–94 *continued*

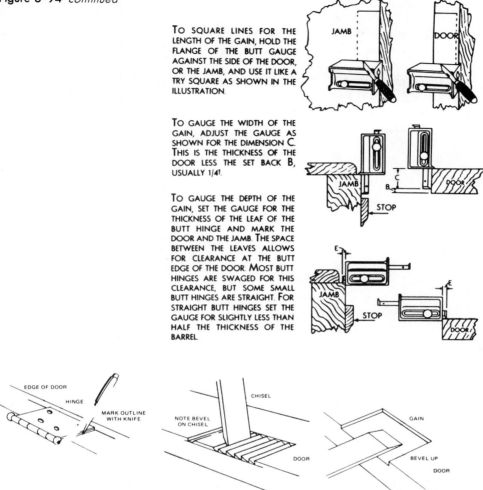

TO SQUARE LINES FOR THE LENGTH OF THE GAIN, HOLD THE FLANGE OF THE BUTT GAUGE AGAINST THE SIDE OF THE DOOR, OR THE JAMB, AND USE IT LIKE A TRY SQUARE AS SHOWN IN THE ILLUSTRATION.

TO GAUGE THE WIDTH OF THE GAIN, ADJUST THE GAUGE AS SHOWN FOR THE DIMENSION C. THIS IS THE THICKNESS OF THE DOOR LESS THE SET BACK B, USUALLY 1/4".

TO GAUGE THE DEPTH OF THE GAIN, SET THE GAUGE FOR THE THICKNESS OF THE LEAF OF THE BUTT HINGE AND MARK THE DOOR AND THE JAMB. THE SPACE BETWEEN THE LEAVES ALLOWS FOR CLEARANCE AT THE BUTT EDGE OF THE DOOR. MOST BUTT HINGES ARE SWAGED FOR THIS CLEARANCE, BUT SOME SMALL BUTT HINGES ARE STRAIGHT. FOR STRAIGHT BUTT HINGES SET THE GAUGE FOR SLIGHTLY LESS THAN HALF THE THICKNESS OF THE BARREL.

Figure 8–95 Mortising a butt with a chisel. *Left:* Laying out gain. *Center:* Cutting out gain. *Right:* Paring gain.

Install the hinges with the head of the loose pin at the top, then check the fit of the door. Be sure it swings freely with the proper clearance at the sides and top. If the door binds against the hinge jamb, place a shim under the hinge. If the hinge-jamb clearance is too great, deepen the gain as necessary.

Non–mortise-type hinges with interlocking leaves are available. These are generally used on closet doors. The clearance between the hinge jamb and door is equal to the thickness of the hinge leaf.

Install the door stops with 4d nails. Set the stop on the hinge side with a ¹⁄₁₆-inch space between it and the door. Place the lock-side stop tight against the door. Do not drive the nails home on the lock-side stop, because some adjustments may be required after the lock is installed.

Installing Locks.   The most commonly used residential locks are tubular and cylindrical locksets. They are very popular because of their ease of installation. They require the boring of two holes, which are easily located using paper templates (*Fig. 8–96*). A cylindrical lock and the required boring is shown in *Fig. 8–97*. A tubular lock is shown similarly in *Fig. 8–98*. Boring jigs with special bits insure accuracy and cut down on installation time (*Fig. 8–99*).

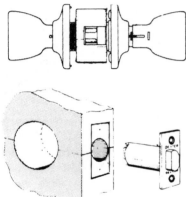

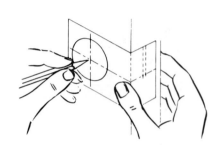

Figure 8–96 Template is used to locate center lines of holes to be bored.

Figure 8–97 Cylindrical locks are easily installed.

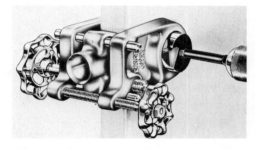

Figure 8–98 Tubular lockset requires two drilled holes and a mortised lock face.

Figure 8–99 Boring jig for door assures accurate work and saves time.

Mortise locks are also available. They provide a greater degree of security, but they are more difficult to install. Deep mortises must be cut into the edge of the door. Special jigs are available, but carpenters often install such locks by boring a series of holes which are then chiseled out to form the mortise. A mortise-type lock is shown in *Fig. 8–100*.

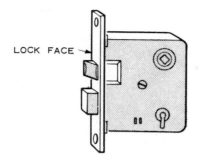

LOCK FACE →

Figure 8–100 A mortise lock.

Some locks can be reversed so they may be used on left or right doors. Others, however, cannot be reversed, and therefore it is important to understand the "hand" of a door.

The hand of any door is determined from the outside. For example, the street side is the outside of an entry door and the corridor is the outside of a room door. In doors that open from room to room, the side where the butt shows is considered the outside. An exception is closet and cupboard doors, where the room side is considered the outside.

When viewed from the outside, if the hinge is at the right side and the door swings inward, it is a right-handed door (RH). If the hinge is at the left and the door swings inward, it is a left-handed door (LH). If the hinge is at the right and the door swings outward, it is a right-hand reverse door (RHR). If the hinge is at the left and the door swings outward (*Fig. 8–101*) it is a left-hand reverse door (LHR). Be sure to order the locks according to the "hand" of the door.

Secure the door firmly in an open position by wedging it at the floor. Mark off the position of the lock on the door edge, usually 36 inches up from the floor level. Square a light line across the edge of the door. Using the template furnished with the lock, fold it according to the instructions, then align it with the mark made on the door edge. Use an awl to mark the centers of the holes to be drilled, then bore the holes and insert the latch bolt. Next, use a knife to mark the gain for the faceplate. Route the gain and install the latch unit.

Locate and mortise the strike on the jamb. Although sizes and details may vary slightly, most tubular and cylindrical locks are installed in this manner.

The procedure for installing mortise locks is illustrated in *Fig. 8–102*. Templates for these locks are also furnished by manufacturers.

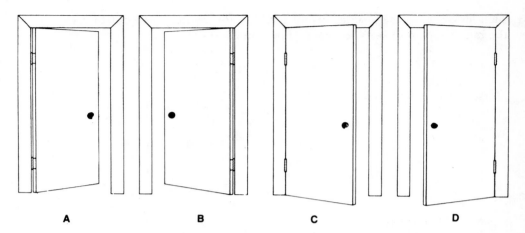

Figure 8–101 Hand of door is determined when facing outside of door: A. Left-hand door. B. Right-hand door. C. Left-hand reverse door. D. Right-hand reverse door.

Figure 8–102 Installing mortise lock: 1. Mark door frame for length of mortise, knob spindle, and keyholes. Bore them. 2. On center line, bore out mortise. 3. Chisel sides of mortise. 4. Insert lock and mark sides of gain with a knife. 5. Chisel out gain for the lock face. 6. Insert lock and fasten.

Thresholds.   Thresholds or saddles are used mainly on exterior doors. In addition to providing swinging clearance and a finished appearance for the floor material at the door, a threshold seals the space between the door bottom and the sill. When properly installed, the threshold will effectively keep rainwater from entering. The outside face of the door should over-hang the beveled portion of the threshold (*Fig. 8–103*).

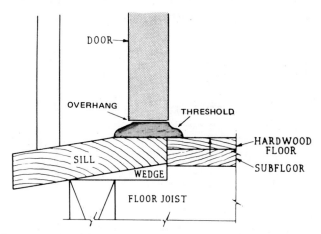

Figure 8–103 Lower edge of door should overhang threshold bevel.

The threshold should be installed after the door has been marked for cutting. Set a divider to the thickness of the threshold, then, with the door hung and in the closed position, scribe a line between the sill and the bottom of the door face. Trim the door and install the threshold. Use caulking to seal the joint, then nail or screw the threshold securely.

A number of patented weather seals are available to seal out dust and weather. One type is shown in *Fig. 8–104*.

Thresholds are sometimes used for interior doors to provide a finish when floor materials differ between adjoining rooms.

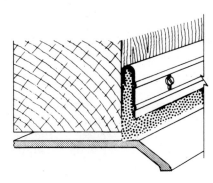

Figure 8–104 Patented weather seal has foam sealing strip.

Pocket Doors.   Sliding pocket doors are useful where swinging doors may pose a problem because of space. These doors are made to slide into an opening in the partition. They are made in various sizes and are furnished complete with partially preassembled frame, track, and hardware. The split-jamb frame must be installed during the rough framing stage, because several studs must be eliminated in the partition. A typical pocket door frame is shown in *Fig. 8–105*. Most units are furnished with adjustable hangers (*Fig. 8–106*) which permit the door to be raised or lowered without cutting.

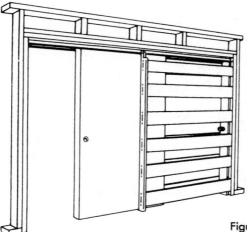

Figure 8–105 Framing for a pocket door.

Figure 8–106 Adjustable mechanism for pocket door.

Prehung Doors.   Prehung single- and double-door units are made for use in residential construction. They are available complete with the door hinged and fastened to the jamb. Doors are bored to fit most common locks. Split jambs allow adjustment for thicknesses of various types of walls (*Fig. 8–107*).

Figure 8–107 Adjustable door jamb. *Left:* Retracted. *Right:* Extended.

The prehung unit is placed directly into the rough opening. Shim and spacer blocks are installed, following the procedure used when installing a conventional door frame. A prehung unit is shown in *Fig. 8–108.*

Figure 8–108 Prefinished prehung door unit is easily installed.

Window Trim. The trim members of a double-hung window consist of the following parts: stool, apron, casings, stop, and mullion (*Fig. 8–109*). The stool must be rabbeted to fit the bevel of the sill, and it extends past

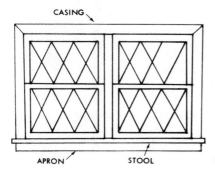

Figure 8–109 Window trim.

the side casings. The top should be level, and provides a surface for the lower end of the side casings. Two rabbeted types are shown in cross section in *Fig. 8–110*.

Figure 8–110 Two types of stool molding.

The apron forms the trim at the lower part of the window opening. It also supports the stool and it usually has the same cross section as the casings.

The casings form the trim at the sides and top of the window opening. They are of the same species and cross section as the door casings.

The stop bead forms one edge of the track in which the lower sash travels. It is fastened to the head and side jambs (*Fig. 8–111*).

Figure 8–111 Apply the stop bead to the jambs with sufficient clearance to allow the lower sash to travel freely.

The mullion is the casing used to frame the division between two windows in the same frame.

The stool must be installed first. Hold it in position, level with the sill and centered from left to right. Mark the position of the side jambs using a square, as shown in *Fig. 8–112*. Also mark a line indicating the wall surface, parallel to the wall, at each end of the stool. Usually this line will be directly above the vertical surface of the rabbet. Cut the corner carefully and check the fit by placing it in position on the sill. It should fit snugly against the wall and jambs.

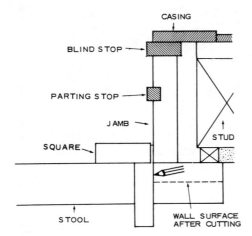

Figure 8–112 Stool layout for double-hung window.

Lower the sash onto the stool and mark a line indicating the edge of the sash. Using a plane, remove $1/16$-inch stock parallel to this line to form the proper clearance between the sash and stool. Trim the length of the stool so that the extension beyond the side casings is equal to the distance from the front of the casing to the front edge of the stool (*Fig. 8–113*). If the front edge of the stool is shaped, the ends should be cut to match. Sand the stool and nail it in place with 8d finishing nails. If required, as in masonry walls, bed it in caulking.

The length of the apron should be the same as the distance from the outside of one side casing to the outside of the casing opposite. Cut the

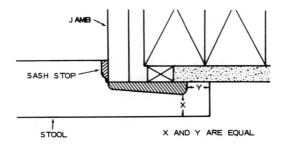

Figure 8–113 Stool position after installation.

apron to length and install. If the apron is molded, it should be coped. To mark the apron for the coped return, use a short piece of the apron stock and draw its outline on the face of the apron (*Fig. 8–114*). Repeat at the other end, then trim with a coping saw. Sand, then install with 8d finishing nails.

Figure 8–114 *Left:* Marking apron molding for return. *Right:* Completed return after sawing and sanding.

The stops should be installed next. Cut the head stop to fit between the side jambs. Raise the lower sash to its top position, then fasten the stop to the head jamb with a $1/16$-inch clearance between it and the sash. Use 4d finishing nails. Cut the side stops straight at the bottom but cope them at the top to fit against the head stop. Fasten these also, allowing $1/16$-inch clearance between the sash and stop.

The casings must be cut and installed last. Cut the bottoms square but miter the tops. To simplify the measuring, cut the top miters first, leaving the square end slightly longer than the finished size. Place the mitered end upside down on the stool, and make a mark for the square cut at the head-stop edge (*Fig. 8–115*). Next, mark the head casing for the miter cuts by holding it in place at the top of the window. Make the marks at the edge of the side stops. Saw the pieces to size and install them with 4d finishing nails at the jambs and 8d finishing nails at the outer edges.

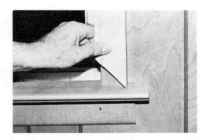

Figure 8–115 Mark length of side casing by placing mitered corner on stool, then scribe a mark at the head stop.

When the plans call for "picture-frame casing," a narrow stool without overhang is used. Four mitered casings are then installed (*Fig. 8–116*).

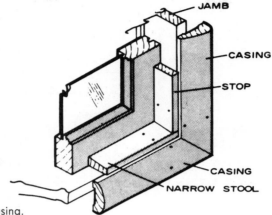

Figure 8–116 Detail of mitered casing.

Base Trim.   Baseboard molding is used to cover the joint between the wall and floor. It is available in a number of sizes and styles. Several are shown in *Fig. 8–117*. Most are made with a slight bevel at the bottom to assure a tight joint at the floor. Base shoe is used to seal the joint between the baseboard and the finish floor. Baseboard and shoe are applied as shown in *Fig. 8–118*. The shoe molding is cut and fitted at the same time as the baseboard, but it is not fastened permanently until the floors have been varnished or otherwise finished.

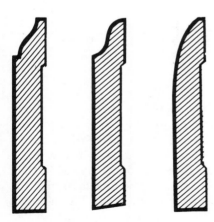

Figure 8–117 Types of base molding.

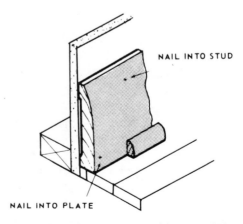

Figure 8–118 Application of base and shoe molding.

The baseboard is applied in a continuous band around the room. It is butted to door casings, as shown in *Fig. 8–119*. Inside corners are coped but outside corners are mitered (*Fig. 8–120*). The coped joint is superior to the mitered joint, because it will not open when the baseboard is nailed. Also, the joint will be less noticeable in the event of shrinkage.

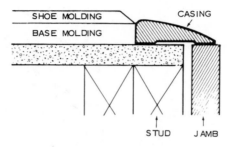

Figure 8–119 Base molding butts against door casing.

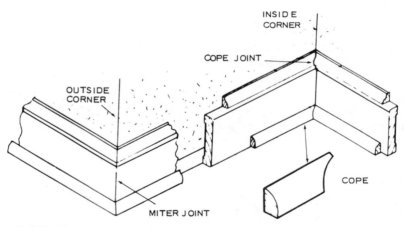

Figure 8–120 Corner treatment for base molding.

In the coped joint, one piece of molding should be shaped at the end so that its contour matches that of the molding to which it is butted (*Fig. 8–121*). To produce a coped joint for a 90-degree corner, place the molding in a miter box in the same position in which it will appear in use. Base moldings are held vertically, but crown and other moldings are held at an angle (*Fig. 8–122*). Make a 45-degree cut (*Fig. 8–123*). The profile formed by this cut is the cutting line to be used to form the cope. Place the molding on sawhorses, then hold the coping saw vertically. Cut away the waste, following the profile. Use the coping saw with the handle up or down,

Figure 8–121 A coped joint.

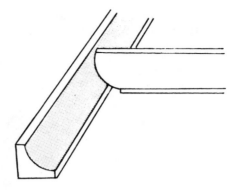

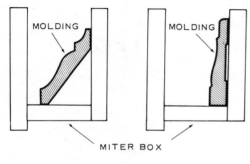

MOLDING

MOLDING

MITER BOX

Figure 8–122 Hold and cut crown moldings in the position in which they will be used.

Figure 8–123 Hold crown molding at its installation position while mitering it.

whichever is more convenient. Be sure the teeth of the blade are pointing downward (*Fig. 8–124*). Some workers like to tilt the saw blade slightly to produce a small undercut. This will produce a tighter joint at the front edge.

Starting on the longest wall, cut the molding to size and fit it carefully into place. If the fit is snug, as it should be, use a block of wood to tap it into place. Nail into place with two 6d or 8d nails at each stud location (*Fig. 8–125*). On plaster walls, a base cap is often used; it forms a better joint if the plaster is not flat (*Fig. 8–126*). Drive the lower nail into the sole plate, the upper one into the studs. Stud locations should be transferred to the floor from markings made earlier when walls were covered.

Figure 8–124 Coping saw being used to cope a crown molding.

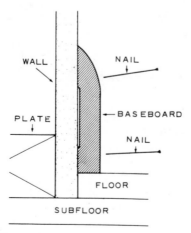

Figure 8–125 Nailing detail for one-piece base molding.

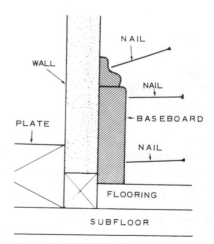

Figure 8–126 Nailing detail for two-piece base molding.

If it becomes necessary to piece moldings, cut a scarfed joint (diagonal), as in *Fig. 8–127*, and be sure to locate the joint over a stud so that both members can be nailed securely.

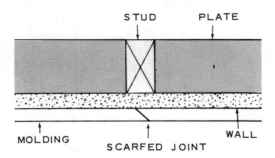

STUD   PLATE

MOLDING   WALL

SCARFED JOINT

Figure 8–127 Joint of pieced molding must occur over a bearing.

The second molding must be butted to the first piece at an inside corner. Make a coped joint as described earlier. If this piece is to go to an outside corner or baseboard, fit it snugly into the corner, then mark the opposite end for cutting. If this second piece is to fit another inside corner, take the measurement from the face of the baseboard to the opposite wall. Cut the outside corners using a miter box or power miter.

Base shoe molding is cut and fitted the same as the baseboards. Miter or round off the ends that terminate at a casing. Nail the base shoe to the baseboard, not to the floor. This will eliminate problems should there be movement of the flooring due to expansion.

Ceiling and Wall Trim. Other moldings used for interior trim include chair rails, picture molds, wainscot, cap, and cornices (*Fig. 8–128*).

Chair-rail moldings are used horizontally at chair-top height to prevent damage to the walls. Used mostly in dining rooms, they are not as popular as they once were. They are installed the same as baseboards with coped internal corners.

Picture moldings are made with a rounded edge at the top to form a recess to accommodate picture hooks. They are sometimes used as part of a ceiling cornice.

In a wainscot wall, the lower section differs from that of the wall above. Usually the lower section is made of plywood or paneling. The top of the wainscot is trimmed with a wainscot cap. Standard caps are made with ¼-inch and ¾-inch rabbets. The cap with the ¼-inch rabbet is also referred to as ply cap molding.

Crown and cove moldings are generally used to trim the joint between the ceiling and the wall. Fitting, cutting, and installation procedures

Figure 8–128 Interior trim moldings.

are the same as for baseboards. When you are coping molding, be sure to use a coping saw fitted with a fine blade (*Fig. 8–129*). A slight recess cut at the top edge of the molding will often improve the appearance of plaster ceilings (*Fig. 8–130*). Finishing nails should be driven into the top plate and, when large moldings are used, into the ceiling joists as well.

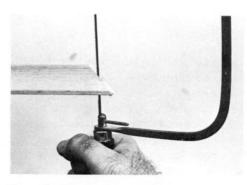

Figure 8–129 Coping a molding.

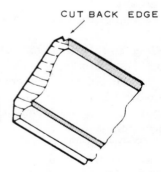

CUT BACK EDGE

Figure 8–130 Cut edge improves appearance at ceiling line.

Mantels.   A mantel is used to decorate the area around the fireplace. Some types are simply shelves or mantelpieces above the masonry; others are very ornate with pilasters and fancy facings (*Fig. 8–131*). Most mantels are factory-made units that are simply installed by the carpenter. Wood is generally used in mantel construction.

Figure 8–131 A traditional fireplace mantel.

Certain precautions are necessary when building or installing a mantel. Flammable materials must not be closer than 3½ inches to the fireplace opening. Also, wood mantelpieces must be at least 12 inches above the opening. A mantel section made with stock moldings is shown in *Fig. 8–132*. It is important that the wood members used in mantel construction be of low moisture content, otherwise joints will open as the wood dries out from the fireplace heat.

Figure 8–132 Mantel made with stock moldings.

Decorative Beams. Decorative beams are often used in residential construction for both walls and ceilings. They may be made of wood, plywood, urethane, or other non-wood materials, and may be distressed or smooth-surfaced.

Molded urethane beams with a realistic hand-hewn appearance are readily available. These are merely cut to size and fastened with nails, screws, or adhesives. Hand-hewn wood beams can easily be made on the job. Use a sharp axe and be sure to keep your fingers at a safe distance (*Fig. 8–133*). Installation details are shown in *Fig. 8–134*.

Figure 8–133 Shaping a solid beam with a hatchet.

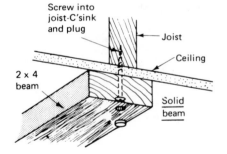

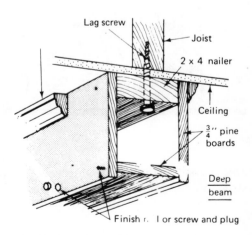

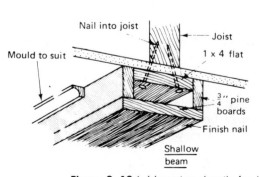

Figure 8–134 Mounting details for beams.

9 Stairs

Stairs are mainly functional, but they should also add to the appearance and beauty of the home. They must be soundly built with the safety and comfort of the user always in mind. When correctly designed, the stairs will have the proper incline, tread size, and riser height. They will be comfortable to ascend and descend and will not cause strain in the leg muscles. To insure safety and reduce accidents, stairs should be built and designed to meet certain standards. A residential stairway is shown in *Fig. 9–1*.

Figure 9–1 A properly designed residential stairway.

The length of the landings should be the same as or greater than the width of the stairway when measured between railings. Winders (pie-shaped steps) should be avoided if possible. In a flight of stairs all the treads should be of equal size. Likewise, all the risers should be of equal height. Exclusive of the nosing, the sum of one tread and one riser should be at least 17 inches and not more than 18 inches. The maximum nosing projection should be 1¾ inches. Lastly, the stairway angle of incline must be not less than 20 degrees and not more than 50 degrees (*Fig. 9–2*).

Figure 9–2 The ideal incline angle for stairs is between 30 and 35 degrees.

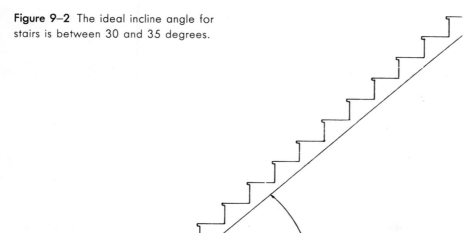

IDEAL ANGLE
30°–35°

Stair Design and Construction

Stair design and building is rather complex and time-consuming. It is customary, therefore, to have main stairways mill-made and delivered to the jobsite ready for installation by the carpenter. Most service stairs, however, are built on the job by carpenters. The principal difference between the two types is in construction details. Most main stairways are made with shaped and routed sections, requiring special tools and equipment. Job-built stairs can be made with ordinary carpenter's tools.

Stairs can be classified as being closed-string, as in *Fig. 9–3*, or open-string, as in *Fig. 9–4*. Stairways without walls on either side or with just one wall are considered to be open. A half-open stairway is a combination of both types.

Figure 9–3 Closed-string stairway.

Figure 9–4 Open-string stairway.

Straight Stairs.   The simplest type of stair is the single-flight straight run which leads from one floor to the next (*Fig. 9–5*). Generally, these stairs have thirteen or fourteen steps, without landings. They often require a long hallway, which could cause problems, especially in a small house.

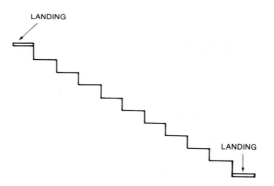

LANDING

LANDING

Figure 9–5 Simple straight stairs.

Stair Types and Landings.   When the flight is longer than fifteen steps, it should be interrupted with a landing at some point in the run (*Fig. 9–6*). This gives the user a chance to pause and is therefore less tiring. Also, landings eliminate the frightening effect that long, uninterrupted stairways can have on youngsters and older persons.

Landings are also used for changing the direction of stairs. An L-shaped stair has two flights placed at right angles to each other with a landing in between (*Fig. 9–7*).

Long L stairs are those in which the landing is close to the upper or lower floor with just a few steps between.

Double L stairs have two turns, one near the top and one near the bottom. Those with the landing near the center of the flight are called wide L stairs.

Figure 9–6 Long flights of stairs should be interrupted with a landing.

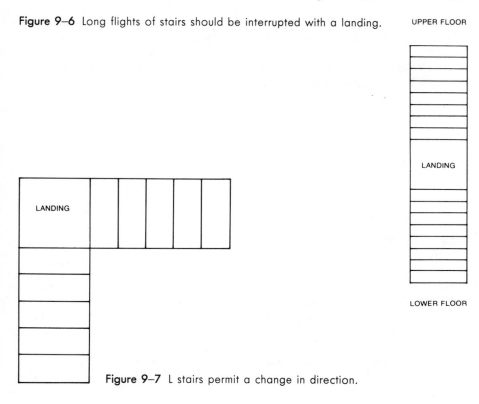

UPPER FLOOR

LANDING

LOWER FLOOR

Figure 9–7 L stairs permit a change in direction.

When space is limited in both length and width, a narrow U stair is often used. It is laid out so that the lower part will go up to a landing at some intermediate point and then double back on itself (*Fig. 9–8*).

The wide U is used in place of the narrow U when space permits. It has two landings with a short flight of steps between the two (*Fig. 9–9*).

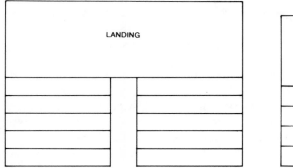

Figure 9–8 Narrow U stairs.

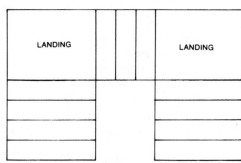

Figure 9–9 Wide U stairs with two landings.

Winders utilize pie-shaped treads to bring about a change in direction (*Fig. 9–10*). They are usually used where space is limited. Falls are a particularly great hazard on winders, and their use should be avoided if possible.

Circular and elliptical stairs are referred to as geometrical stairs (sometimes they are called Hollywood stairs). They are difficult to build and very costly.

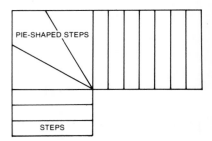

Figure 9–10 Winders permit change in direction when space is limited.

Stair Parts. Basically, stairs consist of risers and trends supported by stringers (*Fig. 9–11*). The following terms apply to stair construction:

Riser: The vertical member of the step. If a step lacks a riser, it is described as an open riser.

Tread: The horizontal member of a step.

Nosing: The projection of the tread beyond the face of the riser.

Unit Rise: The vertical distance from the top of one tread to the top of the next.

Total Rise: The sum of all the risers.

Unit Run: The horizontal distance from riser to riser.

Total Run: The horizontal distance covered by a flight of stairs including the intermediate landings.

Headroom: The clear space between the front edge of the tread to the ceiling above a stairway.

Stringer (or String): An inclined member, parallel to the slope of the stair, to which the treads and risers are joined.

Housed Stringer: A stringer routed to receive the ends of both treads and risers.

Open Stringer: A stringer with the top edge notched to follow the riser and tread.

Closed Stringer: A stringer with parallel edges. The top edge extends above the ends of the stairs.

Carriage: Rough timbers cut to fit below and to support the treads and risers. Also called rough stringers.

Winder: A step with a tread narrower at one end than the other.

Stair Widths. Main stairways should be wide enough to permit two people to pass comfortably. The accepted minimum width for main stairways

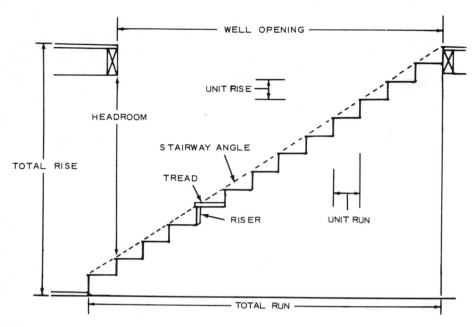

Figure 9–11 Illustration of stair layout terms.

is 3 feet, but 3 feet 6 inches is preferred. The minimum width for service stairs is 2 feet 6 inches.

Consideration must be given to the passage of furniture when stairs are planned. Stairs open on one side offer the best opportunity for moving large pieces of furniture. Those with closed string or narrow U design cause the most difficulty in this respect.

Rail Heights.   The handrail height found to be the most comfortable is 34 inches on the rake and 30 inches on the landing (*Fig. 9–12*). The handrail should be continuous from floor to floor whenever possible and it should be smooth, without sharp edges and free of splinters.

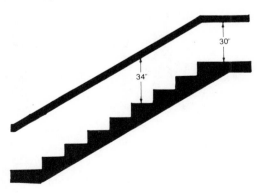

Figure 9–12 Preferred handrail heights.

Headroom. Stairway headroom should be sufficient to permit a person to ascend or descend with ample clearance between the user's head and the soffit or ceiling above. Studies reveal that the headroom should be a minimum of 6 feet 6 inches (*Fig. 9–13*). However, a space of 6 feet 8 inches to 7 feet 7 inches is ideal for the average person.

Figure 9–13 Minimum headroom requirements for main stairs: 6'8''; for basement stairs: 6'6''.

Rise and Run. Stairs that are too steep or too shallow are dangerous and uncomfortable. Those with too short a riser and too wide a tread will cause the user to lean backward while ascending. Too steep an incline will cause the user to lean too far forward. These tendencies are reversed when the user descends such stairs (*Fig. 9–14*). At A the treads and risers are of the improper size; at B the angle is too steep. Properly designed stairs are shown at C.

The riser-to-tread ratio has been established for use in stair design. When stairs conform to the rules set forth below, the user will be able to

Figure 9–14 Properly designed stairs will be easy to ascend and descend.

ascend or descend safely and without undue strain. The rules are as follows:

- *Rule 1*. The sum of two risers and one tread should be 24 to 25 inches.
- *Rule 2*. The sum of one riser and one tread should be from 17 inches to 18 inches.
- *Rule 3*. The tread width multiplied by the unit rise should be from 72 to 75 inches.

For example, applying the rules above, a step with an 8-inch unit rise and a 9-inch tread will meet the requirements of all three rules.

In *Rule 1*, 8 plus 8 plus 9 equals 25.
In *Rule 2*, 8 plus 9 equals 17.
In *Rule 3*, 9 times 8 equals 72.

All three results are within the guidelines of the rules stated above.

Main stair risers are normally kept between 7 inches and 7⅝ inches. An exception is made for basement stairs, which usually have lower ceiling heights. Here, 8-inch risers are permitted. As the riser height increases, the tread width decreases and vice versa. This is clearly illustrated in *Fig. 9–15*, where three different tread/riser dimensions are used, all complying with Rule 1. The desired slope for stairs falls between 30 and 35 degrees above the horizontal.

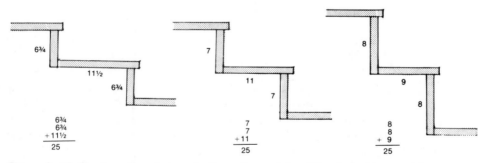

Figure 9–15 The treads and risers in these three stairs differ, but all are satisfactory because they meet requirements of Rule 1.

As mentioned earlier, it is most important that all the risers in one flight of stairs be the same dimension and all the treads be of equal width in that same flight. Although not conscious of it, the user ascends and descends the steps rhythmically. If the steps are "off," tripping could result.

Stairway Framing. The openings for stairways are framed during the floor-framing stage. The long dimension of the opening may run either parallel to or at right angles to the joists. Construction is greatly simplified when the two are parallel (*Fig. 9–16*). If the span is greater than 4 feet, headers and trimmers must be doubled. Headers 6 feet long or longer must be secured with framing anchors unless they are supported by a beam, post, or partition. Their length should be limited to 10 feet unless supported by a wall below.

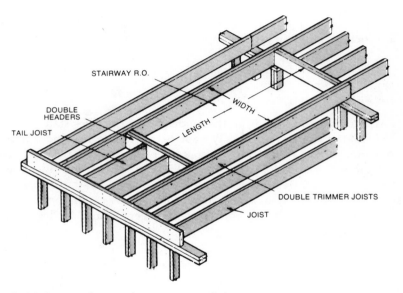

Figure 9–16 Framing for rough opening parallel to joists.

The rough opening for basement stairs is usually 9 feet long by 2 feet 8 inches wide. The minimum rough opening for main stairs is 10 feet long by 3 feet wide. A rough opening detail is shown in *Fig. 9–17*. Here the rough stringers have been installed for two flights, one above the other. *Fig. 9–18* shows the rough framing for a long L stairway. Note how one stringer and the landing are supported by a corner post.

Stringers. Since stringers (also called carriages) must support the load on the stairs, it is important that they be well made. They must be carefully laid out and cut to accept the risers and treads. The lumber used should be 2 × 10 or 2 × 12 for plain stringers. Housed stringers are made from $1^1/_{16}$-inch or thicker material. A $3^1/_2$-inch minimum of stock must remain on the underside of the stringer after the notches are cut. Also, a 4-inch bearing is required at the top where the stringer bears against the header (*Fig. 9–19*).

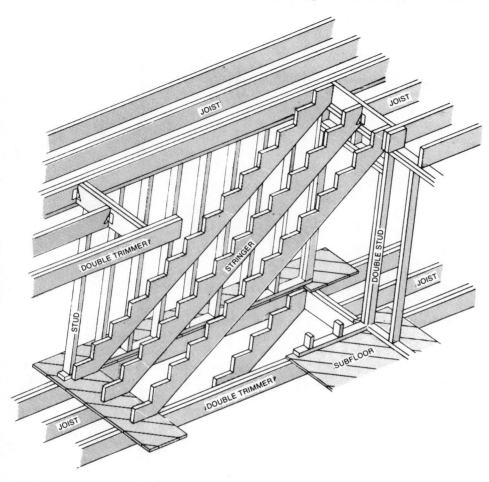

Figure 9–17 Rough framing detail for straight flight, one over one.

When treads are 1¹/₁₆ inches thick and wider than 2 feet 6 inches, three stringers are required. Three stringers are also needed when treads are 1⅝ inches thick and wider than 3 feet.

Calculations for Riser Height and Tread Width. The tread and riser sizes, the angle of slope, and the rough opening dimensions are interrelated. Usually the plans will specify the stairwell opening sizes. Some will also include the number of treads and risers.

When the opening size is known, the carpenter can calculate the tread and riser sizes using the stair ratio principle. If the opening size is not given, it can be determined after the tread and riser sizes have been calculated.

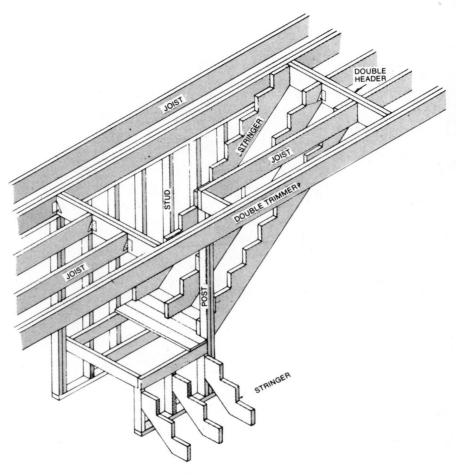

Figure 9–18 Rough framing detail for stair with landing.

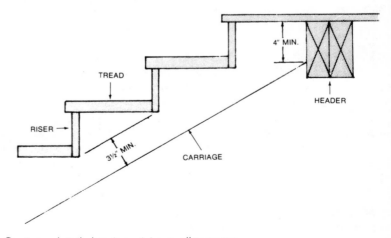

Figure 9–19 Carriage detail showing minimum allowances.

A scale drawing on graph paper can be helpful in calculating and checking the stair layout as well as the length of the stairwell opening. The drawing should show the floor-to-floor height (total rise), two diagonals representing the slope limits, and the steps, determined by using the stair ratio (*Fig. 9–20*). Select the best tread/riser combination as follows:

Find the number and size of the risers by dividing the total rise in inches by 7. For example, assume that the total rise (finish floor to finish floor) is 8 feet 5 inches or 101 inches. Dividing 101 by 7 equals 14.428. Since you must have a whole number of risers, you have a choice of using fourteen or fifteen. If you choose fourteen risers, then divide 101 by 14 and you will get 7.214 (the unit riser), which is approximately $7^3/_{16}$ inches. Referring to Rule 2: the sum of one tread and one riser must equal 17 or 18; if 17 is used, then 17 minus $7^3/_{16}$ inches equals $9^{13}/_{16}$ inches (the tread less nosing). If 18 is used for calculations, then 18 minus $7^3/_{16}$ inches equals $10^{13}/_{16}$ inches. Take an average between the two results and make it $10^5/_{16}$ inches.

Thus the riser height is $7^3/_{16}$ inches and the tread width (less nosing) is $10^5/_{16}$ inches. Both fall within the range of desired angles, as shown in the table in *Fig. 9–21*.

The number of treads in a stair is always one less than the number of risers (*Fig. 9–22*). Therefore, since the number of risers in the example is fourteen, the number of treads will be thirteen. To find the total run, multiply the number of treads by the tread width: $10^5/_{16} \times 13 = 134.06$ inches or 11 feet 2 inches (fractions are dropped). The total run is 11 feet 2 inches.

Using the figures obtained above, lay out the profile of the stairs on the scale drawing. From the layout you will then be able to determine the stairwell opening size. You can also check the slope angle and headroom clearance before laying out and cutting the stringers.

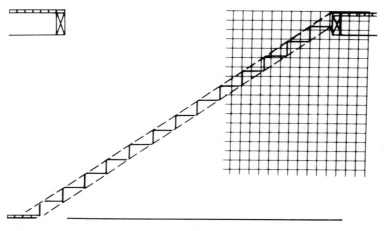

Figure 9–20 A scale drawing on graph paper will help in laying out the stairs.

Angles, Risers and Treads for Stairs

ANGLE WITH HORIZONTAL		RISER IN INCHES	TREAD IN INCHES	
Degrees	Minutes			
22	00	5	12½	
23	14	5¼	12¼	
24	38	5½	12	
26	00	5¾	11¾	
27	33	6	11½	
29	03	6¼	11¼	
30	35	6½	11	
32	08	6¾	10¾	Preferred
33	41	7	10½	
35	16	7¼	10¼	
36	52	7½	10	
38	29	7¾	9¾	
40	08	8	9½	
41	44	8¼	9¼	
43	22	8½	9	
45	00	8¾	8¾	
46	38	9	8½	
48	16	9¼	8¼	
49	54	9½	8	

Figure 9–21 Table of preferred angle for stairs where tread plus riser equals 17½ inches.

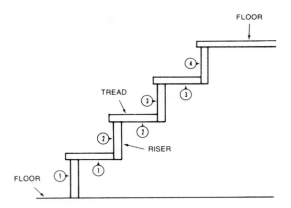

Figure 9–22 The number of treads in a stairway is always one less than the number of risers.

The stringer length can also be determined from the layout. You can double-check your calculations by using a story pole. Carefully place the pole vertically on the finished floor below and mark the top of the finished floor above. Set a divider to the calculated riser height and step off the number of risers, fourteen in our example (*Fig. 9–23*). If the last step-off does not fall on the mark, adjust and step off again, repeating the procedure until the fourteen divisions are equal. This gives the actual riser height. Use this measurement on the tongue of the square and the tread dimension on the blade to lay out the stringer. Use stair gauges or blocks clamped to the square to insure accuracy during the layout (*Fig. 9–24*).

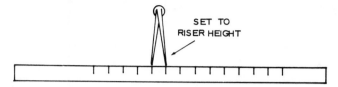

Figure 9–23 Laying out the riser height.

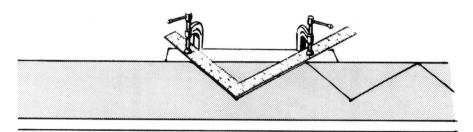

Figure 9–24 Layout of stringers with the aid of a framing square.

Select a sound 2 × 12 about 2 feet longer than the stringer length. Lay it on a pair of sawhorses and check it to make sure it is free of cracks or other defects that may impair its strength.

Place the square on the plank near one end and draw a line along the body and tongue (*Fig. 9–25*). Remove the square and extend the body or tread line to the back edge of the plank. This represents the floor cut (*Fig. 9–26*). Replace the square, then mark and move it along the edge until the required number of steps have been drawn. When the last riser is drawn, extend the line to the back edge of the plank (*Fig. 9–27*). This part of the stringer will bear against the header. For added strength, the stringer may be notched to fit around the header joist.

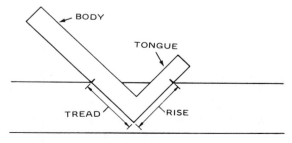

Figure 9–25 Layout of floor cut. Mark stock along the body of square.

Figure 9–26 Extension of tread line forms the floor cut line.

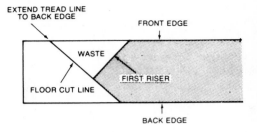

Figure 9–27 Stepped-off stringer.

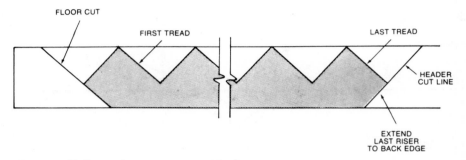

The stringer must be "dropped" at the bottom to allow for the tread thickness, which was not included in the layout. Draw a line equal to the tread thickness and parallel to the floor line, as shown in *Fig. 9–28.*

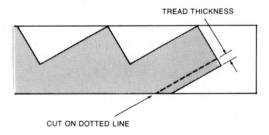

Figure 9–28 Tread allowance at bottom of stringer.

The stringers must be cut with great care. If a portable saw is used, bear in mind that the major part of the saw base is to one side of the blade. Arrange the cutting so that all the outboard cuts are made first. This will permit maximum support for the saw. During the second or notching cut the base is well supported by the remaining stringer stock (*Fig. 9–29*). Use a handsaw to complete the cut and drop out the waste.

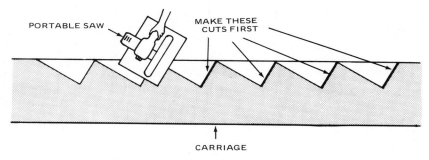

PORTABLE SAW

MAKE THESE
CUTS FIRST

CARRIAGE

Figure 9–29 Cutting sequence is important when notching stringers.

Tread and Riser Details. The treads for main stairways are usually $1\frac{1}{16}$ inches or $1\frac{1}{8}$ inches thick. Both hard- and softwoods may be used. Softwoods must have a vertical grain or, if flat-grained, they must be covered with a suitable floor finish material.

Risers are usually made of ¾-inch stock of the same species as the treads. If they are to be covered, however, less expensive lumber may be used. Plywood can be used when the stairs are to be carpeted. It is ideally suited because of its dimensional stability, flatness and favorable cost.

The front edge of the tread is the nosing. It projects beyond the front of the riser on which it rests (*Fig. 9–30*).

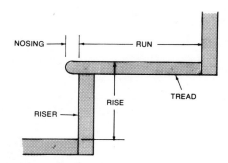

NOSING

RUN

TREAD

RISE

RISER

Figure 9–30 Nosing detail.

Fasten treads and risers with finishing nails at the edges. Drive nails or screws into the treads at the rear through the lower part of the riser. Use glue blocks at the intersection of the rear surface of the tread and riser (*Fig. 9–31*).

Stringers. Stairs with open risers are often used for basement stairways. They can be dadoed or cleated (*Fig. 9–32*). The dadoed stair takes more time to build than the cleated type, but it has a more pleasing appearance. The depth of the dado should be one-third the stringer thickness. Cut the

Figure 9–31 Preferred method of fastening treads and risers.

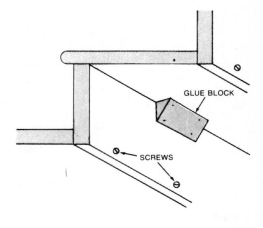

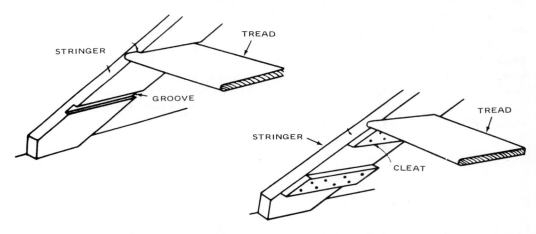

Figure 9–32 Treads on open riser stairs may be installed in grooved stringers as at left. In a simpler version, the tread rests on cleats, as at right.

dadoes so the tread line on the stringer becomes the top edge of the dado cuts. Fasten with nails driven through the stringer into the ends of the treads. Use glue for an improved joint.

For a cleated stringer, draw a line below and parallel to the tread mark on the stringer. Make the space between the two lines equal to the tread thickness. The cleat must be long enough to fully support the tread. Bevel all exposed edges and fasten the cleats securely to the stringers. Drive nails through the stringers into the ends of the treads.

If the side walls or floor are masonry, some means must be used to secure the stringers. Use a suitable masonry anchor. In some construction, wood blocks are embedded in the masonry for this purpose.

Finish Stringers.   Either full stringers or notched stringers can be used at the sides of the stairway. When full stringers are used, they should be fastened directly to the side walls. Then the rough stringer is fastened to the finish stringer and the treads and risers cut to fit between the finish stringers (*Fig. 9–33*).

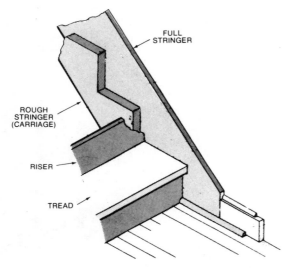

Figure 9–33 Enclosed stairs with full stringer. The rough carriage is applied against the stringer.

A notched stringer is shown in *Fig. 9–34*. Here the rough stringer is fastened directly to the wall. The finish stringer is then notched to match the steps of the rough stringer. Treads and risers are then cut and installed.

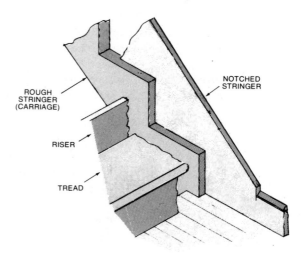

Figure 9–34 Enclosed stairs with notched stringer. The treads and risers are fitted between the stringers.

A low-cost stairway with metal stringers is shown in *Fig. 9–35*. The rough stringers are securely fastened to the wall with masonry fasteners. The metal stringers are then applied and secured with screws. The treads are held by driving nails into the holes provided.

Figure 9–35 A low-cost basement stairway.

Housed Stringers. Although housed stringers are generally mill-made, they can be made on the job by the carpenter. Special stair templates are used to guide the router for making the grooves to receive the treads, risers, and wedges. When one of the walls of the stairway is open, a mitered stringer is used. *Fig. 9–36* shows details of both the mitered and housed stringers.

Winders. As explained earlier, winders are dangerous because of the tapered treads. Unfortunately, occasionally they are required because of space limitations. Some degree of safety can be introduced by bringing the center of convergence to a point out from the corner, as in *Fig. 9–37*. The converging point should be located in a manner that will provide a tread width equal to that of the regular treads along the line of travel. This imaginary line is about 18 inches from the inside stair edge.

The winder can be calculated mathematically, but most carpenters prefer to make a full-size plan view layout showing positions, lengths, and angle cuts of the various components. Riser heights must be the same as those of the straight stairs.

Spiral Stairs. The spiral stairway is a form of winder, but although the treads do taper, they are all uniform and the tendency to stumble is less likely than on a winder. Spiral stairs are used mostly for decorative effect.

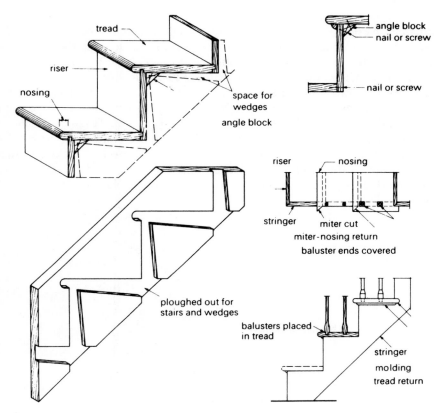

Figure 9–36 Details of housed and mitered stringers.

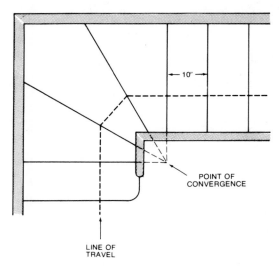

Figure 9–37 Winder details. Point of convergence should be away from corner of stairs.

Balusters, Newels, and Rails.   The open section of stairs must be enclosed with a handrail which is inclined and parallel to the slope of the stair. The rail is supported by balusters and ends at the newel post, to which it is fastened. The balusters may be square or turned spindles that are fastened between the handrail and tread, or between the handrail and a lower rail.

The complete assembly, which includes the newels, balusters, and rails, is called a balustrade and is illustrated in *Fig. 9–38*. These components are mill-made and delivered to the jobsite where they are installed by the carpenter. The rails are fastened to the walls with special brackets which are screwed to the wall frame. Nails should not be used, because they might work loose. Spacing for the brackets should be no greater than 10 feet.

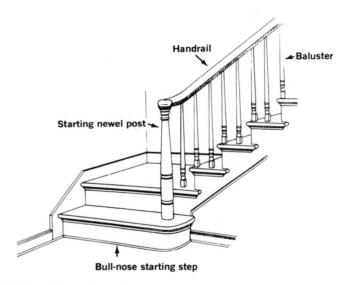

Figure 9–38 Parts of a balustrade.

10 Cabinetry

Not too long ago, most kitchen cabinets were custom-made. However, during the housing boom following World War II the demand for cabinets was too great for the small custom shops. They just could not keep up with the demand and were replaced by large manufacturing plants which were able to mass-produce thousands of units per day. The advantages of factory-made cabinets (*Fig. 10–1*) are many. Lower cost, better quality, superior finishes, and fast delivery are common. Most cabinets are made of hardwood or wood products such as hardboard, particleboard and plywood, but metal is also used.

Figure 10–1 Low-cost factory-built cabinets.

Cabinet Design and Construction

There are three basic kinds of cabinets: base, wall, and miscellaneous. The last group includes a wide variety of storage facilities such as broom, oven, overhead cabinets, and many others.

Base cabinets are usually 24 inches deep and 36 inches high, with a toe space at the bottom. Wall cabinets are usually 12 inches deep and from 12 to 33 inches high (*Fig. 10–2*). The highest shelf should be from 68 to 72 inches high. Anything higher would be too difficult to reach.

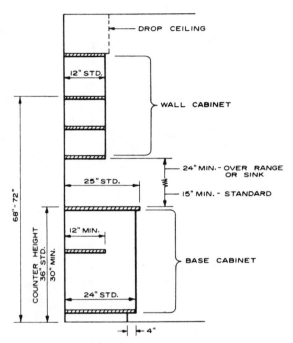

Figure 10–2 Kitchen cabinet specifications.

Kitchen cabinets are delivered to the jobsite ready for installation by the carpenter. Most are made with faces that can be trimmed at the edges to match irregular wall surfaces. The following installation procedure is typical:

Starting at a corner, sweep the floor clean and remove any moldings that may interfere with the installation. Set the cabinet in place and if necessary shim with wedges at the wall and floor until it is level and plumb (*Fig. 10–3*). Scribe the edges that butt against the wall, using a divider, as shown in *Fig. 10–4*. Move the cabinet away from the wall and trim to the scribed lines. Use a plane and undercut the edge slightly to insure a good tight fit on the exposed edge. Replace against the wall and check the fit carefully. Be sure the cabinet is leveled and resting on the shims. Make the necessary adjustments, then fasten it permanently, driving screws through the cabinet backing into the wall framing members. Wall units are installed in much the same manner.

Figure 10–3 Use wedges to level base cabinet.

Figure 10–4 Scribing base cabinet to wall.

When two or more cabinets are butted, be sure to align the front frames so they are flush along the front and bottom edges. Use shims between the cabinet sides at the rear (*Fig. 10–5*). Drill holes through the stiles of adjoining cabinets and secure them with flathead screws set with the heads a trifle below the surface (*Fig. 10–6*).

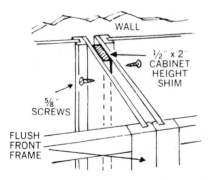

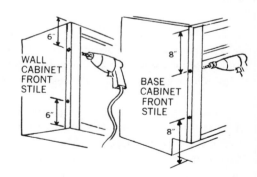

Figure 10–5 Install shims between cabinet sides at the rear.

Figure 10–6 Drilling installation holes in stiles.

Various filler strips are available for use with cabinets used at a corner. A straight filler is used when cabinets are set up with a blind corner, as in *Fig. 10–7*. For open-corner installation, as in *Fig. 10–8*, the corner filler is used. Ceiling-hung units that are installed over a peninsular base should be fastened with lag screws driven into the ceiling joists or to solid framing if the ceiling has been dropped.

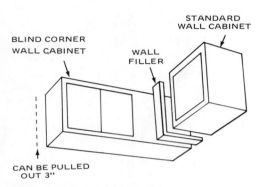

Figure 10–7 Blind-corner filler.

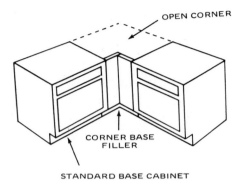

Figure 10–8 Base cabinet with open-corner filler.

Built-Ins.   There are occasions when the carpenter may be called upon to construct built-ins. These include such items as closets, shelves, vanities, room dividers, and many others (*Fig. 10–9*). When factory-built, the units are delivered ready for installation with minor adjustments required for fitting as described previously for kitchen cabinets. When constructed in place on the job, built-ins are usually attached to the structure.

Architects' plans usually give overall dimensions and the appearance of the built-in unit. It is the responsibility of the carpenter to work out the details along with a materials list. Since built-ins are custom-made, sizes and specifications will vary for each job. The space for the built-in should be checked for size to make sure it agrees with the architect's plans and that the unit being made will fit the space. Wall studs and joists should be located and marked to indicate where cabinet members may be fastened. Bear in mind that assembled units that are ceiling height cannot be tipped into place. Walls and corners may be off-square or out of plumb. Check these areas carefully and note all discrepancies. Allowances will have to be made in the cabinet members affected.

A typical storage cabinet is shown in *Fig. 10–10*, together with materials list and cutting diagram. The construction procedure is as follows: Lay out and cut the parts to size, then rabbet and dado the top, bottom, and side members, as shown in the detail. Next, assemble the parts, using glue and nails at all joints. Add the rear panel. Cut and assemble the front frame, using ¼ × ½-inch splines at all joints. Fasten the front frame with

Figure 10–9 Typical built-in cabinet.

glue and finishing nails. Cut and fit the doors, then install, using two hinges on the short doors and three hinges on the third one.

Details for a four-compartment base cabinet are shown in *Fig. 10–11*.

Drawers.   Drawers are of two basic types, flush and overlapping. A simple drawer that can be made by the finish carpenter is shown in *Fig. 10–12*. It has five basic parts: two sides, front, rear, and bottom. The front panels are usually made of ¾-inch stock. Sides and backs are normally made of ½-inch material. Bottoms of ¼-inch plywood or hardboard are common. Various drawer joints are shown in *Fig. 10–13*. Other more complex joints, including dovetails and milled joints, require special tools and are generally made in a cabinet shop.

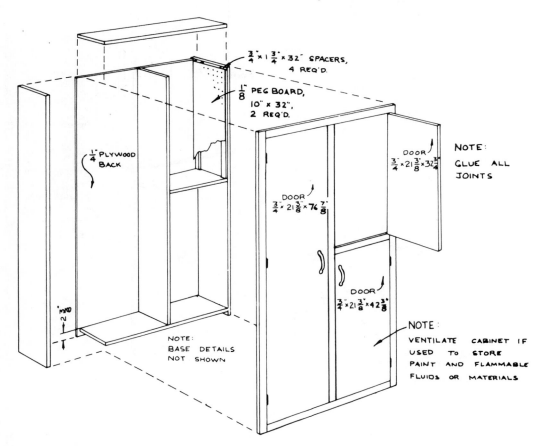

$\frac{3}{4}" \times 1\frac{3}{4}" \times 32"$ SPACERS, 4 REQ'D.

$\frac{1}{8}"$ PEG BOARD, 10" × 32", 2 REQ'D.

$\frac{1}{4}"$ PLYWOOD BACK

DOOR $\frac{3}{4}" \times 21\frac{3}{8}" \times 32\frac{3}{4}"$

DOOR $\frac{3}{4}" \times 21\frac{3}{8}" \times 76\frac{7}{8}"$

NOTE: GLUE ALL JOINTS

DOOR $\frac{3}{4}" \times 21\frac{3}{8}" \times 42\frac{3}{8}"$

$2\frac{1}{2}"$ TYP.

NOTE: BASE DETAILS NOT SHOWN

NOTE: VENTILATE CABINET IF USED TO STORE PAINT AND FLAMMABLE FLUIDS OR MATERIALS

Figure 10–10 Plans for storage cabinet.

Drawer Guides. Drawer guides are the tracks in which drawers slide. Simple guides made of wood are shown in *Fig. 10–14*. Three basic types are widely used. These are the center guide, corner guide, and side guide. In the center guide a grooved strip of wood, fastened to the bottom of the drawer, rides over a runner that is attached to the framing. The corner guide consists of an L-shaped strip which supports both the bottom and sides of the drawer. The side guide consists of grooves cut in the side panels of the drawer. These ride on a runner that is fastened to the framing. The grooved section may be cut in the side of the case, instead, as in *Fig. 10–15*, with the runner attached to the drawer side.

The side-guide drawer does not require a kicker, but the first two do.

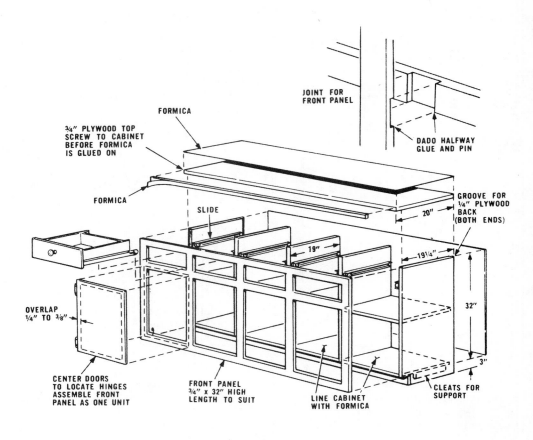

JOINT FOR
FRONT PANEL

DADO HALFWAY
GLUE AND PIN

¾" PLYWOOD TOP
SCREW TO CABINET
BEFORE FORMICA
IS GLUED ON

FORMICA

FORMICA

SLIDE

GROOVE FOR
¼" PLYWOOD
BACK
(BOTH ENDS)

20"

19"

19¼"

32"

OVERLAP
¼" TO ⅜"

3"

CENTER DOORS
TO LOCATE HINGES
ASSEMBLE FRONT
PANEL AS ONE UNIT

FRONT PANEL
¾" x 32" HIGH
LENGTH TO SUIT

LINE CABINET
WITH FORMICA

CLEATS FOR
SUPPORT

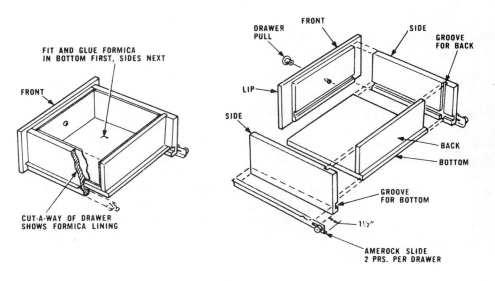

FIT AND GLUE FORMICA
IN BOTTOM FIRST, SIDES NEXT

FRONT

DRAWER
PULL

FRONT

SIDE

GROOVE
FOR BACK

LIP

SIDE

BACK

BOTTOM

CUT-A-WAY OF DRAWER
SHOWS FORMICA LINING

GROOVE
FOR BOTTOM

1½"

AMEROCK SLIDE
2 PRS. PER DRAWER

424   Figure 10–11 Construction details of a base cabinet.

The kicker is a piece of wood mounted above the drawer to prevent it from tipping as it is extended.

Commercial drawer slides are available in wood, metal, and plastic or in a combination of various materials. Most commercial guides do not require kickers.

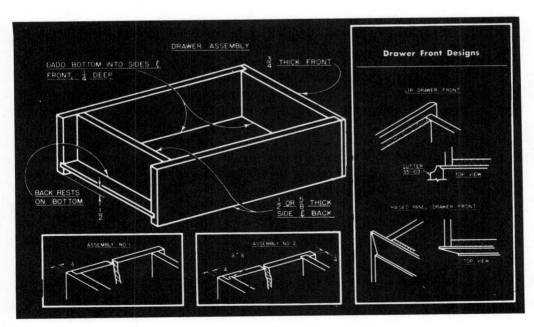

Figure 10–12 Details of a simple drawer.

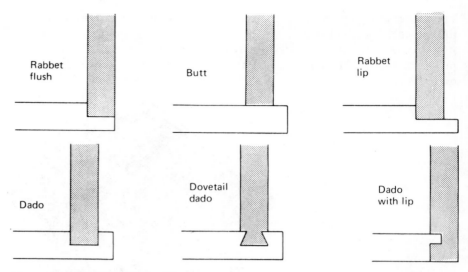

Figure 10–13 Various drawer joints.

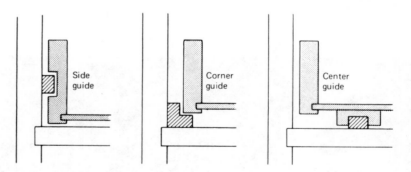

Figure 10–14 Wood-type drawer guide.

Cabinet Doors. There are two basic types of cabinet doors: sliding and hinged. Sliding doors may be made of wood, hardboard, plastic, or glass. They ride in tracks that are mounted at the top and bottom of the opening. Tracks may be of wood, plastic, or metal. Tracks may also be cut in the framing member, as shown in *Fig. 10–16*.

Another type of track is shown in *Fig. 10–17*. Grooves cut in the top and bottom edges of the doors ride on the plastic track. Many other types are available. To insure smooth operation, plastic buttons are often used to reduce friction between the door and track. These buttons are inserted into the lower edge of the doors, one near each end.

Figure 10–15 Drawer rides in groove cut in side.

Figure 10–16 Sliding door tracks cut in framing.

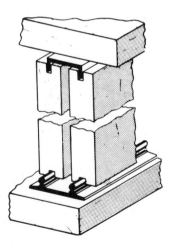

Figure 10–17 Sliding doors riding on plastic track.

The tracks, whether commercially made or constructed on the job, must be made with the upper grooves deeper than the lower ones. To be installed, the doors must be pushed upward into the upper groove, then dropped into the lower grooves or tracks (*Fig. 10–18*). A simple sliding-door

Figure 10–18 Sliding doors are installed in upper track first.

setup can be made with three pieces of wood used for each track (*Fig. 10–19*). Make the lower strips ¼ inch high and those for the top ½ inch high. The thickness of the strips depends on the size of the doors. The center-strip thickness determines the amount of bypass clearance. Finger cups in various sizes and shapes are made for sliding doors. They should be as deep as the doors' stock thickness.

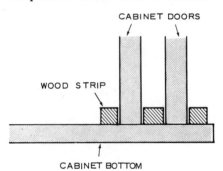

Figure 10–19 A simple sliding-door track.

Hinged Doors. Cabinet doors are usually purchased ready-made; however, the simpler types may be constructed on the job by the carpenter. Either lumber-core material or plywood is generally used for this purpose. Solid lumber requires doweled or shaped glued joints which are not practical to make outside the shop. The doors used for cabinets may be made with lipped edges or square edges. The lipped door is shown in *Fig. 10–20*. The lipped door has a rabbeted edge which overhangs the frame. The size of the rabbet depends on the hinge used. Generally, semiconcealed hinges are used on lip doors, but surface-mounted hinges are also available.

Figure 10–20 Installing semiconcealed hinge on a lipped door.

To determine the width of the lipped door, add twice the amount of the lip to the width of the opening. Do the same for the height. After the lip has been cut, round off the outer edges. Install the hinges on the door and then fasten them to the frame.

The square-edged door may be installed flush or with a full overlap (*Fig. 10–21*). Flush-mounted doors can be hung with butt, pivot, or decorative hinges.

Figure 10–21 Flush-mounted square-edged door.

Butt hinges are installed after the door has been carefully checked for fit. The usual clearance for flush-mounted cabinet doors is $\frac{1}{16}$ inch all around, except for hinges that are not gained; these require more clearance. If the opening is found to be off-square, the door must be trimmed accordingly. After trimming the door, bevel the opening edge slightly so that the inside corner will clear the edge of the frame (*Fig. 10–22*). Cut gains of equal depth in the door and frame, following the same procedure used in hanging interior doors.

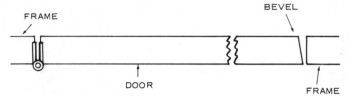

Figure 10–22 Bevel the opening edge of door to clear frame.

Doors that fully overlap the frame are often installed with pivot hinges or pin hinges. Pin hinges are useful on frame and panel doors if screws will hold firmly on the edge. For plywood and particleboard doors, the pivot hinge is best. It is semiconcealed and mounted to the face of the frame and to the rear of the door (*Fig. 10–23*). An angular recess must be cut at the top and bottom edges of the door. A third hinge may be required on heavy doors. Slotted holes in each leaf permit adjustments to be made after the door is hung.

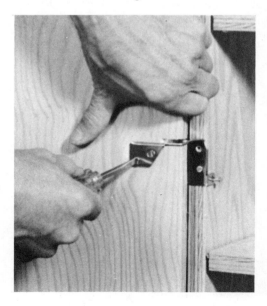

Figure 10–23 Flush door mounted with concealed hinge.

Pulls and Catches. Many types of catches and pulls are made for cabinet doors. Among the catches, the magnetic types are very popular. Most are surface-mounted, but flush catches are also made. They have a round body which is inserted into a hole bored into the edge of the frame. One very popular catch is made with an adjustable magnet. Various catches and pulls are shown in *Fig. 10–24*.

Applying Plastic Laminates. Plastic laminates are widely used in the modern home for furniture, walls, and cabinets, and for surfacing sink and

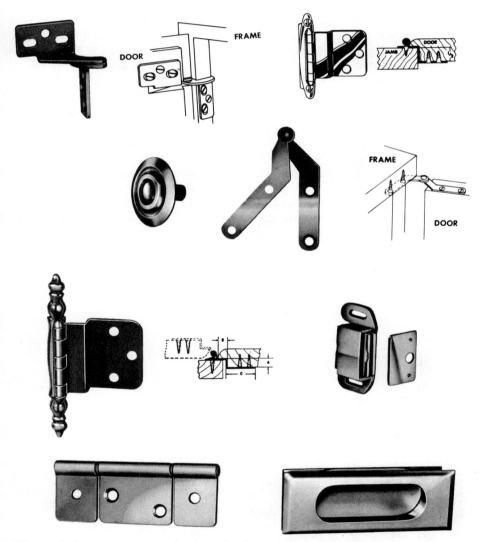

Figure 10–24 Various types of cabinet hardware.

counter tops in the kitchen and bathroom. The durable and heat-resistant qualities of the material make it well suited for these applications. This ideal surfacing material is made in many colors, patterns, and textures. Standard thicknesses are $\frac{1}{32}$ inch and $\frac{1}{16}$ inch. Widths range from 24 inches to 60 inches and lengths from 72 inches to 144 inches.

Backing grades are available in nondecorative form. These are used as a moisture barrier to prevent warping. Cabinet-liner grades are also made. These are similar to the backing grades but are better looking. Both are less expensive than the surfacing grade.

Cutting Laminates.   Laminated plastics may be cut with a handsaw, a circular saw, or tin shears. If you use a handsaw, make sure it is a fine-tooth saw with a low angle stroke, so you will avoid chipping the edge. Saw with the decorative side up. If you use a circular electric saw, cut with decorative side down and hold the saw firmly against the laminated sheet. If you use tin shears, cut with the decorative side up and cut oversize to allow for any chipping that may occur along the cut edge. Always firmly support the laminate sheet as close to the line of cut as possible, and cut $\frac{1}{8}$ inch to $\frac{1}{4}$ inch oversize. Trim off the excess after the sheet is firmly glued to the base material. If you are applying laminate to the edge (edge-banding), apply the edge band before gluing the sheet to the top.

Applying the Adhesive.   Use a clean paintbrush or roller to apply an even coat of contact cement to the back of the laminate sheet. Also apply an even coat to the wood surface to which the laminate is to be glued (*Fig. 10–25*). Allow the cement to dry, following the instructions of the manufacturer. A very porous wood surface will need two coats of contact cement. Apply the second coat after the first coat has dried completely.

Figure 10–25 Application of cement to wood surface and to laminate.

Applying the Laminate to the Surface.   Once the two cemented surfaces touch, the contact cement bonds immediately and no further adjustment of position is possible. Therefore it is extremely important to position the sheet properly before the glue surfaces touch. This is the best method to follow: Cut a piece of heavy brown paper into several strips; place these strips on the glued wood surface so they overlap and cover the surface completely. Now place the laminate sheet on the brown paper and position it for perfect fit. Move the first piece of paper out a few inches at a time, gently pressing the laminate to the wood surface as you do this. Then remove the remaining paper sections, one at a time, slowly sliding them out and pressing the laminate down into position.

Applying Pressure.   Once the sheet is in place, heavy pressure must be applied to the surface. A 3-inch hand roller is recommended. Start in the center and roll toward the edges. For hard-to-reach areas, hold a smooth block of wood on the surface and apply pressure with a hammer (*Fig. 10–26*). Remove excess cement from surface by rubbing with your fingers.

Figure 10–26 Apply pressure with hammer and small block of wood.

Finishing the Edges.   Edge-banding is the preferred method for finishing flat, vertical edges such as those on sinks or counter tops. The edge band should be put on before you apply the sheet to the top surface. For best appearance, build up the thickness of the exposed edge by nailing strips of particleboard or plywood to the underside to make a thickness of 1½ inches (*Fig. 10–27*). Cut a strip of laminate ¼ inch wider than the edge to be covered. Coat the wood edge twice and the back of the laminate strip once with the adhesive. Apply the strip and roll with firm pressure. Use a router to trim the excess laminate flush with the top and bottom of the wood surface. After the edge band has been applied, add the laminated

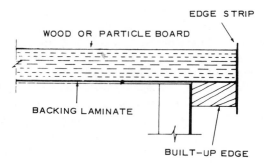

Figure 10–27 Built-up counter-top edge detail.

plastic sheet to the top of the counter as explained previously. Trim off the excess with router fitted with a 22½-degree bevel-trimming bit to bevel the edge joint (*Fig. 10–28*).

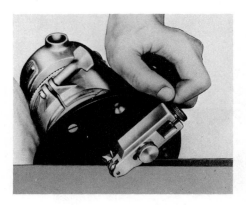

Figure 10–28 Trimming excess laminate.

Glossary

Acoustical Board: Any material used to control sound; also used to prevent the passage of sound from one room to another.

Acoustical Materials: Those materials especially designed to absorb sound.

Acoustics: The science of sound, including its production, transmission, and effects.

Actual Size of Lumber: The size of lumber after seasoning and dressing.

Admixture: A material added to concrete before or while it is mixed to add color or to control its strength or setting time.

Aerated Concrete: A specially made concrete with a cellular structure; it is lightweight and often used in subfloors.

Aggregate: Hard, inert material mixed with Portland cement and water to form concrete.

Air-Dried Lumber: Lumber that has been stored outdoors and stacked so that air can circulate around each piece to facilitate drying (*Fig. G–1*).

Figure G–1 Lumber being air-dried outdoors.

Air-Entrained Concrete: Concrete with millions of tiny air bubbles throughout its mass, making it highly resistant to freeze–thaw cycles.

Anchor: A metal form used to fasten masonry or timber.

Anchor Block: Wood block embedded into masonry to receive fasteners.

Anchor Bolt: A bolt used to anchor sills or plates to masonry (*Fig. G–2*).

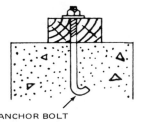

ANCHOR BOLT

Figure G–2 Anchor bolt in foundation wall.

Apron: A flat piece of trim placed on the inside of a window beneath the stool.

Arch: A curved or pointed structure supported at the ends; usually found in doors, passageways, and fireplaces (*Fig. G–3*).

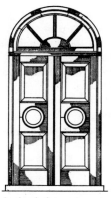

Figure G–3 Arched doorway.

437

Asbestos Shingle: A fireproof shingle made with particles of asbestos.

Astragal: A molding used to cover the joint between a pair of doors or windows (*Fig. G–4*).

Figure G–4 Astragal molding.

Backfill: Coarse dirt, broken stone, or other material used to replace earth after excavating (*Fig. G–5*).

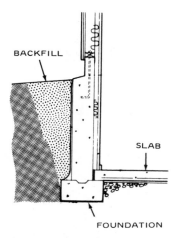

Figure G–5 Backfill at foundation wall.

Balloon Framing: A method of construction in which the wall studs extend in one piece from foundation to roof, and all floor joists are attached to the wall studs.

Baluster: One of a series of vertical supports for a stair rail (*Fig. G–6*).

Balustrade: A railing consisting of a series of balusters, resting on a base.

Bar Number: A number designating the size of a reinforcing rod; each digit equals ⅛ inch in diameter, thus a #3 bar is ⅜ inch in diameter.

Base (or Baseboard): A board placed around the perimeter of a room between floor and wall.

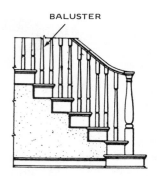

Figure G–6 Baluster.

Base Molding: Molding used to trim the upper edge of interior baseboard.

Base Shoe: Narrow molding used around the perimeter of a room where the baseboard meets the finish floor.

Batten: A thin narrow piece of plywood or lumber used to seal or reinforce a joint; also used as decorative vertical trim over panels.

Batter Boards: Horizontal boards on corner posts which assist in the accurate layout of foundation and excavation lines.

Bay Window: A window projecting from a wall in a square, rectangular, curved, or polygonal shape; usually supported on a foundation extending beyond the main wall (*Fig. G–7*).

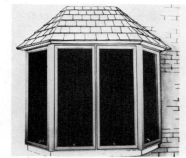

Figure G–7 A bay window.

Beam: A structural member of wood, steel, or concrete, used to support loads between walls or posts.

Beam Ceiling: Exposed beams in a ceiling used for decorative effect (*Fig. G–8*).

Figure G–8 Beam ceiling.

Beam Hanger: A device, usually of steel, used to support beams and other framing members (*Fig. G–9*).

Figure G–9 Beam hanger.

Beam Pocket: An opening in a structural member into which a beam is to fit (*Fig. G–10*).

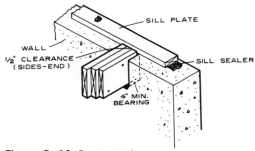

Figure G–10 Beam pocket.

Bearer: A small supporting member.

Bearing Partition: A partition that supports a vertical load in addition to its own weight.

Bearing Plate: A plate used to distribute excessive loads on beams, girders, or columns; the plate is placed under the load-bearing member.

Bearing Wall: A wall that supports a vertical load in addition to its own weight.

Benchmark: A reference mark on a permanent object on the jobsite from which all land measurements and elevations are taken.

Bird's Mouth: A notch consisting of a combination plumb and level cut, at the lower end of a rafter where it meets the top plate of the building.

Blockout: A block added to a concrete form to create a notch to accept a beam or girder at the top of the concrete wall.

Board Measure: A system for measuring lumber, using 1 board foot as the standard; the board foot measures $12 \times 12 \times 1$ inches (nominal) or $12 \times 6 \times 2$ inches (*Fig. G–11*).

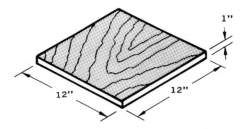

Figure G–11 Example of a board foot.

Boston Ridge: A method of applying shingles at the ridge or hips of a roof as a finish.

Bow: See *Warp.*

Box Beam: A hollow beam formed like a box (*Fig. G–12*).

Figure G–12 A box beam.

Box Cornice: A cornice that is completely enclosed (*Fig. G–13*).

Box Sill: Formed when a header is nailed to the joist ends which rest on a wall plate.

Brace: A piece of framing lumber applied to wall or floor to stiffen the structure; sometimes used only temporarily during framing of structure.

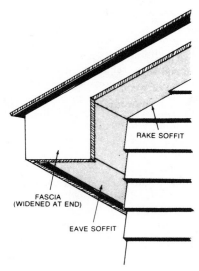

Figure G–13 Box cornice.

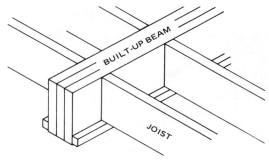

Figure G–14 Built-up beam.

Bridging: Wood or metal pieces placed between floor joists or wall studs to stiffen them.

British Thermal Unit (Btu): A symbol representing the amount of heat needed to raise the temperature of 1 pound of water 1 degree Fahrenheit.

Buck: A boxlike form set in concrete to create an opening for a door or window.

Building Line: An imaginary line beyond which a city prohibits the erection of a structure; the term is also used to denote the line on a building site which designates the outer walls of the building.

Building Paper: A building material, made of paper or felt, used in wall and roof construction as a protection against the passage of air and moisture.

Built-Up Beam: A beam made by fastening two or more planks to increase the strength of the timber (*Fig. G–14*).

Built-Up Roof: A flat or low-pitched roof waterproofed by the application of overlapping layers of roofing material.

Built-Up Timber: Timber that is increased in size by the fastening of two or more boards together.

Bulkhead: A blocking inserted into a concrete form to close that part of the form.

Butt Joint: A joint formed when two parts are fastened together without overlapping.

C: A symbol representing the conductance value of a material, and indicating the amount of heat (Btu's) that will flow through 1 square foot of material in 1 hour with a 1-degree difference between both its surfaces.

Camber: Slight arch in a beam or other horizontal member.

Cant Strip: Triangular piece of wood used under shingles at gable ends or under edges of roofing on flat decks, to divert water away from the structure.

Cantilever: A structural member such as a beam or truss which projects beyond a support; also a bracket-shaped support for a balcony or cornice.

Casing: Moldings of various widths and thicknesses used to trim door and window openings at the jambs (*Fig. G–15*).

Figure G–15 Section through casing molding.

Caulking: The sealing of open joints in a structure with a waterproof compound; applied by hand in strip form or as a continuous bead from a caulking gun (*Fig. G–16*).

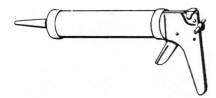

Figure G–16 Caulking gun.

Chamfer: A beveled edge formed on wood or masonry, usually at a 45-degree angle.

Chimney Tiles: Round or rectangular tiles placed within a chimney to protect the brickwork; the glazed surface of the tiles is not affected by smoke or gas.

Cleat: A narrow strip of wood used to support a shelf or other object; also used on door frames diagonally to temporarily reinforce the frame.

Collar Beam: A horizontal tie between roof rafters, placed well above the wall plate (*Fig. G–17*).

Column: An upright supporting structural member, circular or rectangular in cross section; it can also be ornamental.

Common Rafters: Those members extending from top plate to ridge, at right angles to both.

Concrete: A composite material made of Portland cement, water, aggregates, and sometimes admixtures to accelerate or retard setting of the concrete.

Continuous Beam: A beam that spans more than two supporting members.

Continuous Header: A top plate consisting of 2 × 6s on edge and running around the entire building instead of the customary 2 × 4s laid flat; acts as a lintel over openings and eliminates the need for some headers.

Contour Lines: The lines on a contour map which connect all points of the same elevation.

Control Joint: A joint tooled in flat concrete work, permitting the concrete to crack along the joint instead of the finish surface; the cracking usually takes place as the concrete begins to harden.

Coped Joint: In moldings, cutting the end of one piece to fit the molded face of the other at an interior angle.

Corbel: In brickwork, the stepping out of courses to form a ledge (*Fig. G–18*).

Corner Bead: Metal reinforcement placed on corners before plastering.

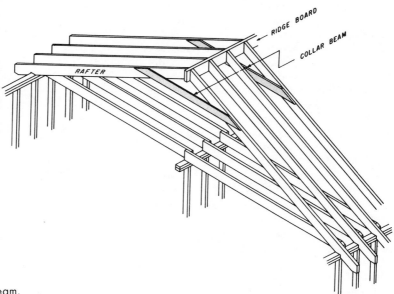

Figure G–17 Collar beam.

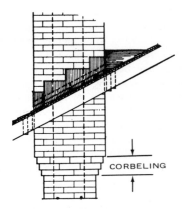

Figure G–18 Corbel in a brick chimney.

Decibel: The unit of measurement of sound.
Dog: Also called pinch dog; a steel bar with ends bent at right angles which is used to fasten timbers (*Fig. G–19*).

Figure G–19 Pinch dogs.

Cornice: The exterior finish and trim at the point where the roof projections and side walls meet; usually it consists of fascia board, soffit, and moldings.
Cornice Return: An extension of the cornice trim around the gable end; it usually extends about 2 feet onto the gable wall.
Counterflashing: A flashing used on chimneys at the roof line to cover shingle flashing and to prevent moisture entry.
Crawl Space: A shallow space between the first floor and ground in houses without a basement.
Cricket: A small rooflike structure which acts as a water shed; usually built behind chimneys or other projections.
Cripple Studs: Shortened studs at door and window openings.
Crook: See *Warp.*
Cross-Cutting: Sawing wood across its grain.
Crown: The top side of a bowed joist.
Cup: See *Warp.*
Curing Concrete: The process of keeping concrete moist for an extended period after placement to insure proper hardening and strength.
d: A nail size; see *Penny.*
Dado: A rectangular groove across the width of a board or plank.
Dead Load: The permanent weight of a structure.

Door Frame: The wood parts, consisting of head and side jambs, which form a support for the door.
Door Trim: The casing or trim used around interior doors to conceal the break between the door frame and the wall.
Dormer: A projection built out from a sloping roof to house a vertical window.
Double Header: The joining of two or more structural members when additional strength is needed (*Fig. G–20*).

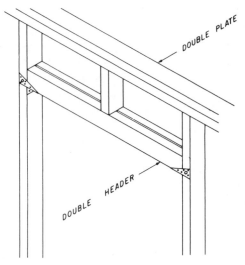

Figure G–20 Double header.

Downpipe: Also called downspout; the vertical pipe connected to rain gutters to carry rainwater from the roof to the drain.

Drain Tile: Clay or concrete pipes, about 4 inches in diameter, laid with open joints and covered with asphalt paper, and placed outside the footings to drain water away from the foundation.

Dressed and Matched: Boards or planks machined on all four edges to form mating tongues and grooves.

Dressed Size of Lumber: Size of lumber after surfacing.

Drip Cap: A molding placed over doors and windows, designed with a groove to let water drip to the outside of the structure (*Fig. G–21*).

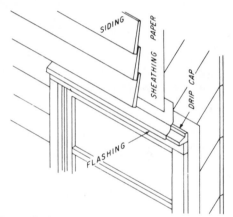

Figure G–21 Drip cap.

Drop Siding: Usually ¾-inch-thick stock made in various patterns with tongue-and-groove or shiplap joints.

Drywall: An interior wall finish consisting of gypsum boards and taped joints.

Eaves: The overhang of a roof projecting beyond the walls.

Elevation: An architectural drawing showing a head-on view of a vertical surface or wall.

Essex Board Measure: A table found on the framing square for rapid calculation of board feet.

Facade: The exterior front face of a building.

Face-Nail: To drive a nail perpendicular to the surface of a piece.

Fascia: Wood or plywood trim used along the eave or gable end of a structure; the outer face of a cornice (*Fig. G–22*).

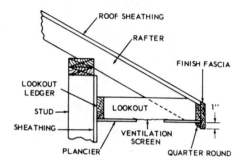

Figure G–22 Fascia.

Fire Cut: The angular cut in a joist that terminates in a masonry wall; designed so that in the event of a fire, the interior of the structure will collapse without bringing the masonry wall down with it (*Fig. G–23*).

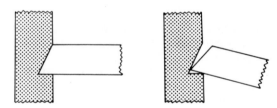

Figure G–23 Fire cut.

Firestop: A block or stop used in the wall of a building between studs to prevent the spread of fire or smoke through the air space.

Flashing: Sheet metal or other material used in roof and wall construction to prevent water penetration.

Floor Load: The total weight on a floor, including the dead weight of the floor plus any live load placed on it; the floor load is usually expressed in pounds per square foot.

Flue: The space in a chimney through which the smoke and gases rise.

Fly Rafter: End rafters of the gable overhang, supported by roof sheathing and lookouts.

Footing: A masonry section, usually concrete, supporting a foundation wall or pier.

Form: A temporary structure erected to contain concrete during placing and initial hardening.

Foundation: The supporting portion of a structure below the first-floor construction or below grade; it includes the footings.

Frieze: The member that connects the top row of the siding to the soffit of the cornice or roof sheathing (*Fig. G–24*).

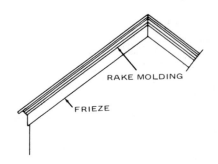

Figure G–24 Frieze.

Frost Line: The depth of frost penetration in soil. Footing should always be placed below the frost line.

Furring: Strips of wood or metal applied to a wall or other surface to even it and to serve as a fastening base for finish materials.

Gable: The triangular end of a double-sloped roof.

Gable Roof: A roof with two sloping surfaces meeting at a ridge and forming a gable at the ends.

Gambrel Roof: A roof with two surfaces having a double slope, the bottom slope being much steeper than the top slope.

Girder: A large horizontal beam used to support interior walls or joists.

Grade: The finished level of earth around a building.

Green Lumber: Unseasoned lumber that has not been exposed to air or kiln-drying and has moisture content over 19 percent.

Gusset: A plate of steel or plywood used to stiffen a corner (*Fig. G–25*).

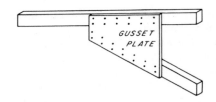

Figure G–25 Gusset.

Gutter: A channel of wood, metal, or plastic, based at the edge of a roof below the eaves to carry off rainwater.

Hardwood: Wood from a tree that has broad leaves.

Head Jamb: The top member of a door or window frame.

Header: A horizontal load-bearing support over an opening such as a door or window; also a supporting member placed at the head and foot of an opening.

Headroom: The vertical space in a doorway and in a stairway.

Hip Roof: A roof with four sloping sides.

Housed Stringer: A stair stringer grooved to take the ends of risers and treads (*Fig. G–26*).

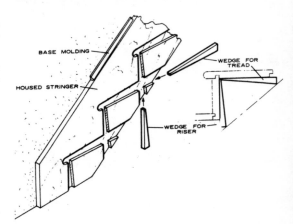

Figure G–26 Housed stringer.

I-Beam: A steel beam with a cross section similar to a capital letter I; now called an S-beam.

Jack Rafter: A short rafter framing from top plate to hip rafter, or from ridgeboard to hip or valley rafter.

Jack Stud: Shortened stud above door and window opening.

Jamb: The inner surface lining the sides and top of door and window frames.

Joist: A horizontal timber laid on edge to support floor and ceiling loads.

Joist Hanger: A metal bracket used to support joists (*Fig. G–27*).

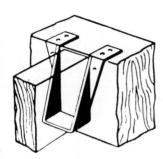

Figure G–27 Joist hanger.

k: A symbol representing thermal conductivity or the amount of heat (Btu's) that passes through 1 square foot of material 1 inch thick in 1 hour when there is 1 degree Fahrenheit difference between both its surfaces.

Kerf: A cut or groove made by a saw.

Keyway: An interlocking groove or channel made in wood or concrete joints for reinforcement.

Kicker: In cabinetry, a low, flat member used above a drawer on the underside of the cabinet, to keep the drawer from tipping downward when extended.

Kiln-Dried: A term applied to wood that has been dried in ovens (kilns) by controlled heat and humidity to specified limits of moisture content.

Lally Column: A metal column or post used to support beams and girders and usually filled with concrete.

Lath: A building material of wood, metal, gypsum, or insulating board, that is fastened to the frame of a building to act as a plaster base.

Leader: Another name for a downspout that carries rainwater from roof to ground.

Ledger: A strip of lumber attached to vertical framing or structural members to support joists or other horizontal framing. Similar to the ribbon strip.

Light: A single windowpane.

Lintel: A horizontal support used over doors, windows, and fireplace openings, often made of wood, steel, or reinforced concrete.

Live Load: The total weight of all moving and variable loads placed on a building.

Lookout: A short structural member which supports the overhang of a roof (*Fig. G–28*).

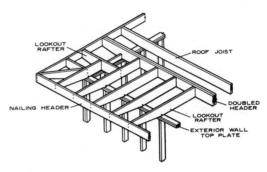

Figure G–28 Lookout.

Lumber: The product of the sawmill which entails sawing, resawing, planing, and cross-cutting; it is available in various sizes as follows:

Boards: Lumber up to 2 inches thick and 2 inches or more in width.

Dimension Lumber: Lumber 2 inches to 4 inches thick and 2 inches or more in width.

Factory and Shop Lumber: Lumber intended to be cut up for use in further manufacture.

Joists and Planks: Lumber 2 inches to 4 inches thick and 4 inches or more in width.

Light Framing Lumber: Lumber 2 inches to 4 inches thick and 2 inches to 4 inches in width.

Matched Lumber: Lumber that is edge-dressed and shaped with a close-fitting tongue-and-groove joint at the edges or ends.

Posts and Timbers: Lumber that is 5 inches square or heavier and having a width not more than 2 inches greater than its thickness.

Yard Lumber: Any lumber intended for general building purposes.

Mansard Roof: A roof having two slopes on each of its four sides; the lower slope is steep and the upper slope is very shallow.

Miter: To cut the ends of two pieces of wood at any angle other than 90 degrees.

Nominal Size of Lumber: A rough-sawed commercial size by which lumber is known and sold.

Mullion: The piece that divides multiple windows.

Muntin: The vertical and horizontal sash bars separating the panes in a window.

Nailing Clip: A notched metal plate with nail holes used to connect joists to steel beams (*Fig. G–29*).

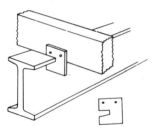

Figure G–29 Nailing clip.

Neat Cement: Pure cement without sand or gravel.

Newell: The upright post at the top and bottom of a stairway which supports the handrail.

Non–Load-Bearing Wall: A wall supporting no load other than its own weight.

Nosing: The part of a stair tread which extends beyond the face of the riser (*Fig. G–30*).

On Center (OC): The measurement of spacing for joists, rafters, studs, etc. in a building.

Open Cornice: A cornice in which the rafter ends are exposed.

Parapet: A low wall or railing at the edge of a roof, balcony, or bridge.

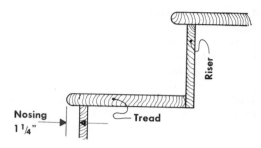

Figure G–30 Nosing.

Parging: Application of a thin coat of plaster cement to a masonry wall to form a smooth or decorative surface.

Particleboard: A highly compressed panel made of wood chips and resin.

Partition: An interior wall separating one room of a house or building from another.

Penny: A nail size, also expressed as the letter d.

Pier: A masonry column used to support other structural members.

Pilaster: A part of a wall that projects inside or out to add strength to the structure (*Fig. G–31*).

Figure G–31 Pilaster.

Piling: A structure of heavy timber, steel beams, or concrete shafts driven into the earth to serve as a foundation.

Pitch of Roof: Denotes the slope; it is the ratio of the total rise divided by the span of the roof.

Plancier: The underside of a cornice.

Plaster: A wall covering consisting of lime, cement, and sand.

Plaster Ground: A strip of wood used as a thickness gauge when a wall is being plastered; usually placed at the floor and around windows and doors.

Plat: A map of surveyed land, showing the location and boundaries and dimensions of the parcel.

Plate: The horizontal structural member of a wall that supports the attic joists and roof rafters.

Platform Framing: A method of construction in which each floor is framed independently; the joists of the floor above rest on the top plate of the floor below.

Plumb: Exactly perpendicular; vertical.

Plumb Line: A weighted cord used for locating a point on a surface from another point above; also for marking vertical lines.

Plywood: A sheet construction material made of three or more layers of wood with adjacent layers cross-grained; the number of layers or plies is always an odd number.

Portland Cement: A pulverized cement or binding agent, made from clay and crushed limestone.

Prefabricated Construction: A type of construction in which most of the construction and assembly is done in a factory; this leaves a minimum amount of assemblage to be done on the jobsite.

Purlin: Horizontal framing members supporting the rafters or spanning between trusses to support the roof covering.

R: A symbol representing the measured resistance of a material to the flow of heat.

Rafter: One of a series of structural roof members which support the roof covering (*Fig. G–32*).

Rail: The horizontal member of a door, window, or other assembly.

Rake: The trim members at the gable end of a roof which form the finish between the roof and wall.

Rebar: Another name for steel reinforcing bars.

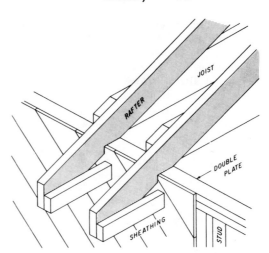

Figure G–32 Rafter.

Reinforced Concrete: Concrete that has been strengthened by the addition of embedded rebars.

Retaining Wall: A wall consisting of concrete, wood, or any other material erected to hold back or support a bank of earth.

Return: The continuation in a different direction of a building member such as a cornice, molding, etc.

Ribbon: A narrow board attached to studding or other vertical members of a frame that adds support to joists or other horizontal members.

Ridgeboard: The horizontal member at the top of the roof to which the rafters are fastened.

Ripping: Sawing wood with or along its grain direction.

Rise: The vertical distance between the plate and ridge in a roof and in stairs; the total height of the stair.

Rod: Also called a story rod or story pole; a long strip of wood, used by carpenters, which contains markings for window heights, stairs, siding, and shingle courses.

Roof Pitch: The slope of a roof, usually indicated in fractions, such as 1/4 pitch, 1/3 pitch, etc.; it is determined by dividing the rise of the roof by its span.

Roof Slope: The rise in inches per foot of run of a roof.

Rough Lumber: Lumber that has been cut to size with saws, but not yet surfaced (dressed).
Run: The horizontal distance from the outer wall to the ridge of a roof; in stairs, it is the horizontal distance from the face of the first riser to the face of the last one (*Fig. G–33*).

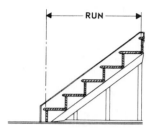

Figure **G–33** Run of stairway.

S-Beam: A structural iron beam shaped like the letter I in cross section; formerly called an I-beam.
Saddle: Small gable-type roof placed in back of a chimney on a sloping roof to direct water away from the chimney. (See *Cricket*)
Sash: The frame of a window into which the glass is set.
Scaffold (or Stage): A work platform erected above ground level.
Scarf Joint: An angled or beveled joint in which pieces are spliced together. The length of the scarf is five to twelve times its thickness.
Screeding: Leveling poured-concrete surfaces with a board or plank.
Seasoning: Removal of moisture from wood by air- or kiln-drying.
Sheathing: Material that is used to cover the outer surface of a wood building's superstructure.
Sheathing Paper: A building material used on floors, walls, and roofs to resist the penetration of air; it is applied between the sheathing and outer covering.
Shed Roof: A roof with only one sloped surface (*Fig. G–34*).
Shim: A thin strip of wood, sometimes wedge-shaped, for plumbing or leveling wood members, such as in door and window framing.

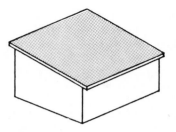

Figure **G–34** Shed roof.

Shiplap: Lumber with rabbeted edges which form a lap joint with adjoining pieces.
Shoring: The use of timbers at an excavation to prevent earth slides; also used as a bracing for temporary support.
Siding: The finish material used to cover the outer walls of a frame house.
Sill: The lowest part of a structure, resting on the foundation; also, the lowest member of a window or door.
Slab: A flat concrete area usually placed on a bed of crushed rock.
Sleeper: A wood strip laid on or in concrete to support a subfloor or finish flooring (*Fig. G–35*).

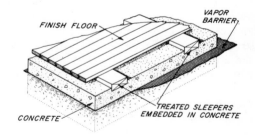

Figure **G–35** Sleepers in concrete.

Slope: The incline of a roof determined by the vertical rise per foot of horizontal run; for example, a roof with a 4-inch rise for each 12 inches of run has a slope of 4 in 12.
Soffit: The underside of a roof overhang; a plancier.
Softwood: Wood produced by trees having needlelike leaves and bearing cones. All evergreens are softwood trees. (This has no reference to actual hardness of the wood.)

Soil Stack: The vent piping for the waste system in a structure (*Fig. G–36*).

Sole Plate: The lowest horizontal member of a wall; supports the studs.

Sound Transmission Class (STC): A number rating that classifies the values of various materials for reducing sound transmission.

Span of Roof: Distance between two opposite walls; measured from the outside of the top plates.

Splice Plate: A piece of wood or plywood used to attach lumber members end to end.

Spreader: In formwork, a wood or steel member temporarily inserted between form walls to keep them evenly apart.

Square: A unit of measure, 100 square feet, usually applied to roofing materials.

Stack Partition: The partition wall that carries the soil pipe in a structure; because of the pipe diameter, the stack wall must be constructed of 2 × 6 or 2 × 8 studding.

Stairway: A series of steps leading from one floor to the next.

Stairwell: The opening in a structure which receives the stairs.

Stile: The vertical outer member of a door or window.

Stool: A flat molding fitted over the window-sill between the jambs.

Stops: Narrow wood strips which keep doors and double-hung windows in place.

Story Pole: A long rod or strip of wood used to lay out and transfer measurements for door and window openings, stairs, and siding and shingle courses.

String: The inclined support for risers and treads of a stair.

Stringer: A supporting member for a series of cross-members; frequently applied to stair supports.

Strip Flooring: Flooring of narrow strips usually 2¼ inches wide and tongue-and-grooved.

Stud: Vertical framing members of a wall.

Stud Pattern: A straight piece of 1 × 4 or 2 × 4 stock which is marked off with floor levels, ceiling heights, and door and window heights.

Subfloor: Boards or plywood laid on joists to serve as a base for the finish flooring.

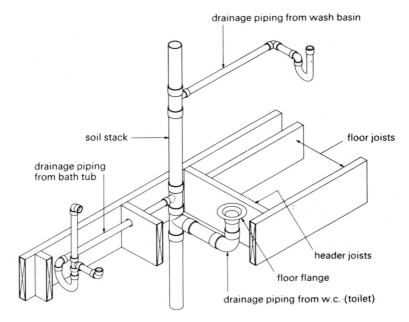

Figure G–36 Soil stack.

Surfaced Lumber: Lumber that has been dressed by being passed through a planer.

Tail of Rafter: That part of rafter that extends beyond the building line.

Tail Joists: Short joists between headers and wall or beam.

Termite Shield: A thin metal strip placed between foundation wall and sill to prevent entry of termites; also used around pipes and other projections.

Threshold: A strip of wood or metal with beveled edges covering the joint between the finished floor and sill of a doorway.

Toenail: To drive a nail in at a slant with the initial surface to permit it to penetrate the second member.

Tongued-and-Grooved (T&G): Shapes cut into edges of boards; the tongue is a projection which fits into the groove or rectangular channel of the mating piece.

Transit Level: An instrument for measuring and laying out vertical and horizontal lines and angles.

Transmission Loss (TL): Measured in decibels, the sound-insulating efficiency and non-transmitting quality of various constructions.

Trimmer: A beam or joist into which a header is framed.

Trimmer Studs: These are adjacent to regular studs at door and window openings and serve to stiffen the sides of the opening and bear the weight of the headers.

Truss: A combination of members usually arranged in triangular units to form a rigid framework for supporting loads over a span.

U: A symbol representing the heat loss through a building section in Btu's per hour.

Underlayment: The material on which a finished floor is placed.

Valley: The internal angle formed at the point at which two sloping roofs meet (*Fig. G–37*).

Valley Rafter: The rafter placed at the intersection where two inclined roofs meet.

Vapor Barrier: A material such as plastic film, used to control moisture transmission through walls and floors. Often used in com-

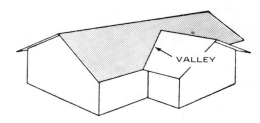

Figure G–37 Roof valley.

bination with insulation to control condensation.

Veneered Wall: The non–load-bearing masonry facing, such as brick, on a frame structure.

Vent Stack: A vertical pipe installed vertically and connected with vent pipes to carry foul air and gases from a building safely into the atmosphere.

Vermiculite: A lightweight mineral with excellent insulating qualities used in pellet form as bulk insulation.

Vernier: A scale used to accurately measure fractions of a degree in an angle.

Wainscot: The wooden lining of the lower part of an interior wall.

Walers: Horizontal timbers used in concrete form construction to brace the section.

Wall Tie: A device, usually a flat metal strip or wire, used to bind the tiers of masonry walls; also used to secure brick veneer to a wood-frame wall.

Warp: Distortion or variation from a true surface; in carpentry, the warp may appear as a bow, crook, cup, or twist. A board is said to be bowed when the face is concave or convex longitudinally. A crook in a board means that its edge is curved. A board with a transverse curvature is said to be cupped. In a board with a twist, all four corners are no longer in the same plane.

Water Table: A specifically milled item made from weather-resistant lumber, such as redwood, with its top surface slanted slightly downward to facilitate the runoff of water away from the siding or foundation of a building (*Fig. G–38*).

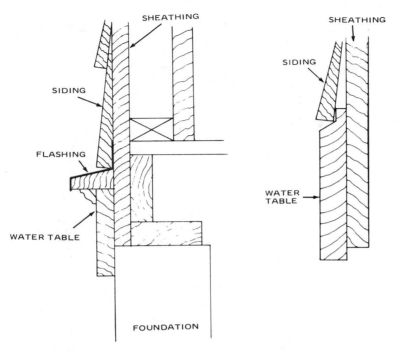

Figure **G–38** Water table.

Weatherstripping: A narrow piece of wood, rubber, plastic, or other material used around doors and windows to prevent air and moisture infiltration.

Weep Holes: Holes in brick or masonry veneer walls to allow the escape of moisture to the exterior of the building.

Winder: A type of stair construction in which the steps are carried around a curve, the treads being wider at one end than the other (*Fig. G–39*).

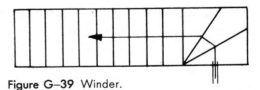

Figure **G–39** Winder.

Window Apron: A trim piece used under the window stool to conceal the rough edge of sheetrock or plaster.

Bibliography

Badzinski, Stanley, *Carpentry in Residential Construction*. Englewood Cliffs, N.J.: Prentice-Hall, Inc., 1972.

Badzinski, Stanley, *Stair Layout*. Chicago, Ill.: American Technical Society, 1971.

Burbank, Nelson, and Herman Pfister, *House Construction Details*. New York: Simmons-Boardman Co., 1968.

Dietz, Albert G. H., *Dwelling House Construction*. Cambridge, Mass.: Massachusetts Institute of Technology Press, 1971.

Durbahn, Walter Edward, *Fundamentals of Carpentry*. Vol. 2, 4th ed. Chicago, Ill.: American Technical Society, 1969.

Huntington, Whitney Clark, and Robert Mickadeit, *Building Construction*. New York: John Wiley & Sons.

Hurd, M. K., *Formwork for Concrete*. Detroit, Mich.: American Concrete Institute, 1973.

Townsend, Gilbert, *The Steel Square*. Chicago, Ill.: American Technical Society, 1939.

U.S. Army, Technical Manual TM5-551B, *Carpenter*. Washington, D.C.: U.S. Government Printing Office, 1971.

U.S. Department of Agriculture, Forest Products Laboratory Handbook #73, *Wood-Frame House Construction*. Washington, D.C.: U.S. Government Printing Office, 1970.

Wagner, Willis, *Modern Carpentry*. South Hooland, Ill.: Goodheart-Willcox Co., 1973.

Wilson, John Douglas, and S. O. Werner, *Simplified Roof Framing*. New York: McGraw-Hill Book Co., 1973.

Index

Anchors, sill 96
Angles, measuring 52–55
Asbestos shingles 301–303
Asphalt roofing 260–273
Asphalt shingles 260–273
Assembling tools 6–8
Attic ventilation 323–325
Attics 223–225, 323–325
Awning windows 245

Backsaws 2–3
Balloon framing 107, 114, 139, 141
Balustrades 416
Baseboard moldings 387–391
Basement stairs 403–404
Basement windows 246
Bathroom floor framing 124
Batter boards 74–76
Beams 112–113, 393–394
Benchmark 77
Bird's mouth cut 160, 170
Block flooring 361–362
Board measure 21–24, 57–58
Board subflooring 128
Boring tools 5–6
Brace table 58–60
Bracing walls 142–143
Brick veneer walls 306–307
Bridging joists 121–122
Bucks 93
Builder's level 42–55
Building code 41
Building plans 33–41
Built-ins 421–422

Cabinet doors 426–431
Cabinetry 418–434
    applying plastic laminates 431–434
    built-ins 421–422

doors, hinged and sliding 426–430
drawers 422–425
installing cabinets 419–421
types of cabinets 418–419
Cantilever joists 124
Casement windows 245
Ceiling insulation 320
Ceiling tiles 347–352
Ceilings 151–155, 347–352, 391–392
Chimney flashing 266–268
Chimney saddles 203–206
Chisels 5
Circular saw 9
Collar beams 176, 202
Common rafters 158–161, 164–177
Concrete 77–101, 319–320
    curbs, walks and driveways 99–101
    footings 82–86
    form materials 80–82
    foundations 77–99
    slab construction 96–98
    slab insulation 319–320
    stairs 99
    wall-form construction 87–95
Condensation 309, 323
Coping saw 3
Corner wall construction 136
Cornices 249–257
    closed 251–253
    open 254
    rake 254–255
Crosscut saw 2
Curbs, concrete 99–101
Cutting tools 2–3

Diagonal bracing 142–143
Door frames 239–241, 366–369
    exterior 239–241
    interior 366–369

455

Doors   370–383, 426–430
Dormer roofs   197–199
Double-hung window   242–244
Downspouts   282–285
Drainage, roof   282–285
Drawers   422–425
Drawings, architectural   35–41
Dressed lumber   20
Drills   5–6
Driveways, concrete   99–101

Eaves   249
Elevations, grade   77
Elevations, plan   34
Ellipses   65–66
Essex board measure   21–24, 57–58
Excavations   76–77
Exterior finish carpentry   238–307
    cornices   249–257
    roofs   257–285
    walls   285–307
    windows   241–249

Fasteners   27–33
Fiberboard sheathing   150
Finish carpentry   238–308, 332–394
    ceilings   347–352
    cornices   249–257
    door frames   239–241, 366–369
    doors   370–383
    exterior walls   285–306
    exterior windows   241–249
    floors   353–365
    interior trim   365–394
    interior walls   332–347
    interior windows   383–387
    roofs   257–285
Finishing tools   5
Fire safety   26
Fixed windows   246
Flashing, chimney and roof   224, 263–268,
277, 312
Flat roof   157, 201
Floor finishing   353–365
    resilient   363
    underlayments   363–365
        hardboard   364–365

    particleboard   365
    plywood   363–364
Floor framing   103–131
    balloon type   107
    bathroom   124
    beams and girders   112–114
    glued floors   131
    joists   114–127
    overhangs   124
    platform type   105
    post-and-beam type   108
    sill construction   111–112
    subflooring   128–130
    termite protection   110–111
Floor plans   34–35
Floor projections   124
Floors, wood   353–365
    block and parquet   361–362
    plank   361
    prefinished   362–363
    strip   353–361
Footings   82–86
Formwork   80–101
    curbs, walks and driveways   99–101
    footings   82–85
    monolithic forms   86
    stairs, concrete   99
    wall forms   87–95
Foundation layout   71–77
Foundations, concrete   71–101
Framing anchors   31
Framing square   55–60, 164–165

Gable dormers   199
Gable roofs   157, 175–179
Gambrel roof   157, 199–201
Girders   112–115
Glued floors   131
Grades of lumber   20–21
Gutters   282–285
Gypsum drywall   337–341
Gypsum lath   332
Gypsum sheathing   150–151

Hammers   6
Handsaws   2–3
Hand tools   1–9

Hardboard formwork   81
Hardboard siding   300–301
Hardboard wall panels   345
Headers   134–136
Hip rafter   179–188
Hip roof   157, 179–188
Horizontal angles   52–54
Horizontal sliding windows   246
Hundredths scale   56

Insulation, thermal   313–323
Interior finish carpentry   332–394
    baseboards   387–391
    ceilings   347–352
    doors   366–383
    floors   353–365
    moldings   387–392
    walls   332–346
    windows   383–387
Intersecting roofs   188–196
Intersecting walls   137

Jack rafters   158, 189–196
Jalousies   246
Joists   114–127, 151–155
    cantilever   124
    ceiling   151–155
    floor   114–127

Ladders   16–19
Lally columns   113
Laminates, plastic   431–434
Let-in bracing   142–143
Leveling instruments   42–55
Levels   4, 42–55
Lintels   95
Locks, installation   378–379
Lumber   19–26, 104
    board measure   24, 25
    grades   20, 21
    moisture control   24, 104–105
    seasoning   24
    sizes   20

Mansard roof   158
Mantel   392–393

Marking tools   4–5
Masonry walls   306–307
Measuring tools   4–5
Metal Lath   334
Metal siding   304
Metric system   60–64
Mineral fiber shingles   301–303
Moisture in wood   24, 104–105
Moldings   387–392
Monolithic forms   86

Nails   27–30
Nominal size of lumber   20

Octagon scale   56–57
Overhang layout   172

Particleboard underlayment   365
Partitions   145
Pitch of roof   162
Planes   5
Plank-and-beam framing   225–236
Plank flooring   361
Plans and specifications   33–41
Plaster   332–341
Plasterboard   337–341; see also gypsum
    drywall
Plastic laminates   431–434
Plastic siding   304
Platform framing   105, 114, 139, 141
Plumb bob   4–5, 47
Plumbing a column   52
Plywood   80–81, 128–131, 149–150, 220–222,
    296–297, 343–345, 363–364
    formwork   80–81
    sheathing   149–150, 220–222
    siding   296–297
    subflooring   128–131
    underlayment   363–364
    wall paneling   343–345
Pocket doors   382
Polygon table   60
Portable saw   9–11
Power tools   9–11
Prefinished flooring   362–363
Prehung doors   382–383
Pythagorean theorem   72

"R" value   314–315
Rafter layout   167–174
Rafter table   164–165
Rafters, common   158–159, 164–176
Rafters, hip   179–188
Rafters, jack   158, 189–196
Rafters, valley   188–189
Rake cornice   254–255
Ripsaw   2
Roof finish carpentry   257–285
   asphaltic products   260–273
   drainage   282–285
   flashing   263–268
   metal   282
   mineral fiber shingles   281
   slate   282
   terminology   259
   tile   282
   underlayments   260
   wood shakes   279–281
   wood shingles   273–279
Roof flashing   263–268
Roof framing   157–222
   chimney saddles   203–206
   collar beams   202
   common rafter layout   164–175
   dormer roof   197–199
   flat roof   201
   framing members   158–160
   gable roof   175–179
   gambrel roof   199–201
   hip roof   179–188
   intersecting roofs   188–196
   jack rafters   189–196
   plank-and-beam framing   225–236
   sheathing   220–222
   shed roof   196–197
   styles   157–158
   terminology   158–162
   trusses   215–219
   unequal pitch roofs   206–214
   valley rafters   188–189
Roof, metal   282
Roof, slate   282
Roof, tile   282

STC (sound transmission class)   327–328
Saber saw   11

Safety rules   9, 10, 11, 16, 18, 19, 26, 66–69
Saws   2–3
Scaffolds   11–16
Screwdrivers   6, 8
Screws   30–31
Seasoning lumber   24
Shakes, wood   279–281
Sheathing, roof   220–222
Sheathing, wall   149–151
Shed dormer   197–198
Shed roof   157, 196–197
Shingles
   asphalt   260–273
   mineral fiber   281, 301–303
   wood   273–279, 294–296
Shrinkage of wood   24, 104–107
Siding   287–306
   hardboard   300–301
   metal   304
   plastic   304, 306
   plywood   296–299
   wood   287–296
Sill construction   111–112
Skylights   246
Slab construction   96–97
Slab insulation   319
Slate roof   282
Slope of roof   162
Soffits   249–257; *see also* cornices
Sole plate layout   134, 139–140
Sound control   326–330
Specifications   33
Squares   4
Stairs   99, 396–416
Step-off rafter layout   168–172
Stepped footings   83–84
Story pole   141, 290
Straightedge   4
Stringers   408–414
Strip flooring   353–361
Stud pattern   141
Subflooring   128–130
Suspended ceiling   352

T-bevel   4
Termites   110–111
Thermal insulation   313–323
Thresholds   381

Tile roofing   282
Tools   1–11
    hand   1–9
    power   9–11
Transit level   42–55
Trusses, roof   215–219

"U" value   315
Underlayments, floor   363–365
Underlayments, roof   260–262
Unequal pitch roofs   206–214

Valley flashing   263–265
Valley rafters   188–189
Vapor barriers   310–312, 321
Veneer walls   306–307
Ventilation, attic & roof   323–325
Vernier scale   52–55
Vertical angle measurement   54

Walks, concrete   99–101
Walls, exterior   133–151, 285–307
Walls, finishing   285–307
    brick veneer   306–307
    sheathing   285
    shingles, mineral fiber   301–303
    shingles, wood   294–296
    siding   287–293, 296–301, 304–306
        hardboard   300–301
        metal   304
        plastic   304–306
        plywood   296–299
        wood   287–293
Walls, foundation   77–99
Walls, framing   133–151
    balloon   139–141

bracing   142–143
corner construction   136–137
headers   134–136
intersecting walls   137–138
partitions   145
platform framing   139, 141
sheathing   149–151
sole and top plates   134, 139–143
stud pattern   141
studs   134, 139–144
wall projections   124
Walls, insulation   321–323
Walls, interior   332–346
Walls, paneling   342–346
    hardboard   345
    plastic   346
    plywood   343–345
    wood boards   342
Walls, plastering   332–336
Walls, sheathing   149–151, 285
Winders   400
Windows, exterior   241–249
    awning   245
    basement   246
    casement   245
    double-hung   242–244
    fixed   246
    horizontal sliding   246
    installing   248–249
    jalousies   246
    skylights   246
Windows, interior   383–386
Wood   19–33, 104–105, 149, 287–296; see also lumber
    fasteners   27–33
    lumber   19–26, 104–105
    sheathing   149
    shingles   294–296
    siding   287–293

APPENDIX I

Termite Control
by Forest Products Laboratory
U.S. Department of Agriculture

Wood is subject to attack by termites and some other insects. Termites can be grouped into two main classes — *subterranean* and *dry-wood*. Subterranean termites are important in the northernmost States where serious damage is confined to scattered, localized areas of infestation. Buildings may be fully protected against subterranean termites by incorporating comparatively inexpensive protection measures during construction. The Formosan subterranean termite has recently (1966) been discovered in several locations in the South. It is a serious pest because its colonies contain large numbers of the worker caste and cause damage rapidly. Though presently in localized areas, it could spread to other areas. Controls are similar to those for other subterranean species. Dry-wood termites are found principally in Florida, southern California, and the Gulf Coast States. They are more difficult to control, but the damage is less serious than that caused by subterranean termites.

Damage from decay and termites has been small in proportion to the total value of wood in residential structures, but it has been a troublesome problem to many home-owners. With changes in building-design features and use of new building materials, it becomes pertinent to restate the basic safeguards to protect buildings against termites.

Subterranean Termites

Subterranean termites are the most destructive of the insects that infest wood in houses. The chance of infestation is great enough to justify preventive measures in the design and construction of buildings in areas where termites are common.

Subterranean termites are common throughout the southern two-thirds of the United States except in mountainous and extremely dry areas.

One of the requirements for subterranean-termite life is the moisture available in the soil. These termites become most numerous in moist, warm soil containing an abundant supply of food in the form of wood or other cellulosic material. In their search for additional food (wood), they build earthlike shelter tubes over foundation walls or in cracks in the walls, or on pipes or supports leading from the soil to the house. These tubes are from ¼ to ½ inch or more in width and flattened, and serve to protect the termites in their travels between food and shelter.

Since subterranean termites eat the interior of the wood, they may cause

461

much damage before they are discovered. They honeycomb the wood with definite tunnels that are separated by thin layers of sound wood.

Dry-wood Termites

Dry-wood termites fly directly to and bore into the wood instead of building tunnels from the ground as do the subterranean termites. Dry-wood termites are common in the tropics, and damage has been recorded in the United States in a narrow strip along the Atlantic Coast from Cape Henry, Va., to the Florida Keys, and westward along the coast of the Gulf of Mexico to the Pacific Coast as far as northern California (fig. 178). Serious damage has been noted in southern California and in localities around Tampa, Miami, and Key West, Fla. Infestations may be found in structural timber and other woodwork in buildings, and also in furniture, particularly where the surface is not adequately protected by paint or other finishes.

Dry-wood termites cut across the grain of the wood and excavate broad pockets, or chambers, connected by tunnels about the diameter of the termite's body. They destroy both springwood and the usually harder summerwood, whereas subterranean termites principally attack spring-wood. Dry-wood termites remain hidden in the wood and are seldom seen, except when they make dispersal flights.

Safeguards Against Termites

The best time to provide protection against termites is during the planning and construction of the building. The first requirement is to remove all woody debris like stumps and discarded form boards from the soil at the building site before and after construction. Steps should also be taken to keep the soil under the house as dry as possible.

Next, the foundation should be made impervious to subterranean termites to prevent them from crawling up through hidden cracks to the wood in the building above. Properly reinforced concrete makes the best foundation, but unit-construction walls or piers capped with at least 4 inches of reinforced concrete are also satisfactory. No wood member of the structural part of the house should be in contact with the soil.

The best protection against subterranean termites is by treating the soil near the foundation or under an entire slab foundation. The effective soil treatments are water emulsions of aldrin (0.5 pct.), chlordane (1.0 pct.), dieldrin (0.5 pct.), and heptachlor (0.5 pct.). The rate of application is 4 gallons per 10 linear feet at the edge and along expansion joints of slabs or along a foundation. For brick or hollow-block foundations, the rate is 4 gallons per 10 linear feet for each foot of depth to the footing. One to 1½ gallons of emulsion per 10 square feet of surface area is recommended for overall treatment before pouring concrete slab foundations. Any wood used in secondary appendages, such as wall extensions, decorative fences, and gates, should be pressure-treated with a good preservative.

In regions where dry-wood termites occur, the following measures should be taken to prevent damage:

1. All lumber, particularly secondhand material, should be carefully inspected before use. If infected, discard the piece.

2. All doors, windows (especially attic windows), and other ventilation openings should be screened with metal wire with not less than 20 meshes to the inch.

3. Preservative treatment in accordance with Federal Specification TT-W-571 ("Wood Preservatives: Treating Practices," available through GSA Regional Offices) can be used to prevent attack in construction timber and lumber.

4. Several coats of house paint will provide considerable protection to exterior woodwork in buildings. All cracks, crevices, and joints between exterior wood members should be filled with a mastic calking or plastic wood before painting.

5. The heartwood of foundation-grade redwood, particularly when painted, is more resistant to attack than most other native commercial species.

Pesticides used improperly can be injurious to man, animals, and plants. Follow the directions and heed all precautions on the labels.

NOTE: Registrations of pesticides are under constant review by the U.S. Department of Agriculture. Use only pesticides that bear the USDA registration number and carry directions for home and garden use.

Care of the New House
by Forest Products Laboratory
U.S. Department of Agriculture

A well constructed house will require comparatively little maintenance if adequate attention was paid to details and to choice of materials. A house may have an outstanding appearance, but if construction details have not been correct, the additional maintenance that might be required would certainly be discouraging to the owner. This may mean only a little attention to some apparently unimportant details. For example, an extra $10 spent on corrosion-resistant nails for siding and trim may save $100 or more annually because of the need for less frequent painting. The use of edge-grained rather than flat-grained siding will provide a longer paint life, and the additional cost of the edge-grained boards then seems justified.

The following sections will outline some factors relating to maintenance of the house and how to reduce or eliminate conditions that may be harmful as well as costly. These suggestions can apply to old as well as new houses.

Basement

The basement of a poured block wall may be damp for some time after a new house has been completed. However, after the heating season begins, most of this dampness from walls and floors will gradually disappear if construction has been correct. If dampness or wet walls and floors persist, the owner should check various areas to eliminate any possibilities for water entry.

Possible sources of trouble:

1. Drainage at the downspouts. The final grade around the house should be away from the building and a splash block or other means provided to drain water away from the foundation wall.

2. Soil settling at the foundation wall and forming pockets in which water may collect. These areas should be filled and tamped so that surface water can drain away.

3. Leaking in a poured concrete wall at the form tie rods. These leaks usually seal themselves, but larger holes should be filled with a cement mortar or other sealer. Clean and slightly dampen the area first for good adhesion.

4. Concrete-block or other masonry walls exposed above grade often show dampness on the interior after a prolonged rainy spell. A number of waterproofing materials on the market will provide good resistance to moisture penetration when applied to the inner face of the basement wall. If

the outside of below-grade basement walls is treated correctly during construction, waterproofing the interior walls is normally not required.

5. There should be at least a 6-inch clearance between the bottom of the siding and the grass. This means that at least 8 inches should be allowed above the finish grade before sod is laid or foundation plantings made. This will minimize the chance of moisture absorption by siding, sill plates, or other adjacent wood parts. Shrubs and foundation plantings should also be kept away from the wall to improve air circulation and drying. In lawn sprinkling, it is poor practice to allow water to spray against the walls of the house.

6. Check areas between the foundation wall and the sill plate. Any openings should be filled with a cement mixture or a calking compound. This filling will decrease heat loss and also prevent entry of insects into the basement, as well as reduce air infiltration.

7. Dampness in the basement in the early summer months is often augmented by opening the windows for ventilation during the day to allow warm, moisture-laden outside air to enter. The lower temperature of the basement will cool the incoming air and frequently cause condensation to collect and drip from cold-water pipes and also collect on colder parts of the masonry walls and floors. To air out the basement, open the windows during the night.

Perhaps the most convenient method of reducing humidity in basement areas is with *dehumidifiers*. A mechanical dehumidifier is moderate in price and does a satisfactory job of removing moisture from the air during periods of high humidity. Basements containing living quarters and without air conditioners may require more than one dehumidifier unit. When they are in operation, all basement windows should be closed.

Roof and Attic

The roof and the attic area of both new and older houses might be inspected with attention to the following:

1. Humps which occur on an asphalt-shingle roof are often caused by movement of roofing nails which have been driven into knots, splits, or unsound wood. Remove such nails, seal the holes, and replace the nails with others driven into sound wood. Blind-nail such replacements so that the upper shingle tab covers the nailhead.

A line of buckled shingles across the roof of a relatively new house is often caused by shrinkage of wide roof boards. It is important to use sheating boards not over 8 inches wide and at a moisture content not exceeding 12 to 15 percent. When moisture content is greater, boards should be allowed to dry out for several days before shingles are applied. Time and hot weather tend to reduce buckling. Plywood sheathing would eliminate this problem altogether.

2. A dirt streak down the gable end of a house with a close rake section can often be attributed to rain entering and running under the edge of the shingles. This results from insufficient shingle overhang or the lack of a metal roof edge. The addition of a flashing strip to form a drip edge will usually minimize this problem.

3. In winters with heavy snows, ice dams may form at the eaves, often resulting in water entering the cornice and walls of the house. The immediate remedy is to remove the snow on the roof for a short distance above the gutters and, if necessary, in the valleys. Additional insulation between heated rooms and roof space, and increased ventilation in the overhanging eaves to lower the general attic temperature, will help to decrease the melting of snow on the roof and thus minimize ice formation. Deep snow in valleys also sometimes forms ice dams that cause water to back up under shingles and valley flashing.

4. Roof leaks are often caused by improper flashing at the valley, ridge, or around the chimney. Observe these areas during a rainy spell to discover the source. Water may travel many feet from the point of entry before it drips off the roof members.

5. The attic ventilators are valuable year round; in summer, to lower the attic temperature and improve comfort conditions in the rooms below; in winter, to remove water vapor that may work through the ceiling and condense in the attic space and to minimize ice dam problems. The ventilators should be open both in winter and summer.

To check for sufficient ventilation during cold weather, examine the attic after a prolonged cold period. If nails protruding from the roof into the attic space are heavily coated with frost, ventilation is usually insufficient. Frost may also collect on the roof sheathing, first appearing near the eaves on the north side of the roof. Increase the size of the ventilators or place additional ones in the soffit area of the cornice. This will improve air movement and circulation.

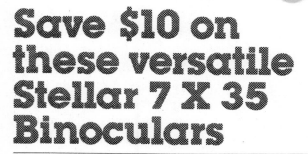